The Patrick Moore Practical Astronomy Series

More information about this series at http://www.springer.com/series/3192

Imaging the Messier Objects Remotely from Your Laptop

Len Adam

Len Adam
Fellow of the Royal Astronomical Society
Leyland, Lancashire, UK

ISSN 1431-9756 ISSN 2197-6562 (electronic)
The Patrick Moore Practical Astronomy Series
ISBN 978-3-319-65384-6 ISBN 978-3-319-65385-3 (eBook)
https://doi.org/10.1007/978-3-319-65385-3

Library of Congress Control Number: 2017964662

Printed on acid-free paper

This Springer imprint is published by the registered company Springer International Publishing AG part of Springer Nature.
The registered company address is: Gewerbestrasse 11, 6330 Cham, Switzerland

Acknowledgements

I would like to thank itelescope.net, the not for profit organization that provides access to telescopes based at sites spread across four locations and three continents, used to take the images of all the Messier objects in this book. In particular, I would like to thank Fernando Abalos at Astrocamp in Spain, part of the itelescope network, for the images of the Nerpio observatories and telescopes that he provided. Thanks also to Takahashi America and to PlaneWave Instruments for the images they provided of some of the telescope models used to capture the Messier images. I would also like to thank Software Bisque for allowing the use of the charts generated through its SkyX software to display the location of the Messier objects. I used this software to plan the imaging program and analyze the resulting images with the aid of its image linking facility to produce astrometric solutions. This helped me to identify field objects such as remote galaxies and minor planets in the images. Much of the supporting data on specific Messier objects was sourced from the SEDS website, so thanks to all the contributors to that very useful resource.

Len Adam

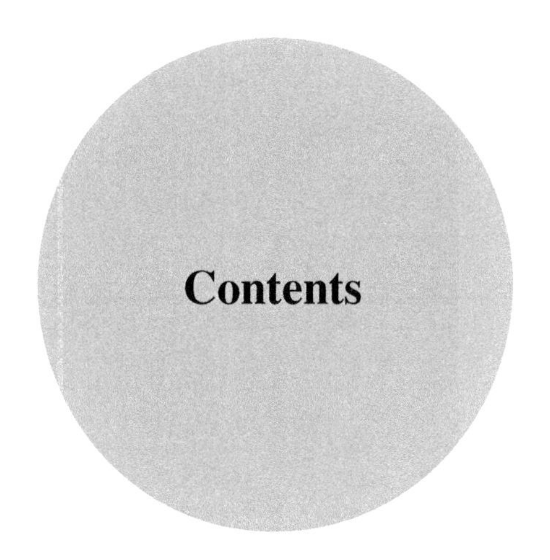

Contents

Part I Background and Resources

Part II The 110 Messier Objects – Images and Corresponding Data

Part I

Background and Resources

Chapter 1

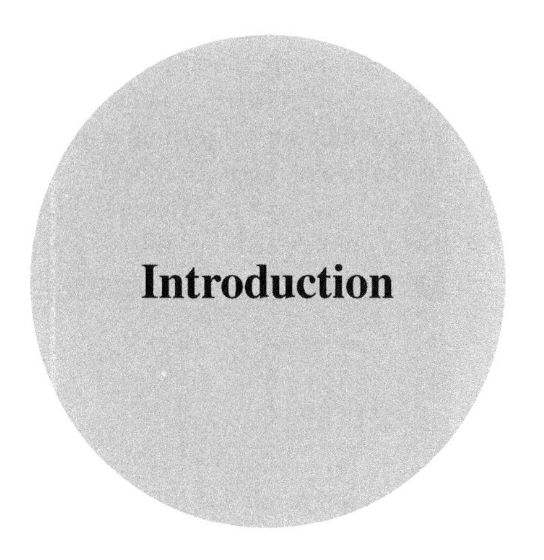

Introduction

The Remote Telescope Messier Imaging Project

The Advantages of Remote Imaging

While you may be interested in astronomy, you might not realize just how easy and straightforward it can be to hire time on high-quality telescopes located across the continents. Perhaps you cannot afford your own telescope or camera to take images. Perhaps you live in an apartment and have nowhere convenient to set up equipment. Maybe you have owned equipment but no longer have the mobility to be able to clamber into that observatory dome. Quite possibly, your once-dark sky site is now compromised by the arrival of a neighboring 24-hour supermarket and its floodlit parking lot.

There are many reasons why you would wish to use remote telescopes, and using this method is much simpler than you think. You do not need to be a professional astronomer, nor do you need an exorbitant amount of money to take advantage of these global 'scopes – in fact, all you really need is Internet access and a basic knowledge of astronomy.

The Messier Objects

The Messier objects have long been a popular target for amateur astronomers. There are many sources explaining how the Messier list came to be. If the term "Messier" is new to you, a quick Internet search will soon bring you up to speed.

L. Adam, *Imaging the Messier Objects Remotely from Your Laptop*, The Patrick Moore Practical Astronomy Series, https://doi.org/10.1007/978-3-319-65385-3_1

Charles Messier, an 18th-century comet hunter, started a list of deep sky objects, such as clusters and nebulae, that could be mistaken for comets. Ironically, this list of objects has become far more famous than his comet discoveries.

The Messier list is an excellent starting point for remote astronomy. There are 110 Messier objects on the list, although M102 is in fact considered by many to be a duplicate of M101. I have taken the route of assuming that M102 is the lenticular galaxy, NGC 5866, so that is the image taken for M102.

The purpose of this book is to show how the Messier objects are quite simple to image and analyze. Of course, a remote telescope allows you to point to any object you like, but the Messier objects are a great and achievable first challenge.

Project Summary

Over the last year, I was able to image all 110 Messier objects using 17 remote telescopes scattered across the world, in places such as New Mexico, California, Southern Spain and Siding Spring in New South Wales. This period coincided with a time when, due to my relocation, my astronomical equipment was locked in a storage unit. For quite some time, I was unable to pack my truck with telescopes, mounts and cameras and drive off for a few months somewhere with clear and dark skies, as I had done before. The image in Figure 1.1 shows a typical location for my normal activity, on top of a mountain, with my telescopes strapped to the mountain itself.

Fig. 1.1 My non-remote telescope

As it was not possible to find time to use my own equipment over the last year, the idea of using a laptop, the Internet and remote telescopes for astronomical imaging became very attractive! While using remote telescopes for my project was not free, I did indeed save on travel and accommodation. Imaging all the Messier objects using excellent telescopes in three continents over the course of a year costs less than the price of a couple of months of accommodation at a dark sky site. Free remote telescopes do exist, but there are few of these, and they are considerably less flexible than the system used for taking images for this book.

There was an additional advantage to this technique -- remote imaging from my laptop meant not having to strain my back to assemble a Paramount ME and lift up a Celestron 14 and other kit onto the mount! Although I will go back to heavy lifting next year when I begin to use my equipment again, I will most certainly make use of the added advantage of also being able to use remote telescopes from any location for future projects.

How the book is organized

The book is split into six chapters. This chapter (Chapter 1) sets the scene for remote imaging, describes the remote telescope network used for the project and explains in general terms how the images in the book were obtained. Although a specific remote telescope network was used, there are alternatives. If you intend to start your own remote imaging project, you will need to get full instructions on how to use a particular network from the website of any remote telescope system. A little Internet research will allow you to choose which remote telescopes would be most appropriate for your requirements.

Additionally, this chapter will explain when remote imaging could be of use and provides an explanation of terms used throughout the book, in case you are not already familiar with these. References are made throughout the book to a number of data sources relating to specific Messier objects and any objects of related interest. These sources are listed at the end of this chapter.

Chapter 2 describes the 17 telescopes used and lists the specific Messier objects imaged with each telescope. It also includes images of some of the actual on-site telescopes or of the general type of telescope used.

Chapters 3, 4 and 5 are the main body of the book. This is where you will find the remotely captured image of each Messier object, together with associated data including exposure time, telescope details, camera details, time, date, altitude, azimuth angle, right ascension, declination, transit time and field of view. An annotated version of the image is included, with details of other objects of interest captured in the frame and a chart showing the sky at the exact time the image was taken. Each chart has the image precisely superimposed on it to scale as a result of plate solving.

Chapter 6 is a Quick Reference Image library, showing the images obtained of all the Messier objects in one place.

Knowledge Level

In this chapter the necessary knowledge that you need to get started is introduced and explained. Terms such as Stellar Magnitude, Right Ascension, Declination, Azimuth and Altitude are explained and illustrated. There is a brief explanation of Spectral Types. These are terms that are used regularly in the details of each Messier object imaged. Much of this will be very familiar to readers and you can skip over these explanations if you already have that knowledge, but I felt that I should try to explain terms that are regularly used throughout the book.

The Telescope Network used to Image the Messier Objects

Network Summary

The remote system used in this book to image Messier objects is called "itelescope.net," although it is by no means the only network you could use. To begin, you must register on the site as a subscribed member. Using this network feels just the same as when you control a telescope with a computer. It is remarkably similar to the way many amateurs operate nowadays, controlling the telescope from inside the house through a cable or wireless connection. In reality, it doesn't matter how far away the telescope is -- 10 yards or 10,000 miles -- the experience is more or less the same.

Start by logging into the remote telescope network website, and then log into the computer connected to the particular telescope that you have selected. The itelescope website will tell you which telescopes are available and when, so you can instantly use a telescope if it is available for the period of your imaging, or you can make an advance reservation so that the telescope will be yours for the duration of the booking. Note that the only thing you pay for is the imaging time, so if you take a 300-second image, you will be charged for that time only, at the rate for that particular telescope. You buy prepaid blocks of points, with each telescope having its own cost per point depending on the size of the telescope. All of this can be checked by reading the guides on the itelescope website.

To start, you can just select the object from a suggested list and the system will do the rest for you, setting an appropriate exposure time and appropriate filter (this is the option titled "One-Click Image"). Later on you can be more adventurous. There are many options associated with each image you take. You may wish to take images using B and V filters to allow you to do some real science. You may wish to track the brightness of a variable star. If you have an AAVSO account (American Association of Variable Star Observers), you may set the option on the particular telescope to automatically send your images to your AAVSO account for you to analyze. When you take an image, you will receive an email within minutes of the exposure finishing, telling you the image is complete – usually accompanied by a preview image. You will then be able to download the full resolution image. The email will also tell you the number of points you have spent.

Whether you plan to use the images as book illustrations, display them on your website, post them on social media platforms or use them in a presentation at your local astronomical society, the images are yours to use as you wish.

It is important to have some form of planetarium software that can identify a date and time that is appropriate for imaging the specific object that you are targeting. Note that while I have taken an approach involving a minimum exposure time (to minimize costs), usually 300 seconds, do not underestimate the quality of the telescope systems available as a result of my simple approach. If you are willing to invest more, the telescopes can produce fantastic images of objects in the Universe with longer exposure times and intensive processing. If you use itelescpe.net, check the user images and the associated details of how the images were taken on the www.itelescope.net website for guidance.

The New Mexico Site

The itelescope network consists of telescopes in four locations on three continents covering the northern and southern hemispheres. The first location is New Mexico, where there are six telescopes that can be used to image Messier objects (at the time of writing). The telescopes are based at the New Mexico Skies site near Mayhill, New Mexico. Some of the observatories at this site are shown in Figure 1.2. All of the sites have additional private telescopes and observatories. For each Messier object image in the book, details of the location and the particular telescope used are included.

For example, one of the telescopes used for this project in New Mexico is a single-shot color, 6-inch telescope (Figure 1.3). This telescope was used for a number of Messier objects, as it allows the user to obtain color images with a relatively low

Fig. 1.2 Observatories at the New Mexico site. Courtesy of itelescope.net

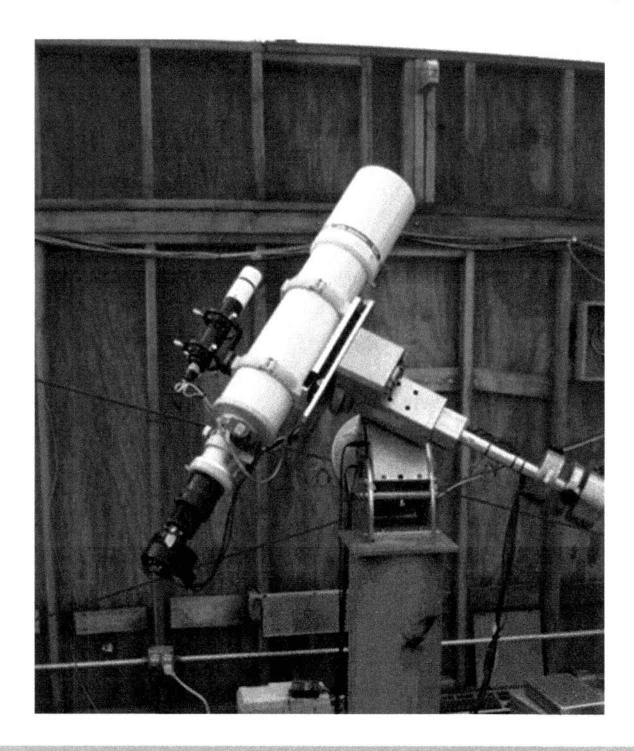

Fig. 1.3 6-inch telescope in New Mexico with single-shot color camera. Courtesy of itelescope.net

exposure time, which in turn means relatively low cost. The other telescopes are capable of color imaging, but you need to take multiple exposures with filters and then add the separate images together later on using your own software. All the Messier object color images in this book were taken with a single-shot color camera.

The largest telescope available in New Mexico has an aperture of 20 inches. All of the mounts are of top quality, as a reliable mount is essential for remote operation.

The California Site

There is only one telescope available in Auberry, California at the Sierra Remote Observatory. However, this is an impressive 24-inch instrument. Part of the observatory is shown in Figure 1.4.

The 24-inch telescope is shown in Figure 1.5.

Fig. 1.4 Runoff roof sheds at the Sierra Remote Observatory. Courtesy of iTelescope.net

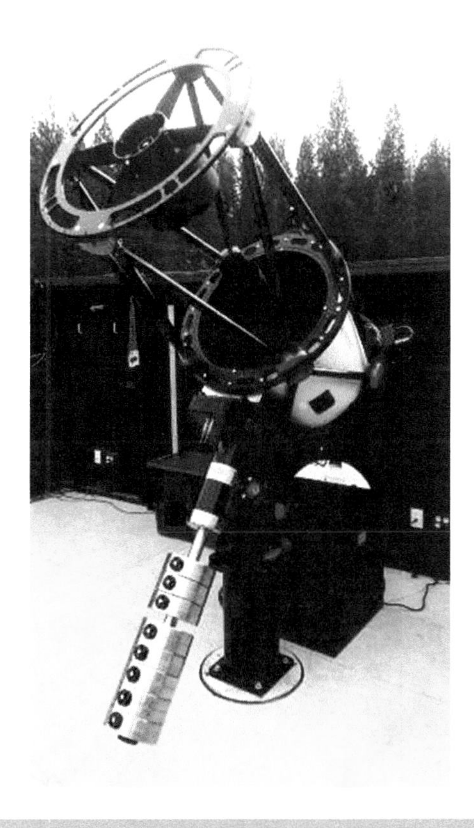

Fig. 1.5 The 24-inch telescope at the Sierra Remote Observatory. Courtesy of iTelescope.net

The Spanish Site

In Spain, three telescopes are available for use. They are based at AstroCamp in Nerpio, Southern Spain. A daylight view of the collection of observatories based at Astrocamp is shown in Figure 1.6.

A stunning view of the Milky Way over the Astrocamp observatories is shown in Figure 1.7. Courtesy of Fernando Abalos.

The Australian Site

The final location is Siding Spring Observatory in New South Wales, Australia. There are 10 telescopes available for use (at the time of writing). Figure 1.8 shows the closed runoff shed that contains all of the telescopes.

The image of the observatory with the roof open (Figure 1.9) shows telescopes that all operate independently, driven by users controlling them from any part of the world. Some of the telescopes are private and not available to public users. If you want to own your own telescope in one of the sites, you can do so with the hosting scheme, details of which are available online.

Fig. 1.6 Daylight view of the AstroCamp site in Nerpio, Spain. Courtesy of Fernando Abalos

Fig. 1.7 The AstroCamp observatories at night

Fig. 1.8 The runoff roof shed housing the 10 available telescopes at Siding Spring. Courtesy of iTelescope.net

There is a small telescope based at Siding Spring that is capable of taking single-shot color images. Shown in Figure 1.10, it is a 3-inch apochromatic refractor capable of taking excellent images.

A larger telescope based at Siding Spring is 17 inches in aperture. This is shown in Figure 1.11.

Fig. 1.9 Siding Spring Remote Observatory. Courtesy of iTelescope.net

Fig. 1.10 3-inch refractor with single-shot color based at Siding Spring. Courtesy of iTelescope.net

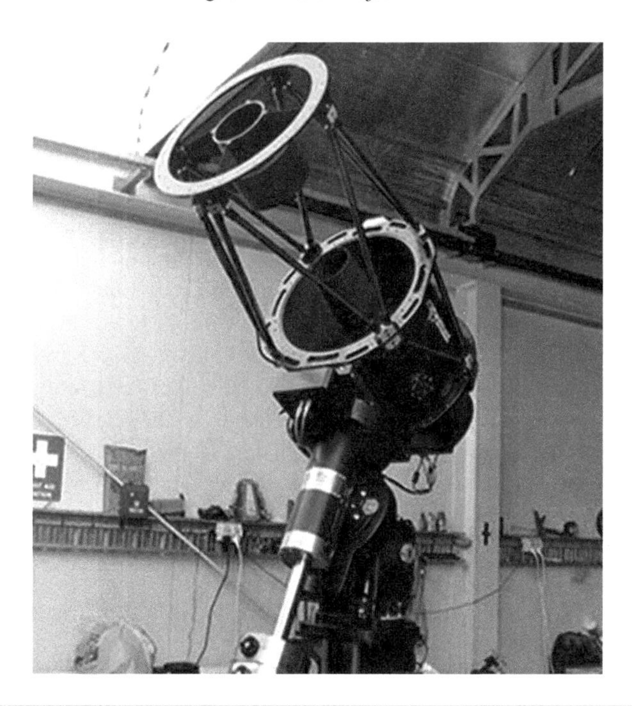

Fig. 1.11 17-inch telescope based at Siding Spring. Courtesy of iTelescope.net

Using Planetarium Software to set the Location

If you are going to use telescopes on other continents, you must be able to predict when particular objects are going to be visible from specific locations and at particular times. A range of "Planetarium" programs that can be found by searching the internet will allow you to do exactly that. If you have an observatory in your backyard, you can set the planetarium program to the Latitude and Longitude of that specific location, but if you are going to be using remote telescopes in three or four different locations you can set the software to produce a version for each location.

All the Messier images for the book involved the use of the SkyX software from Software Bisque, but using a free program such as Stellarium will also allow you to plan your remote imaging. The software allows you to save as many locations as you want, and for the purpose of imaging the Messier objects, I was able to set up a separate document for Mayhill in New Mexico, Auberry in California, Nerpio in Spain and New South Wales, Australia. Once you have the software opened, simply set it to the Latitude, Longitude and Altitude of the location then save it as a document. New Mexico Skies sits at 7300 ft (2225 m), Sierra Remote Observatory at 4600 ft (1405 m), AstroCamp at 5400 ft (1650 m) and Siding Spring at 3700 ft (1122 m). Images in the book taken from a particular observatory include all of the details you need to set that specific location.

The Geometry of the Earth and Sky

Location and Time

To identify the location of a particular place on Earth, we use Latitude and Longitude. We have a grid with coordinates that are based on the equator as a zero point for Latitude, and Greenwich UK as a zero point for Longitude. Going north from the equator, we start with Latitude 0° and move up to 90° at the North Pole. For example, the New Mexico Skies site used for imaging has a Latitude of 32° 54' 11.91" north of the equator. The Siding Spring site in Australia, being in the southern hemisphere, has a Latitude of 31° 16' 24" south of the equator.

The zero point for Longitude is Greenwich, near London UK, and measurements are made east or west of Greenwich. The New Mexico site is 105° 31' 43.32" west of Greenwich. Siding Spring has a Longitude 149° 03' 52" east of Greenwich. The earth rotates from west to east, and a full rotation is 360°, or 24 hours. One hour is equivalent to 15°, so the difference in Longitude between Greenwich and New Mexico of 105° is equivalent to 105 divided by 15, which is 7 hours. Thus, New Mexico time is always 7 hours behind Greenwich time, subject to daylight savings variations.

If you happen to live in the UK in the Greenwich time zone, this means that you could log into a telescope in New Mexico at 9 am (UK time) and immediately start imaging with an exposure start time at 2 am in New Mexico. Be aware of local daylight savings time variations, which will affect this. When you log into a telescope, it will always tell you the local time and the Coordinated Universal Time (UTC or just UT), which is the same as Greenwich Mean Time (GMT).

Siding Spring is 149° to the east of Greenwich, which is equivalent to a roughly 10-hour (149 divided by 15) time difference. Time zones exist so that we are never dealing with differences less than one hour. So, if it is 9h 30m 10s in Greenwich (9:30 am) it will be 19h 30m 10s in the evening of the same day in New South Wales – again, you need to check for daylight savings variations. If you live in the UK, you can start imaging from Siding Spring during UK daylight hours – simply check and make sure that the sun has set in New South Wales. If you live in New York, which has a Longitude of 74° west, there is a time difference of 5 hours (74 divided by 15 to the nearest hour) between New York and Greenwich, and 2 hours between New York and New Mexico Skies. Nerpio, Spain has a Longitude of 2° 19' west but strangely has a time zone of +1 -- a relic of the 20th century military general Francisco Franco's support for Nazi Germany, for which he aligned Spain's time to German time. In reality, it is 2° W and should be in the Greenwich Time Zone. If you live in New York and it is 3 am, then it will be 8 am in Greenwich and 6 pm in Siding Spring. At 9 am in New York, it will be 2 pm in Greenwich and midnight in Siding Spring. At other locations on Earth, you can use the above calculations to determine the local time zone and work out the best times to use those telescopes.

The Celestial Equator and the Zenith

To begin remote telescope imaging, it helps if you have an understanding of the way in which stars, galaxies, nebulae and other objects are located through the celestial equivalent of Longitude and Latitude. If you can imagine Earth's equator extended and projected onto the sky, it will form the celestial equivalent, or the **celestial equator**. If you are at the Latitude of New York, face north, and look up at an angle equal to your Latitude of 40.7°, you will see the star Polaris – the North Star (i.e. its altitude is 40.7°). Figure 1.12 shows the position of Polaris at roughly 41° above the northern horizon. This means that it is 90° - 41° = 49° from Polaris to the **zenith** – the point in the sky over your head. Moving 90° from Polaris straight through the zenith takes you to the celestial equator – again, the earth's equator projected onto the sky. Looking to the south, the angle between the horizon and the celestial equator will be 49° -- that is to say its altitude is 90° - 41° = 49°.

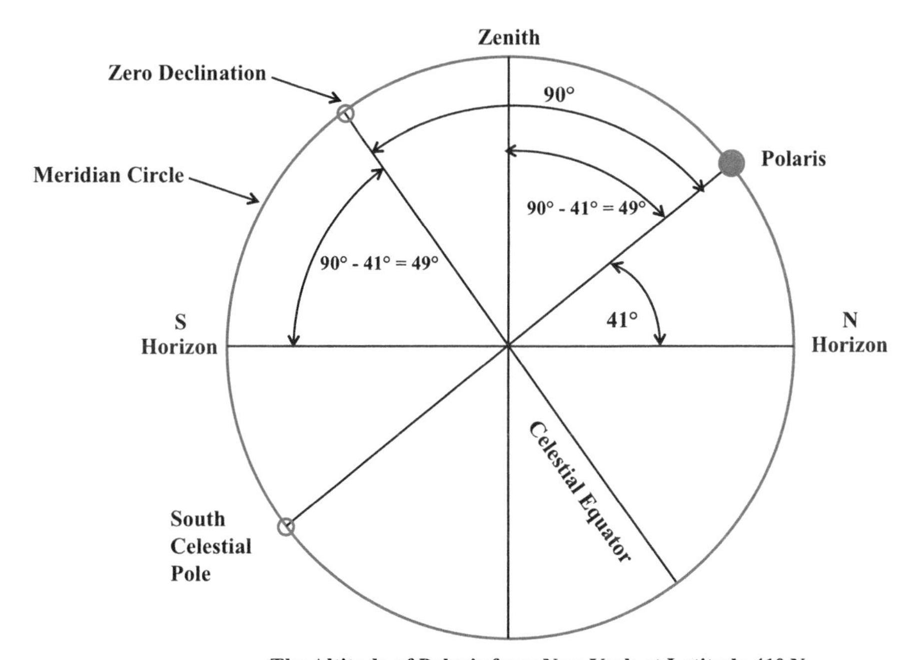

Fig. 1.12 The celestial equator and meridian

The Meridian, Altitude and Azimuth

Pretend that you are in your favorite observing location on a dark, clear night, and that this location is in the northern hemisphere, facing south. Now imagine a hoop that emerges from the southern horizon in front of you, rises vertically in the sky, then curves over your head and down behind you to the northern horizon. This is a fixed imaginary hoop -- the **meridian.** It is always there and does not move. The meridian (Figure 1.12) passes through the north polar point close to the star Polaris.

Still looking due south, as the earth rotates from west to east, the stars in the sky appear to move from east to west, approaching, then crossing this hoop known as the meridian. Any object in the sky will always be at its highest as it crosses the meridian line. The height of a star or other object above the horizon is its **altitude**. The other term that you should be familiar with is **azimuth**. If you stand looking north, a celestial object to the north will have an azimuth angle of 0°. If you turn to the right from your north-facing position, the azimuth angle increases up to 90° as you face east, then goes up to 180° as you face south, 270° as you face west and then back to 360°, or 0°, as you return to north. This is illustrated in Figure 1.13.

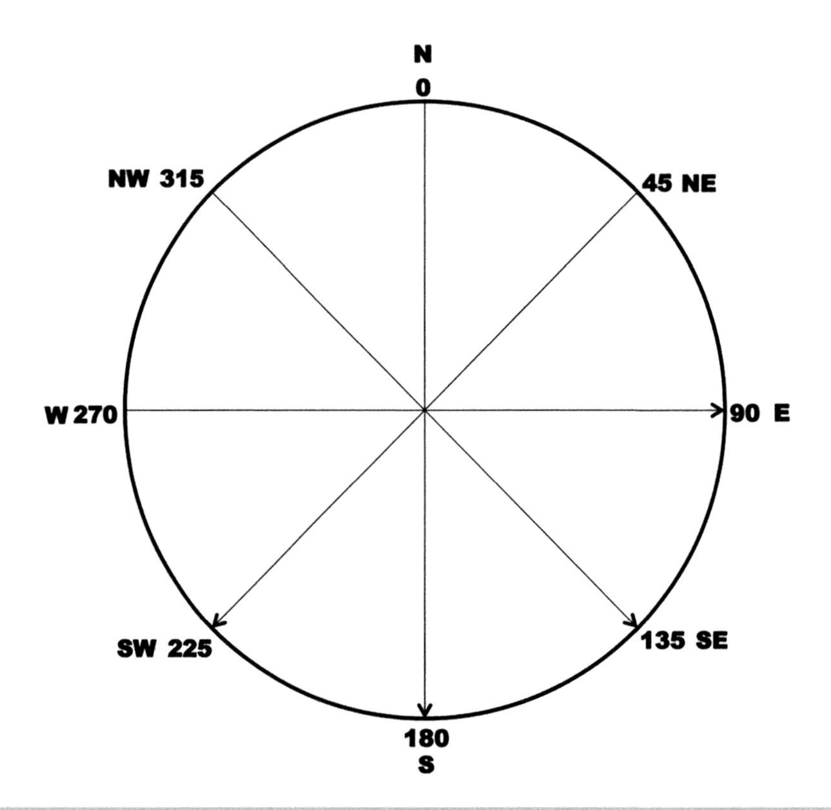

Fig. 1.13 The azimuth circle

Declination

The position of a celestial body can be indicated by means of its azimuth and altitude. The problem with this is that the altitude and azimuth of the star constantly change as the object rises in the east, moves to the west, crosses the meridian at its highest altitude (when its azimuth angle is 180°) and then drops back down in the west. Some objects are circumpolar, so they never set.

On Earth, we have Latitude and Longitude to specify the exact position of a city or town. The equivalent of Latitude in the sky is **declination**. The zero point for Latitude is the earth's equator, and in the sky, the zero point for declination is the celestial equator. Thus, declination is 0° at the celestial equator and goes up to 90° at the north polar point. This direction is regarded as positive, so, for example, the declination of Messier 31 would usually be written +41° 21' 38", or it may be shown as 41° 21' 38" North. If you need to point an equatorial telescope at an object in the sky, this is one value you will need to know. M31 was chosen randomly, but referring back to the New York diagram in Figure 1.12, you will notice that coincidentally, the Latitude of New York and the declination of Messier 31 are both around 41°. When this happens -- that is, the declination and the Latitude of the observing location are the same -- the object will always pass overhead through the zenith as shown in Figure 1.14.

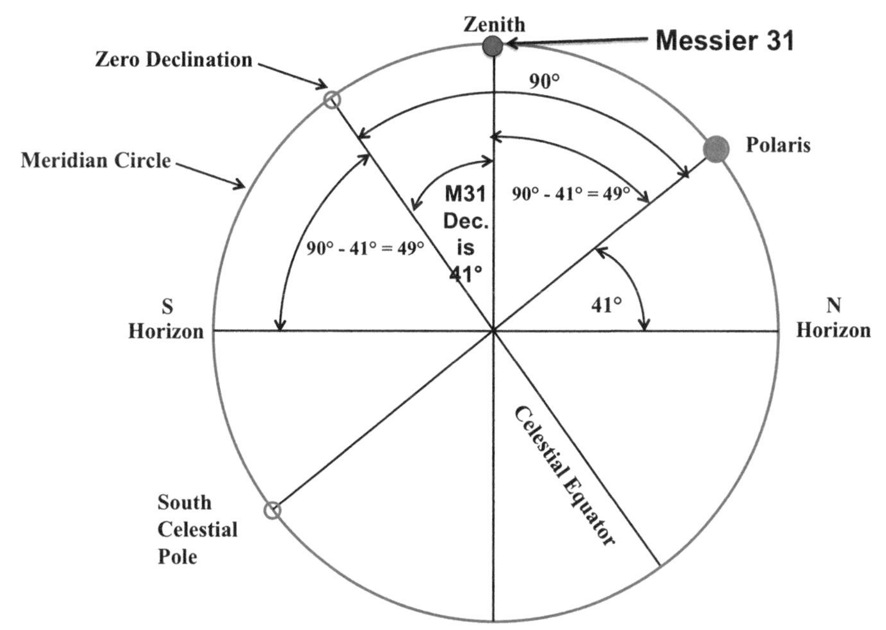

Fig. 1.14 M31 from New York

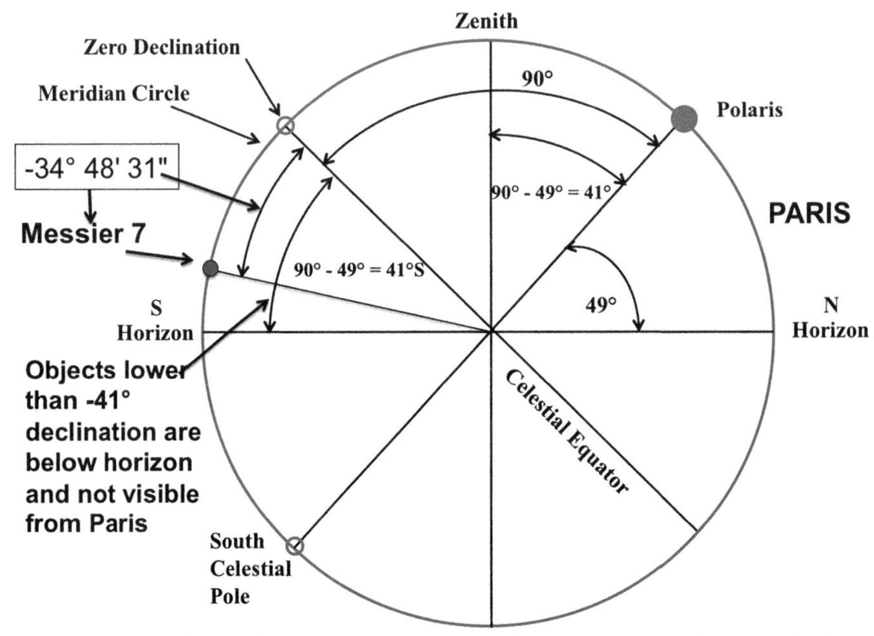

Fig. 1.15 Messier 7 from Paris

Objects that are below the celestial equator are given negative declinations. For example, M7, which is the most southerly of the Messier objects, has a declination of -34° 48' 31", or 34° 48' 31" S. Charles Messier was observing from Paris with a Latitude of 49°, so a large part of the southern sky was not available to him. Of course, in taking the image of M7 for this book, I cheated slightly by using a telescope based in Australia! Figure 1.15 illustrates how M7 is situated for a Paris observer.

So which objects will be circumpolar from a particular location? Using the example of New York, Figure 1.16 illustrates this case. If you visualize yourself standing and facing north, the stars will rotate in a counterclockwise direction around Polaris. Some will make the full rotation and still be visible the entire time, while others will drop down below the northern horizon. It is 41° down from Polaris to the stars just being cut off, and those stars will rotate 41° up 12 hours later. The angle from the Celestial Equator to those stars will be 49°, the minimum declination required to be circumpolar.

The Ecliptic, Equinoxes and Right Ascension

Declination is one thing we need to know when identifying the position of an object in the Sky. On Earth, we refer to the Latitude of a town or city and its **Longitude**. Longitude has a zero point at Greenwich, a value that is now accepted worldwide (it was not always so). What we need is a "Greenwich in the sky."

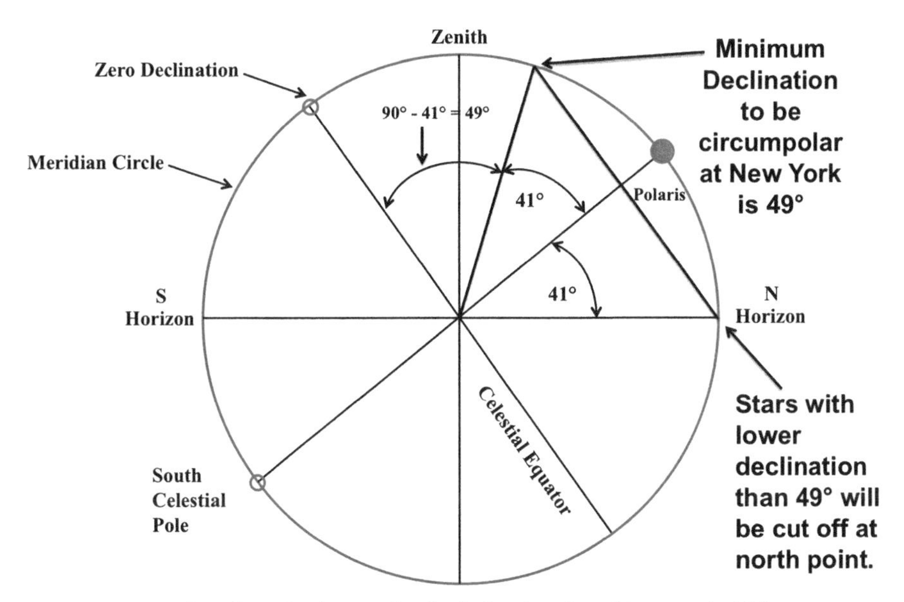

Fig. 1.16 Which objects will be circumpolar?

We have to imagine that we are inside a celestial sphere on which all of the objects in the sky are positioned. The zero point that was chosen for the celestial equivalent of Longitude was the Vernal Equinox – the point where the Sun crosses the celestial equator from south to north in the spring. The path of the Sun and planets is called the **ecliptic**. This crosses the celestial equator twice, corresponding to the Sun crossing it in a northerly direction in spring (**Vernal Equinox**) and a southerly direction in autumn (**Autumnal Equinox**). **Right Ascension** (RA) is the equivalent of Longitude in the sky but uses hours rather than degrees and increases from the Vernal Equinox towards the east. It increases in hourly increments equivalent to 15° per hour until it returns to the start at 24 hours, or 360°.

The First Point of Aries, Precession, Epochs and Local Sidereal Time

When Hipparchus of Rhodes was investigating the equinoxes, the position of the sun's south to north crossing was in the zodiacal constellation of Aries at the time, but he discovered it was slowly shifting westwards. This is because the earth spins on its axis rather like a wobbling top. It goes through one cycle in about 26,000 years. This is called **precession**, so we refer to the "precession of the equinoxes." The zero point for RA is the **First Point of Aries**. Of course, this zero point keeps moving, so RA changes slowly. Declination (Dec) also changes. In terms of objects in the sky, the RA and Dec will change with time. Specific **Epochs** are defined so that a star map can be drawn with RA and Dec axes just like Longitude and Latitude, but *for a specific date*. The current Epoch is for the year 2000 and is called "J2000," with the J referring to the Julian calendar. The "First point of Aries" is now located in the constellation of Pisces! Note that the RA and Dec given for each Messier object in this book is for the date and time that the image was taken. This will differ from J2000 coordinates.

As the earth rotates, the sky moves from east to west and objects cross the meridian. At any particular time, there will be a specific RA on the meridian. This is what is known as the **Local Sidereal Time,** or LST. Knowing the RA of a particular object and the Local Sidereal Time means that you can work out when that object will cross the meridian and be at its highest altitude in the sky. Messier 31 has an RA of 00h 43m 41s, so if the LST is 23h 43m 41s, the RA on the meridian has that value, and it will be 1 hour exactly until M31 crosses the meridian. If the LST is 4h 43m 41s, then M31 crossed the meridian 4 hours ago. Your planetarium program will tell you the LST for the particular local time that it is set to at a particular location.

Figure 1.17 shows the situation in the northern hemisphere with east to the left and west to the right. North is up.

The celestial equator crosses the chart at the vertical center, with its declination of 0° (+00) marked. The chart goes down to roughly -20° declination and up to +20° declination. In the center of the chart is marked the Right Ascension figure of 00h. As referred to previously, RA increases to the east (left) going up to 02h on the chart. On the west (right) side of the chart, you can see the hours of 22h, 23h and then 24h, which is the same as 00h. The Ecliptic crosses the chart diagonally -- imagine the

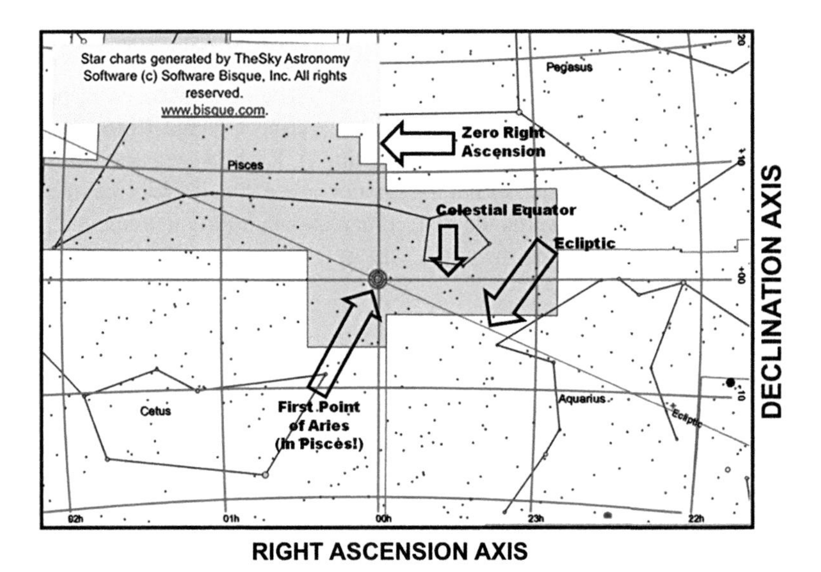

Fig. 1.17 Right ascension and declination

Sun in the sky sliding up the ecliptic, then crossing the center point of the chart in the spring. There will be a gradual westward shift of the point from east to west (left to right) as the years progress.

Practical Considerations

Plate-Solving

Good Planetarium software is essential for planning your remote telescope sessions. Further, software that includes a plate-solving facility adds another very useful dimension to your imaging. The word "plate," which originates from the period when astrophotography involved the use of glass plates, simply refers to the image. You will find the expression "solving the plate" used frequently throughout the book in the descriptions of the imaging and object identification process. Put simply, plate solving involves matching the stars in your image with a star chart through the use of software.

Once a successful match is obtained, the plate is "solved," and the software superimposes the image on top of the star chart. Your image has effectively become a star chart! For example, you can click on that fuzzy blob on your image and the planetarium software might inform you that it is a remote galaxy and even identify which one for you. This can be especially useful when you find an object that was not present on previous images you had taken of the same field. Say you spot an unidentified point-like object or apparent 'star' that happens to be in the field of view of your plate solved image. When you click on it, the software might identify it as a known minor planet or asteroid. (This assumes that the appropriate minor planet database is loaded into your software). This did happen while imaging two Messier objects for this project (See Messier 23 and the minor planet 346 Hermentaria, and Messier 105 and the minor planet MPL 26 or Prosperpina). Of course it could also be a previously unknown minor planet, a new nova, supernova or other 'transient' object that the software would not be able to identify.

Note that for the software to identify an asteroid (for example), you would need to make sure that a minor planet database has been loaded into the software. New Minor Planets are always being discovered so it is important to make sure that you have uploaded the latest verion of the database. All good planetarium software has a facility for loading databases of stars, galaxies, nebulae, double stars, minor planets and comets. In this case, using the SkyX software allowed each of the 110 Messier object images to be plate-solved in about 3 seconds after initiating the command. When an image is plate-solved in SkyX, there is a click and a locked padlock icon in the corner of the image appears, signifying that the image and chart are locked together, meaning that the 'plate' has been solved. Dragging the chart around the screen will now drag the image with it. You can click on the image to identify objects that have been recognized by the plate solution. The chart for each Messier image includes the "padlocked" solved image.

FITS and TIFF Images

All the cameras used to image the Messier objects save images in the FITS (Flexible Image Transport System) format. This is a specialized format for astronomical images, where **metadata** is saved with the image listing the time the image was started, the date, the resolution and other useful information. Thus, once you have the image, you automatically have all the relevant information stored with it.

The itelescope network also offers the option to save images as TIFF files in addition to FITS files. A preview JPEG file is emailed to you very shortly after the image is taken, which you can check immediately. If the image outcome is unsatisfactory, there is a procedure on the website for asking for a points refund. If it is satisfactory, however, you may go to the website and log in to download all the relevant files to your laptop. It is easiest to zip all of them up using the online zip facility before downloading them. For this book, I saved images in FITS and TIFF format, then use Photoshop to open the TIFF image and process it for the main image. Photoshop allows easy cropping of an image when necessary. I would open the FITS image in the SkyX software, and a simple instruction "Set Chart to Photo Time" adjusted the charts to the time the image was started. This means that the chart will automatically show the exact sky view at the time the image was taken, including the location of neighboring minor planets etc. at that time. Once opened, the image could be plate-solved, locking the image to the chart which is set to the time the image was taken.

Camera Position and Position Angle

The way the camera is attached to the telescope varies between different telescopes in the network. The images taken for the book are all displayed in landscape format, but that does not mean north is at the top. **Position angle** is measured clockwise from the north. For example, the image of M1 has a position angle of 88° 11' from the north. This means that the image has to be rotated by 88° 11' counter-clockwise to get north at the top. So, in the original image position, north was towards the right and east was towards the top. When the image is plate-solved, it automatically determines the position angle by checking the image against the chart.

Another example is the image of M77, which has a position angle of 270° 10' from North. This means that the image has to be rotated by 89° 50' clockwise to get north at the top (or by 270° 10' counterclockwise). In the original image position, north was towards the left and west was towards the top. The camera used to take the image of M77 had a rectangular chip so that, when rotated, the embedded image shown on the M77 pages would change to (approximate) portrait format. Many cameras have square chips. The position angle for each image is supplied in the data on the appropriate Messier pages. Planetarium software includes tools to allow rotation of the chart.

Astronomical Considerations

Stellar Magnitudes

Stars are given a magnitude to represent their brightness, with the faintest stars that can be seen with the naked eye having a magnitude of 6 or so, and the brightest stars around magnitude 1. A first magnitude star is defined as being exactly one hundred times as bright as a sixth magnitude star. This means that a difference of one magnitude between any two stars indicates that one star is brighter than the other by roughly two and a half times. For example, the star Altair in the constellation of Aquila has a magnitude of 0.76, which means that it is brighter than a first magnitude star and is thus one of the brightest stars in the sky. Most stars featured in the Messier image pages tend to have magnitudes that are too faint to be seen with the naked eye, usually in the range of magnitude 6 to about magnitude 18.

Spectral Types of Stars

At various times throughout the book I will refer to Spectral Types. Stars have a range of colors and temperatures. For our purposes, think of a sequence of stars ranging from the hottest stars to the coolest. A series of letters is used to represent this, but not as you might imagine as A, B C, etc. For historical reasons that I will not go into, the sequence is O B A F G K M, where O stars are the hottest and M stars are the coolest, representing different ‘spectral types’ to be explained shortly. The standard mnemonic device used by astronomers to remember this order is “Oh Be A Fine Girl/Guy Kiss Me,” but feel free to make up your own!

O stars are the hottest, B stars are still hot but cooler than O stars, and so on down to M stars, which are much cooler. Additionally, each letter in the series represents a different color determined by the star’s temperature. O stars are Blue – for example, Mintaka, the star at the western side of Orion’s Belt. The star Rigel, also in Orion, is a blue-white B-type star. Sirius and Vega are both white A stars. F stars are yellow-white. The star Procyon in the constellation of Canis Minor is an F-type star. G-type stars are yellow. The Sun is a G-type star, and so is Capella in the constellation of Auriga. K stars are orange. Epsilon Eridani and 61 Cygni are examples of a K-type star. M stars are red-orange. The star Gliese 581 in the constellation of Libra is an M-type star.

The colors quoted are approximate. To specify the spectral type more precisely, the basic type is broken down further into a range of 0 to 9. This was established by Annie Jump Cannon, a researcher for Edward C. Pickering at Harvard who worked on a spectral classification project funded by the estate of the (amateur) astronomer Henry Draper. A remote telescope in Siding Spring is dedicated to the memory of Annie Jump Cannon (along with Jocelyn Bell-Burnell and Henrietta Swan Leavitt). This telescope was used to take the images of M46 and M47 that appear in this book. The system she developed designates spectral types from A0 to A9, M0 to M9 etc. An A1 star is cooler than an A0 star.

Table 1.1 The Spectral Classes of Stars

Spectral Class	Color	Temperature (Kelvin)	Example	Color Index B-V
B	Blue-White	10000-28000	Spica	-0.2
A	White	7500-10000	Sirius	0
F	Yellow-White	6000-7500	Procyon	0.4
G	Yellow	5000-6000	Capella	0.8
K	Orange	3500-5000	Arcturus	1.23
M	Red-Orange	2500-3500	Betelgeuse	1.85

You may be wondering why the term 'spectral type' is used. The spectral absorption lines in a star gradually change as you move through the spectral types from O to M and from A0 to A9 etc. Annie Jump Cannon analyzed thousands of stellar spectra to allocate a particular spectral type to each star. She used the strength of the Balmer Absorption Lines to determine a star's spectral class. Table 1.1 shows a range of spectral types matched against temperature and color.

Object Size

When examining the size of the Messier objects that appear throughout the book, some simple calculations can be of use. If you know the apparent size of an object in arcmin and how far away it is in light-years (ly), you can calculate the actual size of the object in light-years with some simple geometry.

As an example, look at the image of M57 in Figure 1.18. My planetarium software has a tool that measures angular separation and size in minutes of arc (') on an image that I have plate solved. With M57, the major axis size was measured at 1.4'. M57 lies at a distance of 2300 ly. Knowing the angle of 1.4' and the distance, the size of M57 can be calculated to be roughly 0.9 ly. The calculation is shown in Figure 1.18.

The Minor Axis has an angular size of 1'. A similar calculation shows the actual size of the Minor Axis to be roughly 0.7 ly. This **small angle formula** is used to calculate size for a number of the Messier objects throughout the book.

An Example Imaging Session using iTelescope.net

Taking an image can be very simple. This is a short example of what you will need to do, up until you receive a preview of the image. Refer to the itelescope website to see the full procedure.

Once you have logged in and found an available telescope, enter the target object and the exposure time. As an example, I logged in and found that the single-shot color camera/telescope T13 in Siding Spring was available, so I put in the necessary information, as you can see in Figure 1.19.

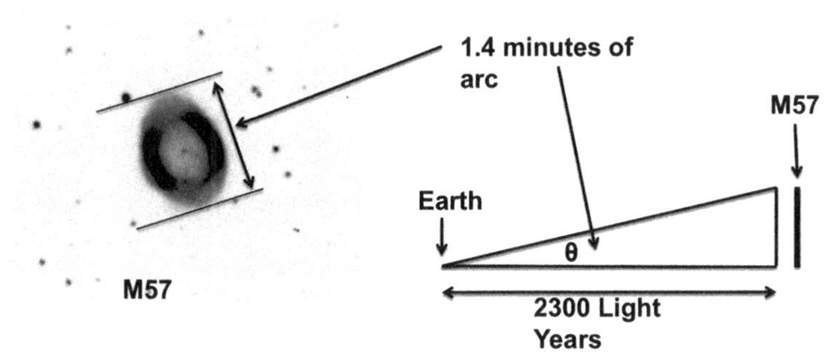

Fig. 1.18 Calculating the size of M57 in light-years

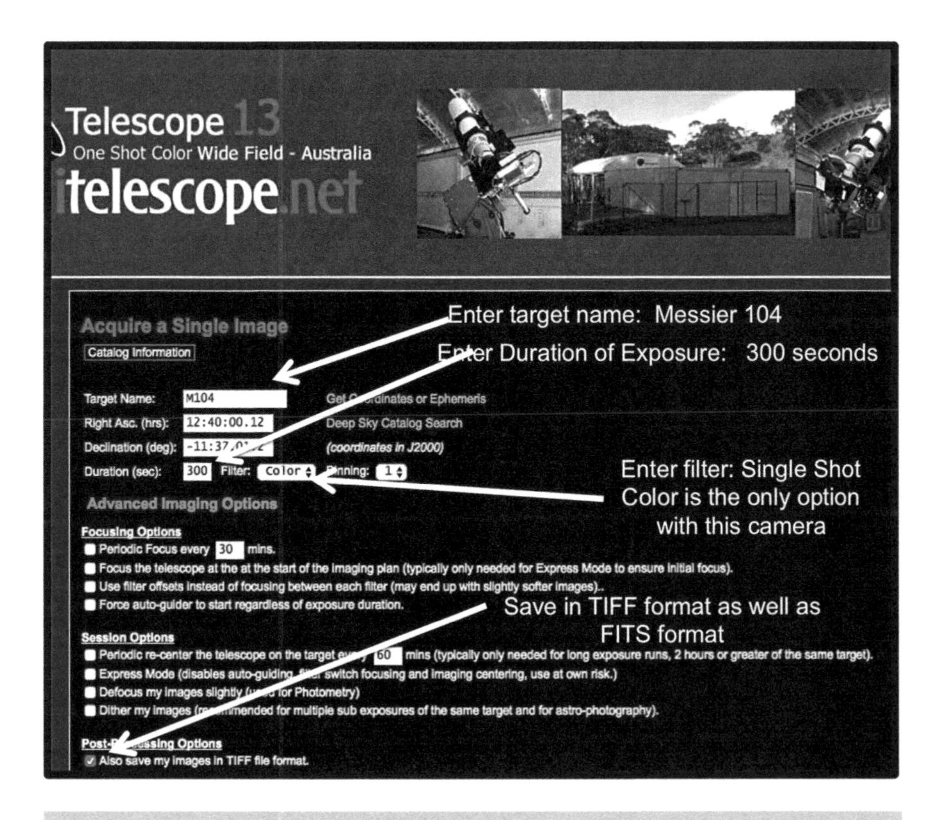

Fig. 1.19 Entering information to take a single image. Courtesy of iTelescope.net

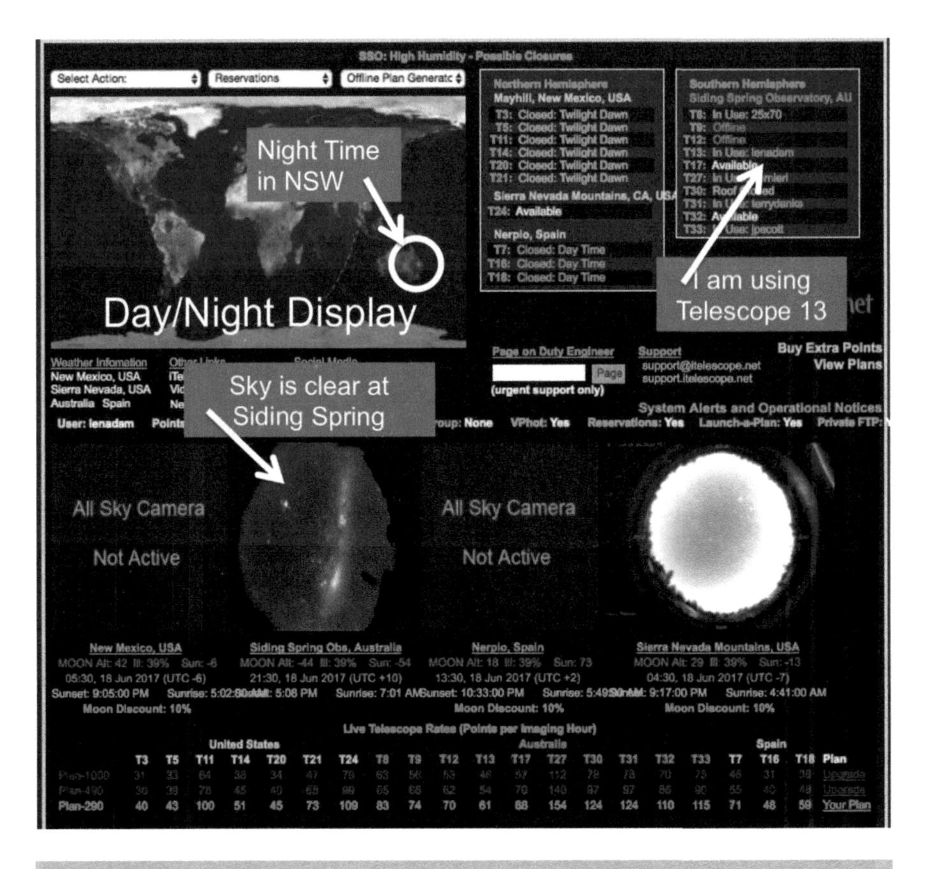

Fig. 1.20 The Launchpad view as I am taking an image. Courtesy of iTelescope.net

I entered a target, Messier 104, clicked the "Get Coordinates" text, then entered the exposure time of 300 seconds and put in a request for it to be saved as a TIFF file as well as a FITS file. I then clicked "Acquire Image" to start the exposure. Then, I checked on the "Launchpad," shown in Figure 1.20.

The Day/Night display shows the current position of the Sun and the location of the telescopes on three continents. Each observatory has a fisheye lens view of the sky so that you can check if it is clear. You can see from the list of telescopes that I am using telescope 13 and that Telescopes 17 and 31 are available for use.

The 5-minute exposure ran and completed, and a few minutes later, I received an email telling me that the image had been taken:

"Images Taken (First 15 Images Only):

Image 1: T13-lenadam-M104-20170618-213536-Color-BIN1-W-300-001.fit (300 seconds)

iTelescope.Net

Advancing Your Horizons in Astronomy"

A JPEG preview image was attached to the email, as shown in Figure 1.21.

Fig. 1.21 Example of Messier 104 preview image

References and Databases

- The SEDS website was used as a major reference for basic information about the Messier objects and their history of discovery. http://messier.seds.org/m/mindex.html
- The Hipparcos Stellar Catalog is integrated into the SkyX software and is useful because it provides stellar magnitudes, spectral types and distances to stars. Hipparcos was a satellite launched in 1989 to obtain Astrometric data. The Wikipedia page is: https://en.wikipedia.org/wiki/Hipparcos
- The Tycho Catalog is also integrated into the SkyX software and was obtained from the same Hipparcos Satellite as a separate project. Wikipedia Page: https://en.wikipedia.org/wiki/Tycho-2_Catalogue . You should be able to download both the Hipparcos and Tycho Catalogs into your planetarium software.
- A number of stars have been identified using the UCAC4 catalog. http://dc.zah.uni-heidelberg.de/ucac4/q/s/info
- The New General Catalog (NGC) and Index Catalog (IC) were used extensively within the SkyX software to obtain information on all of the Messier objects which all have NGC numbers and incidental objects that were in the field of view of the images. https://en.wikipedia.org/wiki/New_General_Catalogue

- A number of distant galaxies were identified in the fields of view of some of the Messier objects by using the integrated SkyX Principal Galaxy Catalog (PGC). https://en.wikipedia.org/wiki/Principal_Galaxies_Catalogue
- Occasional reference was made to this leading source of double and multiple stars. Washington Double Star Catalog – integrated into the SkyX. http://www.usno.navy.mil/USNO/astrometry/optical-IR-prod/wds/WDS
- This is a very useful online source of data on star clusters. WEBDA. https://www.univie.ac.at/webda/

Chapter 2

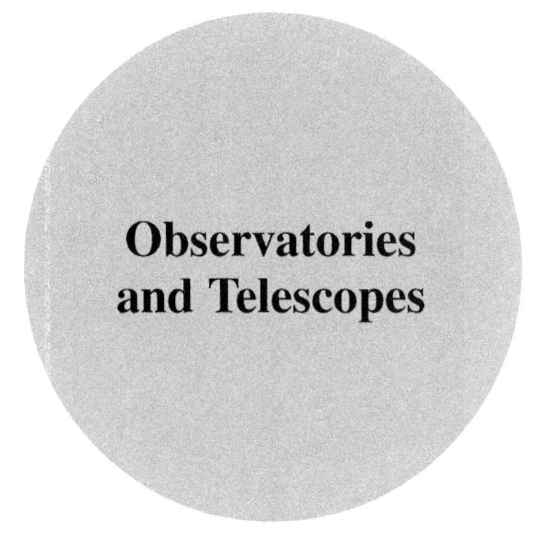

Observatories and Telescopes

Telescope Summary

Three telescopes were used to image the Messier objects in Spain, seven in Australia, six in New Mexico and one in California. The smallest telescope (Siding Spring B, or T13), located in Australia, has an aperture of 3 inches, while the largest (Siding Spring D or T27), also in Australia, has an aperture of 27.5 inches. Ironically, the smallest telescope T13 was used to image the largest number of Messier objects, simply because its single-shot color camera makes it easy to capture color images in a very short time. All 17 telescopes are listed in Table 2.1. Each of the 17 telescopes is detailed with its iTelescope.net reference number and the number of Messier objects imaged for this project using that specific telescope.

Spain

Spain Telescope A (T7): A 17-inch (0.43m) aperture f/6.8 Corrected Dall-Kirkham Astrograph. It carries a large format SBIG STL - 11000M CCD.

Spain Telescope B (T16): A 6-inch (150mm) aperture f/7.3 Apochromatic Refractor. The camera is an SBIG STL-11000M CCD.

Spain Telescope C (T18): A a 12½ -inch (318mm) aperture f/7.9 Corrected Dall-Kirkham Astrograph with a KAF-6303E camera.

L. Adam, *Imaging the Messier Objects Remotely from Your Laptop*, The Patrick Moore Practical Astronomy Series, https://doi.org/10.1007/978-3-319-65385-3_2

Table 2.1 The 17 telescopes used in this book and the number of Messier objects imaged by each

Telescope	iTelescope Reference	Telescope Details	Number of Images	Telescope Manufacturer
SPAIN A	T7	17-inch (0.43 m) aperture f/6.8 Corrected Dall-Kirkham Astrograph	5	Planewave
SPAIN B	T16	6-inch (150 mm) aperture f/7.3 Apochromatic Refractor	2	Takahashi
SPAIN C	T18	12½ -inch (318 mm) aperture f/7.9 Corrected Dall-Kirkham Astrograph	4	Planewave
SIDING SPRING A	T8	4-inch (106 mm) aperture f/5 Petzval Apochromatic Astrograph	2	Takahashi
SIDING SPRING B	T13	3-inch (90 mm) aperture f/5.6 Apochromatic Refractor	23	Takahashi
SIDING SPRING C	T17	17-inch (431 mm) f/6.8 Corrected Dall Kirkham Astrograph	1	Planewave
SIDING SPRING D	T27	27.5–inch (700 mm) f/6.6 Corrected Dall-Kirkham Astrograph	2	Planewave
SIDING SPRING E	T31	20–inch (510 mm) f/4.4 Corrected Dall-Kirkham Astrograph	3	Planewave
SIDING SPRING F	T32	17-inch (431 mm) f/6.8 Corrected Dall-Kirkham Astrograph	8	Planewave
SIDING SPRING G	T33	16-inch (400 mm) f/3.5 Fast Newtonian Astrograph	3	Astro-Systeme Austria
NEW MEXICO A	T3	6-inch (150 mm) aperture f/7.3 Apochromatic Refractor	20	Takahashi
NEW MEXICO B	T5	10-inch (250 mm) aperture f/3.4 Hyperbolic Flat-Field Astrograph	2	Takahashi
NEW MEXICO C	T11	20–inch (510 mm) f/4.5 Corrected Dall-Kirkham Astrograph	14	Planewave
NEW MEXICO D	T14	4-inch (106 mm) aperture f/5 Petzval Apochromatic Astrograph	6	Takahashi
NEW MEXICO E	T20	4-inch (106 mm) aperture f/5 Petzval Apochromatic Astrograph	10	Takahashi
NEW MEXICO F	T21	17-inch (431 mm) aperture f/4.5 Corrected Dall-Kirkham Astrograph	1	Planewave
CALIFORNIA A	T24	24-inch (610 mm) aperture f/6.5 Corrected Dall-Kirkham Astrograph	4	Planewave

Messier Objects imaged using Spain A (T7)

The Milky Way is splendid overhead at the dark sky site, located at over 5,400 ft (1,650 m) in southern Spain. This telescope was used to image five Messier objects in the constellations of Andromeda, Cygnus, Lyra and Virgo. This included two elliptical galaxies, a dwarf elliptical galaxy, a globular cluster and an open cluster. Andromeda has a declination of roughly 38°, Cygnus 42°, Lyra 36° and Virgo -3°. From the Spanish site at AstroCamp with its Latitude of 38° N, all objects in the constellations will be at a reasonable Latitude, with Virgo objects being the lowest in the sky. Figure 2.1 shows the observatory for telescope Spain A (T7) in its spectacular location at Nerpio. Table 2.2 lists the objects imaged using this 17-inch telescope.

Fig. 2.1 The Milky Way over the roll-off roof observatory at AstroCamp in Nerpio, Spain. Courtesy of Fernando Abalos

Table 2.2 Five Messier objects imaged using Telescope Spain A (T7) based at Astrocamp

Messier Object	Object Type	Object Size	Constellation
M29	Open Cluster	10' x 10'	Cygnus
M32	Dwarf Elliptical Galaxy	8' x 6'	Andromeda
M56	Globular Cluster	8.8 x 8.8	Lyra
M59	Elliptical Galaxy	5' x 3.5'	Virgo
M60	Elliptical Galaxy	7' x 6'	Virgo

This telescope has a field of view of roughly 42' x 28' so all of the objects listed in Table 2.2 fit well within the frame. The camera on this telescope has an **Anti-Blooming Gate** (**ABG**), which means that if individual pixels saturate with too much light, the gate will prevent light from spilling over into adjacent pixels. This is good for astrophotographers but bad for scientists, as the measurement of stellar brightness becomes unreliable due to the non-linearity of the relationship between incoming light and brightness on the image when the ABG is introduced. If a camera does not have an ABG, it is called an **NABG** camera. You must limit the exposure time to avoid saturation with an NABG camera. It needs to be stressed that this telescope is designed as a **science platform,** but, because of the ABG feature, exposures must be limited to avoid saturation when being used for **photometry**. Photometry means "measuring light," so if you are trying to measure the magnitude of a star, perhaps using a B or V filter, you need to make sure that it has not saturated. It may appear fine on the image, but if some of the light has been effectively drained off by the ABG and you have entered the non-linear part of the "input light" to "measured light" graph, you will get a false indication of brightness and thus of magnitude.

If you are using filters, for example the B and V filters mentioned above, you would need to take an image with a B filter and then an image with a V filter. The camera used with Spain A has a **filter wheel** with 8 filters. These filters are Luminance, R-Johnson, V-Johnson, B-Johnson, I-Johnson, Hα, OIII and SII. You can specify the filter that you wish to use on any exposure and the filter wheel will rotate to put that filter into the light path.

Calibration files (flats and darks) are taken for you and can be applied automatically or manually to your images. The calibrated and un-calibrated files are provided for you and can be downloaded from the easily accessible web-based FTP server.

As already stated above, telescope Spain A has an aperture of 17 inches. An indicator of the light grasp of a telescope is the area of the mirror or lens, which is proportional to the square of the aperture. The **aperture** is the clear diameter of the mirror or lens. In this case, the square of the aperture is 17 x 17, or 289. By comparing the square of the apertures of each telescope, you can get an idea of its light grasp. The telescope is shown in the closed observatory (center telescope) in Figure 2.2.

Messier Objects imaged using Spain B (T16)

Only two Messier objects, an open cluster and a planetary nebula, were imaged using this telescope, found respectively in the constellations of Taurus and Ursa Major. Taurus has an approximate declination of 19° and Ursa Major 55°. Spain B at a Latitude of 38° was able to capture an image of M45 when it was at an altitude of over 67° and of M97 when it was at an altitude of almost 60°. Figure 2.3 shows the Spain B (T16) telescope in its roll-off roof shed in Spain. The telescope is the

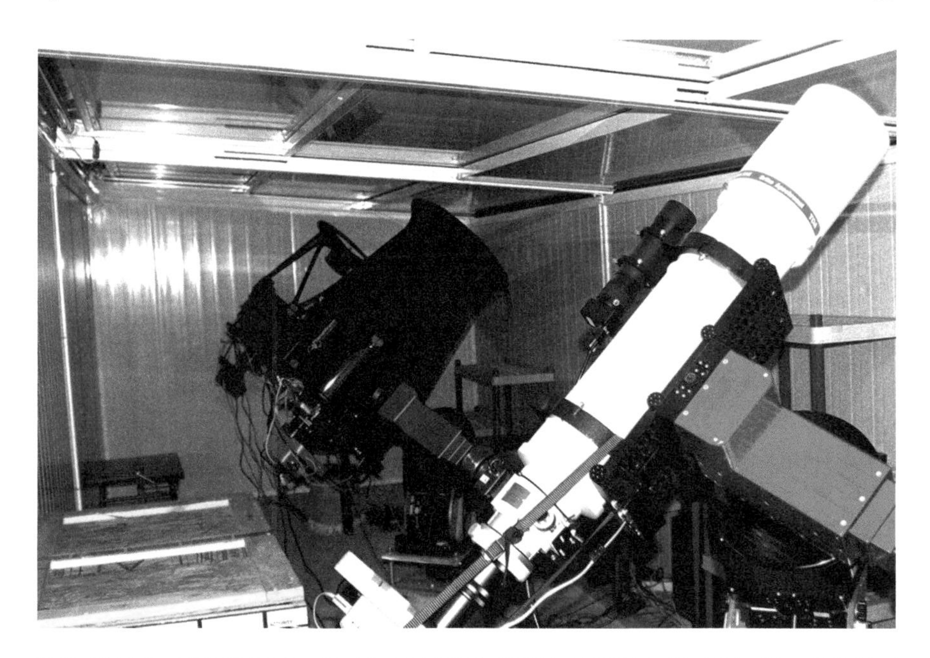

Fig. 2.2 Spain A (T7) in its closed roll-off roof observatory (center telescope). Courtesy of Fernando Abalos

Fig. 2.3 Telescope Spain B (T16) underneath the Milky Way at AstroCamp in Nerpio, Spain. Courtesy of Fernando Abalos

Table 2.3 Two Messier objects imaged using Telescope Spain B (T16) at AstroCamp

Messier Object	Object Type	Object Size	Constellation
M45	Open Cluster	110'	Taurus
M97	Planetary Nebula	3.4' x 3.3'	Ursa Major

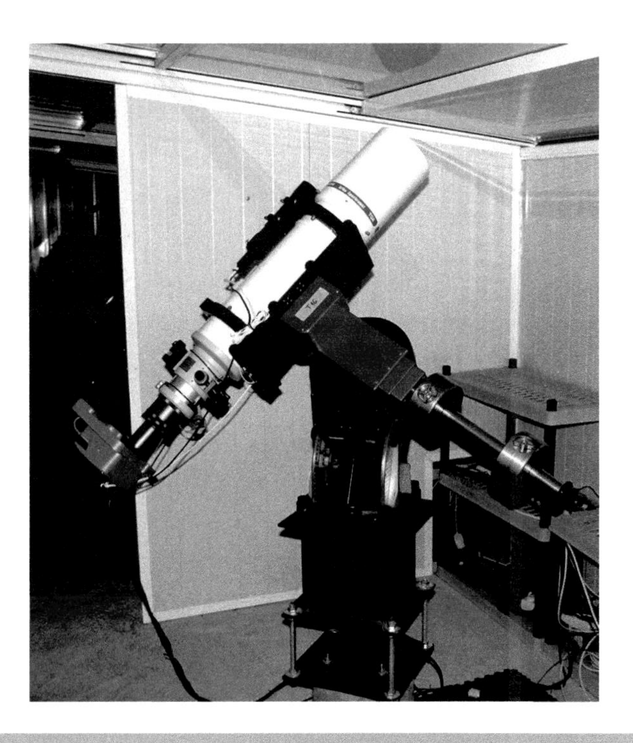

Fig. 2.4 Spain B (T16) in its roll-off roof observatory at AstroCamp with the roof closed. Courtesy of Fernando Abalos

small one just visible to the left. The details of the two Messier objects imaged with this ABG telescope are given in Table 2.3.

Figure 2.4 gives a closer view of this telescope.

The filters available for this telescope are Luminance, R-Johnson, V-Johnson, B-Johnson, I-Johnson, Hα, OIII and SII.

This telescope has an aperture of 6 inches, and the area of the objective lens is proportional to the aperture squared, which is 6 x 6 = 36. If you compare this with Spain A, you will see that Spain A has 289/36 = 8 times the light grasp of Spain B. Thus, Spain A (T7) gathers 8 times the light of Spain B (T16). The major advantage of Spain B is that it has a very wide field of view of roughly 75 x 113 minutes of arc. The comparison is shown in Figure 2.5.

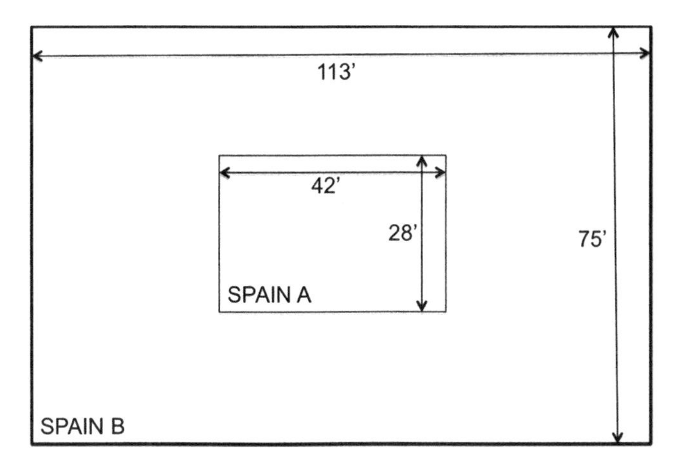

Fig. 2.5 A comparison of the Fields of View of Spain A (T7) and Spain B (T16) Telescopes

Table 2.4 The four Messier objects imaged using Telescope Spain C (T18) at AstroCamp

Messier Object	Object Type	Object Size	Constellation
M13	Globular Cluster	20' x 20'	Hercules
M51	Spiral Galaxy	11' x 7'	Canes Venatici
M57	Planetary Nebula	1.4' x 1.0	Lyra
M71	Globular Cluster	7.2' x 7.2'	Sagitta

Messier Objects imaged using Spain C (T18)

Four Messier objects were imaged using Telescope Spain C (T18). This included two globular clusters, a planetary nebula and a spiral galaxy in the constellations of Hercules, Sagitta, Lyra and Canes Venatici. Hercules has a declination of roughly 30°, Sagitta 18°, Lyra 37° and Canes Venatici 41°. The Messier objects are listed in Table 2.4.

The aperture of Spain C is 12 1/2" (318 mm), and it has a field of view of roughly 37 x 25 minutes of arc. This field of view was an excellent choice for Messier 13 with its size of 20' x 20' but was large for the Ring Nebula (M57) with its diminutive size of 1.4' x 1.0'. Spain C has just over half the light grasp of Spain A and just over 4 times the light grasp of Spain B. Figure 2.6 shows Spain C (T18) in the observatory at Siding Spring, Australia. The filters available for this telescope are AstroDon Series E LRGB, AstroDon 10nm Ha, SII, OIII, and Johnsons-Cousins V. These filters can obtain color images by combining images taken through the LRGB filters and V magnitudes through the photometric V filter.

A diagram illustrating the relative fields of view of all the Spain telescopes is given in Figure 2.7.

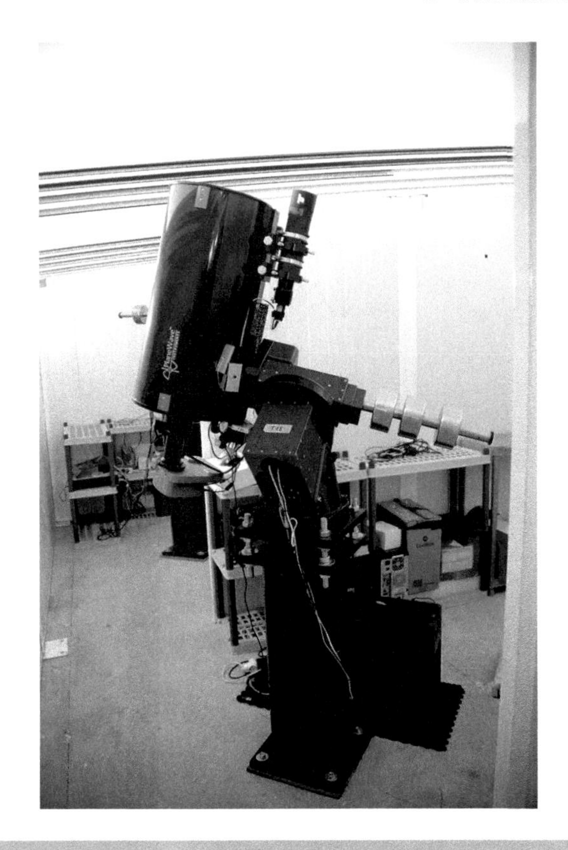

Fig. 2.6 Telescope Spain C (T18) at AstroCamp. Courtesy of Fernando Abalos

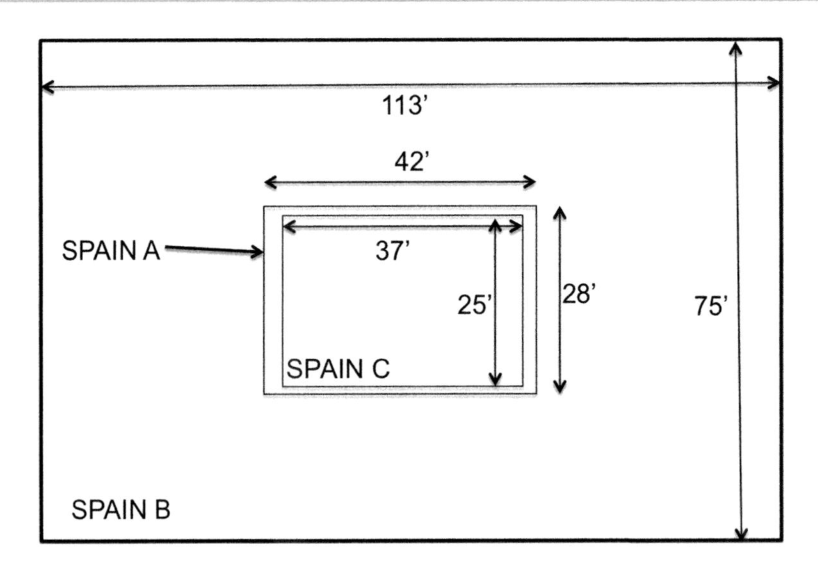

Fig. 2.7 Fields of view of the three telescopes used at Astrocamp

The fields of view of Spain A and Spain C are similar, but Spain A has a higher light grasp with its greater aperture. Bear in mind that the position angle of the camera has not been shown in Figure 2.7, so in reality the actual fields of view will be tilted with respect to north to varying degrees depending on how the camera has been attached to the telescope. In Chapters 2, 3 and 4, each Messier image has associated data that informs you of the actual camera position angle when the image was taken, determined by comparing the image with the actual position of north.

Australia

Seven telescopes were used from the Siding Spring site in New South Wales, Australia. These telescopes are:

Siding Spring A (T8): A 4-inch (106 mm) aperture f/5 Petzval Apochromatic Astrograph with an FLI MicroLine 16803 CCD camera.

Siding Spring B (T13): A 3-inch (90 mm) aperture f/5.6 Apochromatic Refractor with an SBIG ST2000XMC camera.

Siding Spring C (T17): A 17-inch (431 mm) f/6.8 Corrected Dall Kirkham Astrograph with an FLI Proline PL4710 camera.

Siding Spring D (T27): A 27–Inch (700 mm) f/6.6 Corrected Dall-Kirkham Astrograph with an FLI PL09000 CCD camera.

Siding Spring E (T31): A 20–inch (510 mm) f/4.4 Corrected Dall-Kirkham Astrograph with an FLI PL09000 CCD camera.

Siding Spring F (T32): A 17-inch (431 mm) f/6.8 Corrected Dall-Kirkham Astrograph with an FLI Proline 16803 camera.

Siding Spring G (T33): A 16-inch (400 mm) f/3.5 Fast Newtonian Astrograph with an Apogee Aspen CG16070 Class 1 CCD.

Messier Objects imaged using Siding Spring A (T8)

Only two Messier objects were imaged using Siding Spring A (SSA), Messier 24 and Messier 80. Messier 24 is a star cloud and Messier 80 is a globular cluster, which are very different celestial objects, both formed from stars but in very different ways. M24 is in the constellation of Sagittarius and M80 in Scorpius, both southern constellations. Sagittarius has an approximate declination of almost -26° and Scorpius roughly -37°. All of the M objects were observed by Messier from Paris, so these objects were in his far southern sky. Siding Spring, however, with its Latitude of roughly 31° south had no problem with imaging both M24 and M80 above 70° altitude. Details of the objects images with this 4-inch telescope are given in Table 2.5. This telescope uses a camera equipped with an anti-blooming

Table 2.5 Two Messier objects imaged using Telescope SSA (T8) from Siding Spring in New South Wales, Australia

Messier Object	Object Type	Object Size	Constellation
M24	Star Cloud	90'	Sagittarius
M80	Globular Cluster	10' x 10'	Scorpius

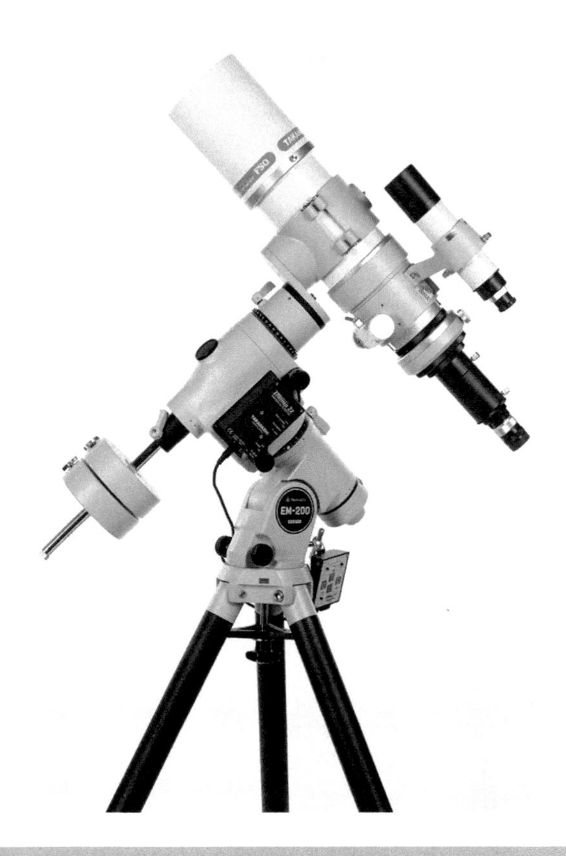

Fig. 2.8 The Takahashi FSQ 106, as used in telescope SSA (T8), set up for visual use. Image Courtesy of Takahashi US

gate. SSA (T8) has an extremely wide field of view of is 238.8' x 238.8'. The Takahashi FSQ telescope is shown in Figure 2.8 in the manufacturer's image. Note that SSA (T8) incorporates one of these telescopes mounted on a Paramount ME, manufactured by Software Bisque.

The filters provided with SSA are AstroDon LRGB, 5nmHA, 5nmSII, 5nmOIII, 5nmNIR, and ExoPlanet Filter.

Messier Objects imaged using Siding Spring B (T13)

As you may have noticed in Table 2.1, Siding Spring B (SSB or T13) was used to image more objects than any other in the project. The reasons are 1) that the telescope is single-shot color, which means it captures all of the information necessary to produce a color image in a single exposure, and 2) it has a wide field of view of 97.6’ x 73.2’. The constellations containing the 23 objects were Auriga at almost +42° declination, Coma Berenices at +22°, Orion at +4.5°, Ophiuchus at -4°, Virgo at -4°, Aquarius at -10°, Cetus at - 11°, Hydra at 19°, Capricornus at -20°, Lepus at -20°, Canis Major at -22°, Sagittarius at -26°, Puppis at -39°, and Scorpius at -37°. Most of these are below the celestial equator apart from Auriga, Coma Berenices and Orion, using declinations at roughly the center of each constellation. Orion, for example, straddles the celestial equator.

This telescope imaged eight globular clusters, seven open clusters, three star-forming nebulae, one elliptical galaxy, three spiral galaxies and one asterism. Details are given in Table 2.6.

Table 2.6 The 23 Messier objects imaged by the single-shot color Telescope SSB

Messier Object	Object Type	Object Size	Constellation
M4	Globular Cluster	36' x 36'	Scorpius
M6	Open Cluster	25' x 25'	Scorpius
M7	Open Cluster	80' x 80'	Scorpius
M8	Star-forming Nebula	90' x 40'	Sagittarius
M20	Star-forming Nebula	28' x 28'	Sagittarius
M21	Open Cluster	13' x 13'	Sagittarius
M25	Open Cluster	32' x 32'	Sagittarius
M28	Globular Cluster	11.2' x 11.2'	Sagittarius
M30	Globular Cluster	12' x 12'	Capricornus
M37	Open Cluster	24' x 24'	Auriga
M41	Open Cluster	38' x 38'	Canis Major
M42	Star-forming Nebula	85' x 60'	Orion
M49	Elliptical Galaxy	9' x 7.5'	Virgo
M61	Spiral Galaxy	6' x 5.5'	Virgo
M62	Globular Cluster	15' x 15'	Ophiuchus
M68	Globular Cluster	11' x 11'	Hydra
M70	Globular Cluster	8' x 8'	Sagittarius
M72	Globular Cluster	6.6' x 6.6'	Aquarius
M73	Asterism	2.8'	Aquarius
M77	Spiral Galaxy	7' x 6'	Cetus
M79	Globular Cluster	9.6' x 9.6'	Lepus
M88	Spiral Galaxy	7' x 4'	Coma Berenices
M93	Open Cluster	22' x 22'	Puppis

Messier Objects imaged using Telescope SSC (T17)

Only one image was taken through this telescope, as you can see in Table 2.7. It is an excellent telescope, but its camera does not have an anti- blooming gate (NABG), as it is designed for scientific work. Exposures need to be limited to 300 seconds. The camera on this telescope has extended red sensitivity (near-infrared). Because of its extreme sensitivity, it is not recommended for general imaging, but it has been used to image some of the faintest and most distant objects ever detected by amateur astronomers, holding multiple world records for this. Dr. Christian Sasse used this telescope to image the distant quasar 'J1148+5251' taking 16 hours of exposures (199 x 300s). The quasar lies at a distance of 12.79 billion years and was the furthest object ever observed when it was originally discovered in 2003. Dr. Sasse points out that, using this telescope, amateurs can reach magnitudes fainter than 23!

The object imaged in this project, Messier 54, is a globular cluster in Sagittarius but surprisingly is not within our own galaxy, as explained in Chapter 4. The field of view of this telescope is 16.1' x 15.7', so M54 fitted well within this. The effect of not having an ABG can be seen in the star at the bottom right of the image obtained, where distinct vertical blooming is evident. (M54 – Chapter 4). The filters available with this telescope are AstroDon Clear, LRGB, UVBRI, 6 nm Ha, OIII, SIII (953.1/10 nm) and Helium HEII (1012.2/10 nm).

A manufacturer's image of the Optical Tube Assembly used in SSC is shown in Figure 2.9.

Table 2.7 The single Messier object imaged with telescope SSC (T17)

Messier Object	Object Type	Object Size	Constellation
M54	Globular Cluster	12' x 12'	Sagittarius

Fig. 2.9 The Planewave 17-inch telescope as used in SSC (T17). Image Courtesy of Planewave Instruments

Messier Objects imaged using Telescope SSD (T27)

Two Messier objects M46 and M47, both open clusters, were imaged using this ABG telescope Both lie in the constellation of Puppis, which has an approximate declination of -39.5°. This is a big telescope with an aperture of 27.5 inches (0.7 m) and is the largest telescope used for this project. It can be used for imaging or for photometry with the filters AstroDon LRGB, Ha, SII, OIII, V, Rs, Ic available in the filter wheel. The field of view of SSD is 27.8 x 27.8 arcmin. M46 just squeezes into the frame, while M47 is not fully imaged with its slighty larger size of 30 arcmin (M46/M47 Chapter 4). Table 2.8 gives the details.

The telescope is shown in a manufacturer's image of the Planewave Optical Tube Assembly and Alt-Azimuth mount used in SSD (T27) in Figure 2.10.

This telescope is dedicated to three famous women astronomer as shown in the representation of the plaque attached to the telescope in Figure 2.11.

Table 2.8 Two Messier objects imaged with telescope SSD

Messier Object	Object Type	Object Size	Constellation
M46	Open Cluster	27' x 27'	Puppis
M47	Open Cluster	30' x 30'	Puppis

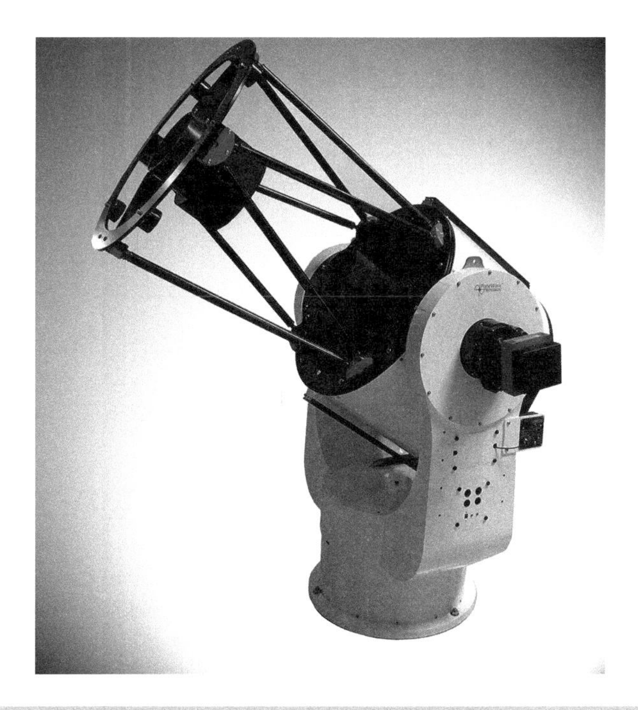

Fig. 2.10 The Planewave 27-inch telescope as used in SSD (T27). Image Courtesy of Planewave Instruments

This instrument was opened by Dr Pamela Gay (SIUE) and Dr Amanda Bauer on 3rd October 2014 and is dedicated to three great women whose contributions to astronomy remain vital to our understanding of the universe.

Dame Jocelyn Bell Burnell
Annie Jump Cannon
Henrietta Swan Leavitt

Hence, using the initials of their surnames, the instrument has been named the "BCL" telescope.

The BCL instrument was provided by the Octans group of companies.

AAO: Australian Astronomical Observatory
SIUE: Southern Illinois University

Fig. 2.11 The wording on the Plaque attached to Telescope SSD (T27) in Siding Spring

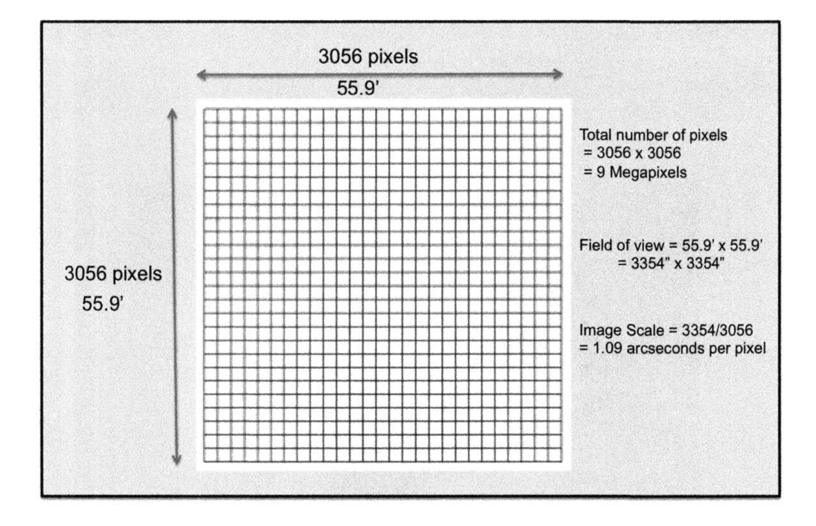

Fig. 2.12 Determining the image scale of the camera used on SSE (T31)

Messier Objects imaged using Telescope SSE (T31)

Three Messier objects were imaged using this telescope: the star-forming nebula M43 and open clusters M50 and M67, in the constellations of Orion, Monoceros and Cancer. Cancer has an approximate (central) declination of +20°, Orion +4.5° and Monoceros -6°. The field of view of this telescope is 55.9' x 55.9', and it has an aperture of 20" (510 mm). The chip on the camera contains 9 Megapixels with 3056 pixels per side. With a field of view of 55.9' x 55.9', this gives a resolution or image scale of 1.09 arcsec per pixel. The calculation is shown in Figure 2.12.

Table 2.9 Three Messier Objects imaged by SSE (T31) at Siding Spring

Messier Object	Object Type	Object Size	Constellation
M43	Star-forming Nebula	20' x 15'	Orion
M50	Open Cluster	16' x 16'	Monoceros
M67	Open Cluster	30' x 30'	Cancer

Fig. 2.13 The Planewave 20-inch telescope as used in SSE (T31). Image Courtesy of Planewave Instruments

The filter set on SSE consists of AstroDon E-Series LRGB, Ha, SII, OIII, and UBVRI, suitable for imaging and science applications. The camera does have an anti-blooming gate. Table 2.9 gives the details of the Messier objects captured with this telescope. A manufacturer's image of the Planewave telescope used in SSE (T31) is shown in Figure 2.13.

Messier Objects imaged using Telescope SSF (T32)

This 17-inch telescope was used to image eight Messier objects in the constellations of Sagittarius, with a rough declination of -26°, Hydra at -19° and Virgo at -4°. The eight objects consisted of two globular clusters, three spiral galaxies, two elliptical galaxies and a single lenticular galaxy. The objects are listed in Table 2.10.

Table 2.10 The 8 Messier objects imaged by SSF at Siding Spring

Messier Object	Object Type	Object Size	Constellation
M55	Globular Cluster	19' x 19'	Sagittarius
M75	Globular Cluster	6.8' x 6.8'	Sagittarius
M83	Spiral Galaxy	11' x 10'	Hydra
M86	Lenticular Galaxy	7.5' x 5.5'	Virgo
M87	Elliptical Galaxy	7' x 7'	Virgo
M89	Elliptical Galaxy	4' x 4'	Virgo
M90	Spiral Galaxy	9.5' x 4.5'	Virgo
M104	Spiral Galaxy	9' x 4'	Virgo

Fig. 2.14 The Planewave 17-inch telescope as used in SSF (T32). Image Courtesy of Planewave Instruments

The field of view of SSF (T32) is 43.9' x 43.9'. All of the objects selected to be imaged with SSF were sized well within that FOV. The filters on this telescope are Astrodon E-Series, Red, Green, Blue, Luminance, Ha, SII, OIII, Johnsons-Cousins V and I, so the telescope is suitable for imaging and science. A manufacturer's image of the Planewave telescope used in SSF (T32) is shown in Figure 2.14.

Messier Objects imaged by Telescope SSG (T33)

Telescope SSG (T33) is a 16-inch aperture fast (f/3.5) Astro Syteme Austria (ASA) Astrograph. Three Messier objects, a spiral galaxy, an open cluster and a globular cluster, were captured with this telescope in the constellations of Leo, with an

Table 2.11 Three Messier objects imaged by Telescope SSG (T33)

Messier Object	Object Type	Object Size	Constellation
M26	Open Cluster	15' x 15'	Scutum
M69	Globular Cluster	9.8' x 9.8'	Sagittarius
M96	Spiral Galaxy	6' x 4'	Leo

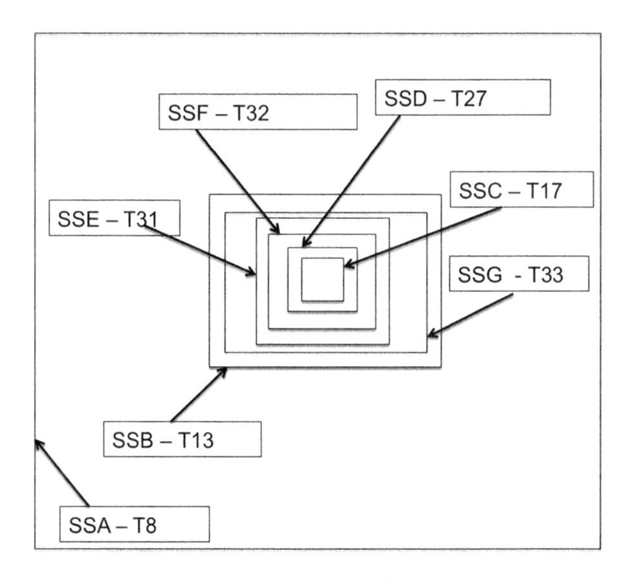

Fig. 2.15 Relative Fields of View of the telescopes used at Siding Spring

approximate central declination of +16°, Scutum at -10° and Sagittarius at -26°. Only Leo lies north of the celestial equator. The field of view is 86.6' x 57.5'. The SSG camera chip is rectangular as opposed to a square chip, as is the case with SSF (T32). The target Messier objects for this telescope, listed in Table 2.11, have relatively small dimensions. The relative fields of view of all the Siding Spring telescopes used for the project are shown in Figure 2.15.

United States - New Mexico and California

Six telescopes were used at New Mexico Skies in Mayhill, New Mexico, which has a Latitude of 32° 54' 11.91". A single telescope was used at Sierra Remote Observatory in Auberry, California, which is at a Latitude of +37° 04' 12". The telescopes used were:

New Mexico A (NMA or T3): A 6-inch (150 mm) aperture f/7.3 Apochromatic Refractor. The camera is an SBIG ST-4000XCM One-Shot Color CCD.

New Mexico B (NMB or T5): A 10-inch (250 mm) aperture f/3.4 Hyperbolic Flat-Field Astrograph. The camera is an SBIG ST-10XME.

New Mexico C (NMC or T11): A 20–inch (510 mm) f/4.5 Corrected Dall-Kirkham Astrograph. The camera is an FLI ProLine PL11002M CCD.

New Mexico D (NMD or T14): A 4-inch (106 mm) aperture f/5 Petzval Apochromatic Astrograph. The camera is an SBIG STL-11000M.

New Mexico E (NME or T20): A 4-inch (106 mm) aperture f/5 Petzval Apochromatic Astrograph. The camera is an SBIG STL-11000M

New Mexico F (NMF or T21): A 17-inch (431 mm) aperture f/4.5 Corrected Dall-Kirkham Astrograph with the FLI-PL6303E camera.

California A (CA or T24): A 24-inch (610 mm) aperture f/6.5 Corrected Dall-Kirkham Astrograph with an FLI-PL09000 camera.

Messier Objects imaged using New Mexico A (T3)

As you can see in Table 2.12, this telescope was used to image 20 Messier objects. The telescope has a single-shot color camera and has an aperture of 6 inches. (150 mm). It has a field of view of 47’ x 47’, so using it to image the Messier 33 galaxy

Table 2.12 The 20 Messier Objects imaged by NMA (T3)

Messier Object	Object Type	Object Size	Constellation
M2	Globular Cluster	16' x 16'	Aquarius
M9	Globular Cluster	12' x 12'	Ophiuchus
M10	Globular Cluster	20' x 20'	Ophiuchus
M11	Open Cluster	14' x 14'	Scutum
M15	Globular Cluster	18' x 18'	Pegasus
M16	Open Cluster and Nebula	7' x 7'	Serpens Cauda
M17	Open Cluster and Nebula	11' x 11'	Sagittarius
M19	Globular Cluster	17' x 17'	Ophiuchus
M27	Planetary Nebula	8.0' x 5.7'	Vulpecula
M33	Spiral Galaxy	73' x 45'	Triangulum
M34	Open Cluster	35' x 35'	Perseus
M36	Open Cluster	12' x 12'	Auriga
M38	Open Cluster	21' x 21'	Auriga
M40	Double Star	0.8'	Ursa Major
M52	Open Cluster	13' x 13'	Cassiopeia
M74	Spiral Galaxy	10.2 x 9.5	Pisces
M76	Planetary Nebula	2.7 x 1.8	Perseus
M81	Spiral Galaxy	21' x 10'	Ursa Major
M82	Irregular Galaxy	9' x 4'	Ursa Major
M109	Spiral Galaxy	7' x 4'	Ursa Major

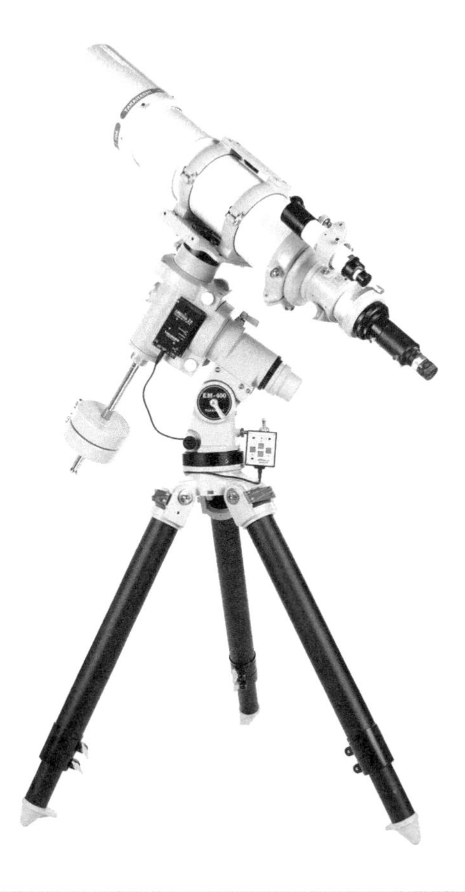

Fig. 2.16 The Six-inch Takahashi TOA -150, as used in telescope NMA (T3), set up for visual use. Image Courtesy of Takahashi US

was a little ambitious, but the essence of the galaxy was captured nevertheless (see M33, Chapter 3). At the other end of the scale is M40, composed of two non-associated stars that happen to appear together in the sky. There is no problem in fitting that "object" into the field of view! (see M40 Chapter 4). The NMA (T3) telescope captured five globular clusters, five open clusters, two planetary nebulae, four spiral galaxies, one irregular galaxy, one double star and two open clusters with nebulae. The 12 constellations containing these objects were Cassiopeia at a central declination of +62°, Ursa Major at +55°, Perseus at +42°, Auriga at +42°, Vulpecula at +25°, Triangulum at +32°, Pegasus at +20°, Pisces at +11°, Ophiuchus at -4°, Aquarius at -10°, Scutum at -10° and Serpens Cauda at -13°. The Takahashi telescope OTA used in NMA (T3) is shown in Figure 2.16 in the manufacturer's image.

Messier Objects imaged using New Mexico B (T5)

Only two Messier objects, listed in Table 2.13, were imaged using New Mexico B (NMB or T5). This 10-inch (250 mm) reflecting telescope is mainly used for science, as the camera is of the non-anti blooming gate type (NABG). The telescope was used to image the open cluster M35 in Gemini and the elliptical galaxy M84 in Virgo. The exposure of the M35 image was only 60 seconds, avoiding any blooming of the stars, but the 600-second exposure of M84 shows distinct blooming on some stars. Gemini has a central declination of roughly +23° and Virgo of -4°. The telescope comes equipped with the filters RGB and Ha, SII, OIII, Clear and Johnsons- Cousins BVI.

Messier Objects imaged using New Mexico C (T11)

This telescope imaged 14 Messier objects consisting of seven globular clusters, two open clusters, three spiral galaxies, a star-forming nebula and a lenticular galaxy. The central constellation declination range was from Draco at 62.5°, through Cassiopeia at +62°, Canes Venatici at +41°, Hercules at +30°, Coma Berenices at +22°, Serpens Caput at +9°, Orion at +4.5°, Ophiuchus at -4°, Virgo at -4°, and Sagittarius at -26°. The field of view of this telescope is 54' 47" x 36' 31". This 20-inch (510 mm) telescope does have an anti-blooming gate, and the mount is so accurate that active guiding is disabled. It can reach magnitude 21.5 with 5- to 10-minute exposures. The filter set consists of AstroDon LRGB, 3nm Ha, SII, OIII and UBVRI. Details of the Messier objects imaged with this telescope are given in Table 2.14.

Messier Objects imaged using New Mexico D (T14)

This telescope (NMD or T14), based in New Mexico at Latitude +33°, was used to image six Messier objects spread across five constellations. This included two open clusters, two elliptical galaxies, a spiral galaxy and a globular cluster. The sequence of declinations of the approximate center of the constellations from north to south is from Andromeda at +39°, Cancer at +20°, Leo at +16°, Ophiuchus at -4° and Sagittarius at -26°. This 4-inch (106 mm) telescope has a very wide field of view at

Table 2.13 Two Messier Objects imaged by NMB (T5)

Messier Object	Object Type	Object Size	Constellation
M35	Open Cluster	28' x 28'	Gemini
M84	Elliptical Galaxy	5' x 5'	Virgo

Table 2.14 The 14 Messier Objects imaged by NMC (T11)

Messier Object	Object Type	Object Size	Constellation
M3	Globular Cluster	18' x 18'	Canes Venatici
M5	Globular Cluster	23' x 23'	Serpens Caput
M12	Globular Cluster	16' x 16'	Ophiuchus
M14	Globular Cluster	11' x 11'	Ophiuchus
M18	Open Cluster	9' x 9'	Sagittarius
M22	Globular Cluster	32' x 32'	Sagittarius
M53	Globular Cluster	13' x 13'	Coma Berenices
M58	Spiral Galaxy	5.5' x 4.5'	Virgo
M64	Spiral Galaxy	9.3' x 5.4'	Coma Berenices
M78	Star-forming Nebula	8' x 6'	Orion
M91	Spiral Galaxy	5.4' x 4.4'	Coma Berenices
M92	Globular Cluster	14' x 14'	Hercules
M102	Lenticular Galaxy	5.2' x 2.3'	Draco
M103	Open Cluster	6' x 6'	Cassiopeia

Table 2.15 Six Messier Objects imaged by NMD (T14)

Messier Object	Object Type	Object Size	Constellation
M23	Open Cluster	27' x 27'	Sagittarius
M31	Spiral Galaxy	178' x 63'	Andromeda
M44	Open Cluster	95' x 95'	Cancer
M105	Elliptical Galaxy	2' x 2'	Leo
M107	Globular Cluster	13' x 13'	Ophiuchus
M110	Elliptical Galaxy	17' x 10'	Andromeda

233.5' x 155.7'. This field of view was determined by plate-solving an image taken with NMD, and measuring the angular dimensions of the result. The wide field of view allowed M31 in Andromeda to be imaged, and the telescope was still able to capture the small elliptical galaxy M105. The image scale of the telescope/camera combination is 3.5 arcsec per pixel at 4008 x 2672 pixels, providing rectangular images. The camera on this telescope has an anti-blooming gate (ABG) and filters LRGB, Ha, SII, OIII and V, so it can be used for imaging and science projects. Details of the Messier objects imaged with NMD are shown in Table 2.15.

Messier Objects imaged using New Mexico E (T20)

Seven Messier objects were imaged by NME (T20), which is basically the same ABG telescope and camera as NMD (T14). There is a slight difference in field of view when I measure the images after plate solution. I get a FOV of 234.2' x 156.1'

Table 2.16 Seven Messier objects imaged by NME (T20)

Messier Object	Object Type	Object Size	Constellation
M63	Spiral Galaxy	10' x 6'	Canes Venatici
M65	Spiral galaxy	8' x1.5'	Leo
M66	Spiral Galaxy	8' x 2.5'	Leo
M85	Lenticular Galaxy	7.1' x 5.2'	Coma Berenices
M95	Spiral Galaxy	4.4' x 3.3'	Leo
M98	Spiral Galaxy	9.5' x 3.2	Coma Berenices
M99	Spiral Galaxy	5.4' x 4.8'	Coma Berenices

Table 2.17 The single Messier object imaged with NMF (T21)

Messier Object	Object Type	Object Size	Constellation
M48	Open Cluster	54' x 54'	Hydra

and an image scale of 3.51 arcsec per pixel for NME. The Messier objects are from the three constellations of Canes Vanatici at a central declination of +41°, Coma Berenices at +22° and Leo at +16°. The Messier objects imaged, listed in Table 2.16, consist of six spiral galaxies and one lenticular galaxy.

Messier Objects imaged using New Mexico F (T21)

This telescope has an aperture of 17" (431 mm) and a field of view of 49' 06" x 32' 44". Only one image, M48 in Hydra, described in Table 2.17, was taken with NMF, as this is an NABG instrument. You will notice some blooming of stars on the M48 image in Chapter 4. The constellation of Hydra has a central declination of -19.5° and the image of the open cluster, taken from New Mexico Skies at a Latitude of 33°, was at an altitude of 37°.

Messier Objects imaged using California A (T24)

The only telescope used in California (T24) is an excellent large aperture (24-inch) instrument, which imaged four Messier objects in four constellations, consisting of two spiral galaxies, an open cluster and a supernova remnant as listed in Table 2.18. The constellations are Ursa Major at a central declination of +55°, Cygnus at +42°, Canes Venatici at 41°, and Taurus at +19°. The field of view is 32' 08" x 32' x 08".

A manufacturer's image of the Planewave telescope used in California A (T24) is shown in Figure 2.17.

The relative fields of view of all the telescopes used in New Mexico and California are shown in Figure 2.18.

Table 2.18 Four Messier objects imaged by telescope California A (T24)

Messier Object	Object Type	Object Size	Constellation
M1	Supernova remnant	6' x 4'	Taurus
M39	Open Cluster	32' x 32'	Cygnus
M94	Spiral Galaxy	7' x 3'	Canes Venatici
M108	Spiral Galaxy	8' x 1'	Ursa Major

Fig. 2.17 The Planewave 24-inch telescope as used in California A (T24). Image Courtesy of Planewave Instruments

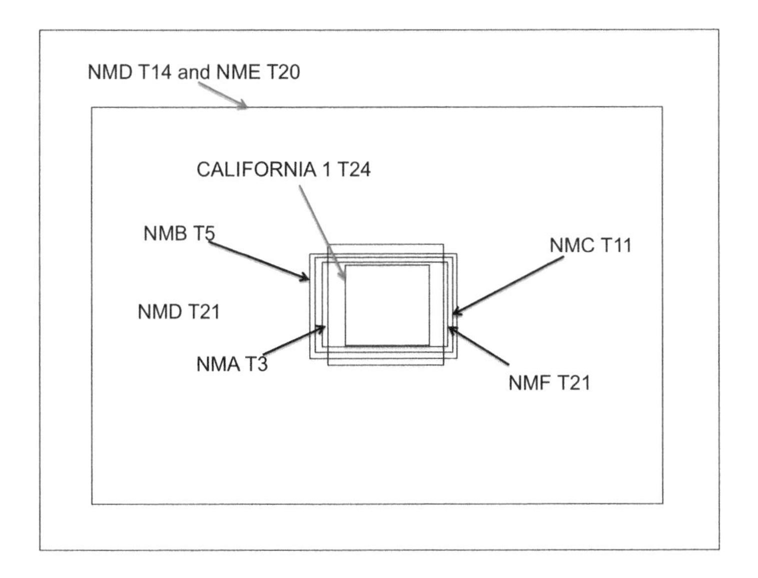

Fig. 2.18 Relative fields of view of the New Mexico and California Telescopes

Part II

The 110 Messier Objects – Images and Corresponding Data

Chapter 3

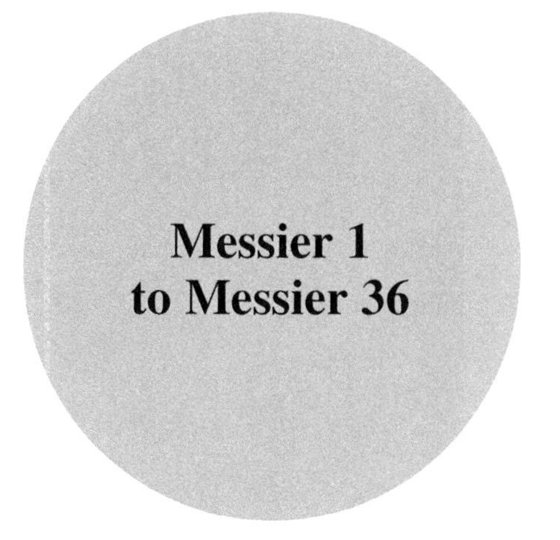

Messier 1

Messier 1 (Figure 3.1) is a very popular target for visual observers and astrophotographers. Of course it is not first in the list because it has greater astronomical significance than the rest of the list, but rather because it was this particular object that inspired Messier to generate a list of objects that could be confused with comets in the first place. His observation of M1 was made on September 12, 1758. He described it as a "Nebula above the southern horn of Taurus, it doesn't contain any star; it is a whitish light, elongated in the shape of a flame of a candle, discovered while observing the comet of 1758" (SEDS).

The large aperture telescope located in Auberry, California in the foothills of the Sierra Nevada Mountains was used to take the original image. The original image was a square measuring just over 38' x 38' of arc. M1 is only 6' x 4' in size, which was quite small on the original. The image was enlarged and cropped via Photoshop to provide the one shown. M1 lies at a distance of 6,300 light-years, a rather small distance compared to the size of the Galaxy, which is about 100,000 ly across. The supernova that was observed by Chinese astronomers on July 4, 1054 signified the birth of this nebula. In the near-1000 years since that explosion, the M1 Nebula has expanded to the size we see now.

The negative image of M1 shown in Figure 3.2 is annotated with a few stars to give an idea of magnitude. For example, the adjacent stars to the right of the nebula are magnitude 13.9 and 13.7. The decimal points are not shown on the image, in case they are confused with stars. The bright star at top left has a magnitude of 10.2. Knowing that the maximum angular diameter of M1 is about 6' of arc and that its

L. Adam, *Imaging the Messier Objects Remotely from Your Laptop*, The Patrick Moore Practical Astronomy Series, https://doi.org/10.1007/978-3-319-65385-3_3

Fig. 3.1 Image of Messier 1 (The Crab Nebula)

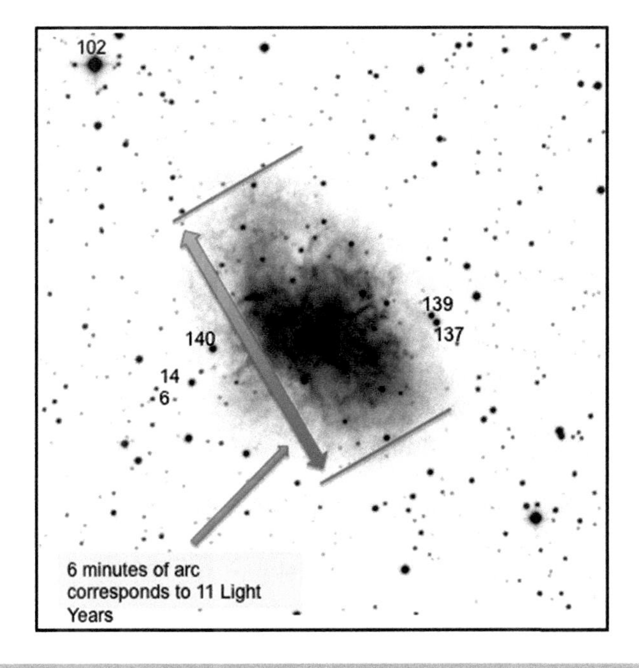

Fig. 3.2 Negative image of Messier 1

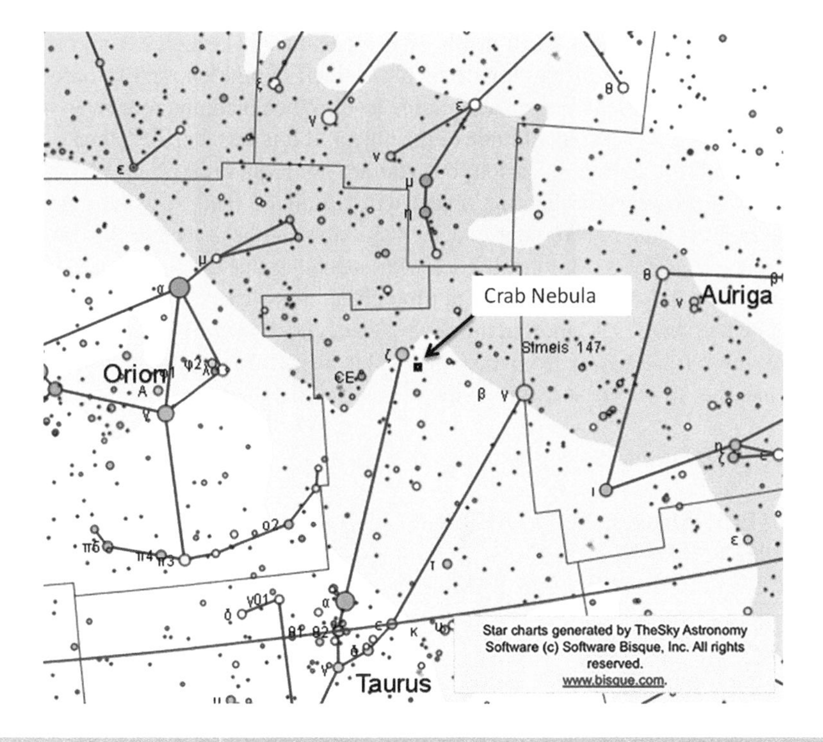

Fig. 3.3 Chart showing the location of the Crab Nebula, M1

distance is 6,300 ly, you can calculate that its maximum diameter is around 11 ly. This shows us that the nebula resulting from the supernova has expanded from 0 to 11 ly in just under 1,000 years. Unseen at the heart of M1 is a pulsar – a neutron star that is rotating 30 times per second.

The Crab Nebula can be found in the constellation of Taurus, lying between the constellations of Orion and Auriga. The chart in Figure 3.3 shows its position. The tiny black square is actually the original image (38 minutes square) precisely superimposed on the chart to give an idea of scale. To the left of M1 is the magnitude 2.97 star Alheka, which is ζ (Zeta) Tauri. To the right is the magnitude 1.65 star Elnath, which is β (Beta) Tauri. You may think I have made a mistake with this designation, as this star sits on the boundary with Auriga. At one stage, it was indeed also known as γ (Gamma) Aurigae, but it is now considered to be in Taurus. The Galactic Equator is not far away from Elnath. Elnath is in fact the brightest star near the Galactic Anti-Center, which lies just into Auriga. If you were to "stand" (on Earth) with your back to the center of the Galaxy and look away from it, you would be looking at the Galactic Anti-Center. These stars, Alheka and Elnath, define the horns of the bull. Messier 1 lies between the horns and can be spotted with binoculars.

The image of M1 was taken at 4h 48m local time in Auberry, California on September 30. M1 was at an altitude of 68°, which is high enough to reduce the thickness of the atmosphere to an acceptable level. When planning to take an image, you need to check the current altitude of the object. If it is less than 45°, then it might be best to wait for another occasion, but you are not required to. Your planetarium software will advise you when the object will culminate (i.e. cross the meridian), offering the best time to take your image. Remember to check the all-sky camera to get an idea of the sky conditions. It goes without saying that the Sun should be well out of the way and that the Moon is not too close. Sometimes good images can be obtained when there is a Moon in the sky – again, check your planetarium software to make sure it is well away from the target and in a less intrusive phase. When doing this, remember to set the software to the location of the telescope!

Messier 1: Data relating to the image in Figure 3.1

REMOTELY IMAGED FROM AUBERRY, CALIFORNIA
NGC 1952 (CRAB NEBULA)
OBJECT TYPE: SUPERNOVA REMNANT

RA: 05h 35m 32s
DEC: +22° 01' 21"
ALTITUDE: +68° 40' 48"
AZIMUTH: 130° 01' 31"
FIELD OF VIEW: 38' X 38'
OBJECT SIZE: 6' X 4'
POSITION ANGLE: 88° 11' from North
EXPOSURE TIME: 600 s
DATE: 30th September
LOCAL TIME: 04h 48m 37s
UNIVERSAL TIME: 11h 48m 37s
SCALE: 0.62 arcsec/pixel
MOON PHASE: 0.28% (waning)

Telescope Optics
OTA: Planewave 24" (0.61 m)
Optical Design: Corrected Dall-Kirkham Astrograph
Aperture: 610 mm
Focal Length: 3962 mm
F/Ratio: f/6.5
Guiding: Active Guiding Disabled
Mount: Planewave Ascension 200HR

Instrument Package
FLI-PL09000 CCD
A/D Gain: e-/ADU
Pixel Size: 12 μm square
Array: 3056 x 3956

Sensor: Frontlit
Cooling: -35°C default
FOV: 38 x 38 arcmin

Location
Observatory: Sierra Remote Observatory – MPC U69
UTC Minus -8.00 (Daylight savings time is observed)
Minimum Target Elevation: Approx 25 Degrees
37.07°N, 119.4W
Elevation: 1405 meters (4610 ft)

Messier 2

Messier 2, shown in Figure 3.4, is a globular cluster found in the constellation of Aquarius. In total, Messier's list contains 29 globulars – so, there are 28 more to go in this long-distance Messier marathon. M2 has a diameter of about 16' of arc

Fig. 3.4 Image of Messier 2, a globular cluster in Aquarius

-- pretty large when you consider that the full moon is 30' across. In reality, it is roughly 175 light-years across, although it is slightly elliptical in shape. It lies at a distance of around 37,500 ly. You can see from the image that it rather dense, with a massive concentration of stars. There are approximately 150,000 stars in this globular. Messier did not discover M2 – it had been observed earlier in 1746 by French-Italian astronomer Giacamo Maraldi when he was observing Comet de Chésaux. Messier notes that M2 is a "nebula without star in the head of Aquarius, its center is brilliant, & the light surrounding it is round; it resembles the beautiful nebula which is situated between the head & the bow of Sagittarius [M22], it is seen very well with a telescope of 2 feet, placed below the parallel [same Dec] of Alpha of Aquarius" (SEDS).

The size of the image is 47' x 47'. To get some perspective on the distance scale of the objects in the M2 image, three foreground stars, A, B and C, are identified on the negative of the image shown in Figure 3.5. Remember that the globular cluster lies at a distance of 37,500 ly. Star A (Tycho 5208:497) has a magnitude of 10.4 and lies at a distance of 105.6 ly, which is about 355 times closer to us than M2. Star B is magnitude 10.26 at a distance of 261 ly, which is about 144 times

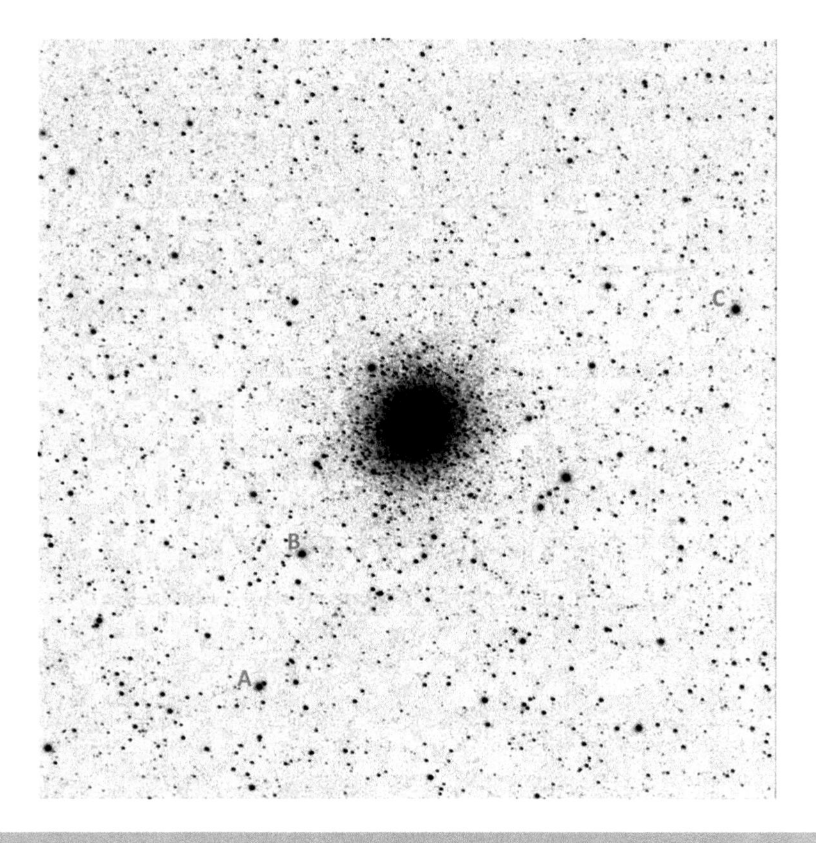

Fig. 3.5 Negative Image of Messier 2

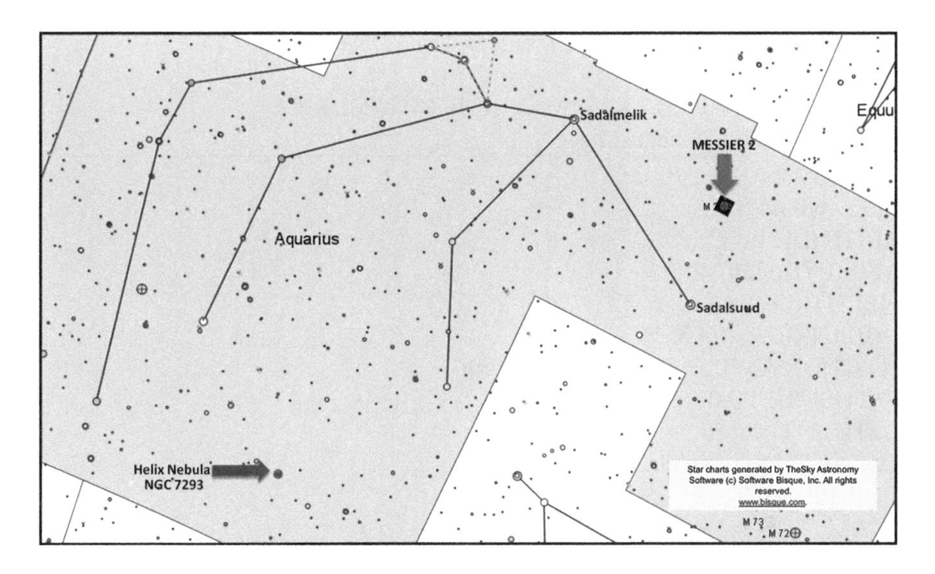

Fig. 3.6 Chart showing the location of M2

nearer than M2. Star C (HIP 106324) is magnitude 9.8 and is 610 ly distant – about 61 times closer to us than the globular. Globular Clusters ring the Galaxy, which itself is 100,000 ly across. M2 is about 1/3 of that size away from us. All the stars in the image are relatively close to us, as the three examples confirm. M2 contains a variable star with a period of 67 days (SEDS) that lies at the eastern edge of the globular cluster – a little way to the north. This was discovered by French amateur astronomer A. Chèvremont. It varies in brightness between magnitude 12.1 and magnitude 14. The star is an orange red giant.

M2 is found in Aquarius, as shown in Figure 3.6. There are two other Messier objects in this constellation, M72, a globular cluster, and M73, which is an insignificant asterism. Messier referred to M2 being on the same declination as Alpha Aquarius. On the chart, you can see M2 to the right of Alpha, which is named Sadalmelik. This star has a magnitude of 2.95. Below M2, you can see the star Sadalsuud, which has a magnitude of 2.9. Another object of interest in Aquarius -- but not a Messier object -- is NGC 7293 (Caldwell 63), otherwise known as the Helix Nebula. This is a planetary nebula that has more recently been nicknamed the "Eye of Sauron" from Tolkien's Middle-Earth story, "The Lord of the Rings."

The image was taken in New Mexico at 23h 38m local time on October 5. At the time the image was taken the globular was at an altitude of 46°. The Right Ascension of M2 is 21h 34m 19s, and its Declination is -00° 44' 42". Thus, it was almost exactly on the Celestial Equator. The Local Sidereal Time was 23h 36m 03s when the image was taken, so M2 had crossed the Meridian roughly two hours previously.

Messier 2: Data relating to the image in Figure 3.4

REMOTELY IMAGED FROM MAYHILL, NEW MEXICO
OBJECT TYPE: Globular Cluster

RA: 21h 34m 19s
DEC: -00° 44' 42"
ALTITUDE: +45° 52' 23"
AZIMUTH: 226° 40' 15"
FIELD OF VIEW: 47' 6"' X 47' 6"
OBJECT SIZE: 16' X 16'
POSITION ANGLE: 358° 17' from North
EXPOSURE TIME: 600 s
DATE: 5th October
LOCAL TIME: 23h 38m 03s
UNIVERSAL TIME: 05h 38m 03s
SCALE: 1.38 arcsec/pixel
MOON PHASE: 23.03% (waxing)

Telescope Optics
OTA: Takahashi TOA-150
Optical Design: Apochromatic Refractor
Aperture: 150 mm
Focal Length: 1095 mm
F/Ratio: f/7.3
Guiding: Internal
Mount: Paramount GTS

Instrument Package
SBIG ST-4000M One-Shot Color CCD
A/D Gain: 0.6e-/ADU
Pixel Size: 7.4um square
Resolution: 1.45 arcsec/pixel
Sensor: Frontlit
Cooling: -20°C Winter (-10°C Summer)
Array: 2048 x 2048 (8.3 Megapixels)
FOV:49.6 x49.6 arcmin

Location
Observatory: New Mexico Skies
UTC Minus 7.00 (Daylight savings time is observed)
Minimum Target Elevation: Approx 25 – 45 Degrees
(N or S) 32° 54' Decimal: 32.9 North
(W or E) 105° 31' Decimal: 105.5 West
Elevation: 2225 meters (7298 feet)

Messier 3

Messier 3, shown in Figure 3.7, is a fine globular cluster that is located in the constellation of Canes Venatici. Its approximate size is 25' of arc across. The size of the frame of the image is 54' 47" on the long side and 36' 31" on the shorter side. The position angle is 271° 12' from north, so in the image, north is to the left. On May 3, 1764 Messier noted that the nebula "discovered between Bootes & one of the Hunting Dogs of Hevelius [Canes Venatici]... doesn't contain any star, its center is brilliant, & its light is gradually fading away, it is round; in a beautiful sky, one can see it in a telescope of 1-foot: It is reported on the chart of the comet observed in 1779. Memoirs of the Academy of the same year. Reviewed on March 29, 1781, always very beautiful (Diam. 3')" (SEDS).

The negative image in Figure 3.8 shows three prominent stars that form an isosceles triangle enclosing M3. Star A is HIP 66890 with magnitude 8.40 lying at a distance of 567.23 ly from us, in comparison with M3 itself, which is 33900 ly distant (SEDS). Star B is HIP 66891 with magnitude 10.56 at a distance of 667 ly. Star C is Tycho 2004:1683, with magnitude 9.8 and a distance of 209 ly.

The chart in Figure 3.9 shows the position of Messier 3 within its parent constellation Canes Venatici. This constellation is home to the spectacular M51 galaxy, as well as the spiral galaxies M106, M63 and M94. As you can see from the

Fig. 3.7 Image of M3, a globular cluster in Canes Venatici

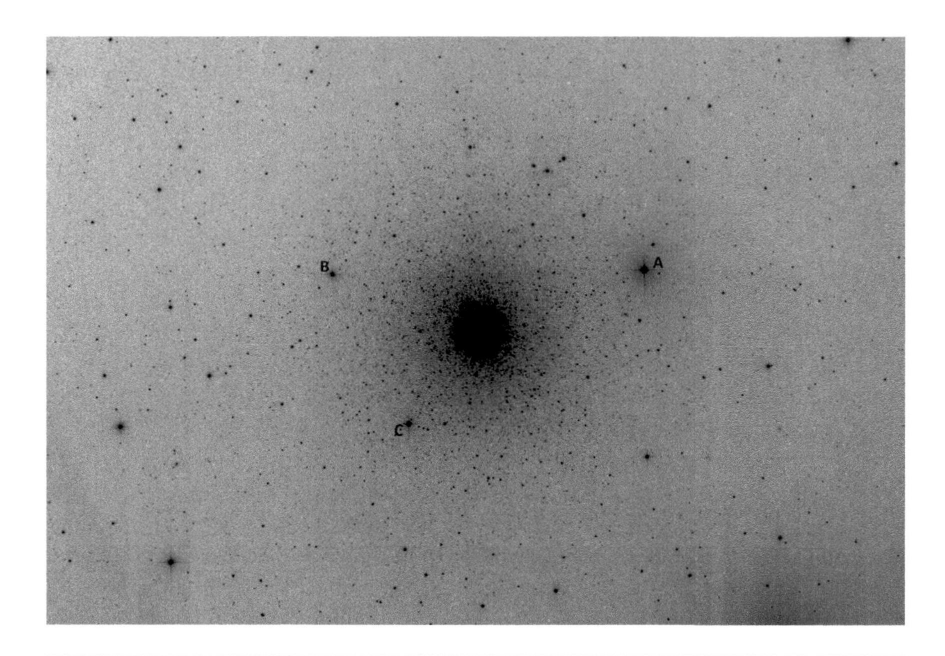

Fig. 3.8 Negative image of Messier 3

chart, M3 is surrounded by the constellations Ursa Major, Bootes and Coma Berenices. At the time the image was taken, the altitude of M3 was almost 85° and had transited the meridian about 13 minutes earlier. You can see the proximity of the meridian to M3 on the chart, where the meridian is represented by the vertical line to the left of M3.

Messier 3: Data relating to the image in Figure 3.7

REMOTELY IMAGED FROM MAYHILL, NEW MEXICO
OBJECT TYPE: Globular Cluster

RA: 13h 42m 57s
DEC: +28° 17' 38"
ALTITUDE: +84° 45' 21"
AZIMUTH: 211° 56' 57"
FIELD OF VIEW: 0° 54' 47" x 0° 36' 31"
OBJECT SIZE: 25' x 25'
POSITION ANGLE: 271° 12' from North
EXPOSURE TIME: 300 s
DATE: 14th April
LOCAL TIME: 01h 27m 12s
UNIVERSAL TIME: 07h 27m 12s
MOON PHASE: 51.63% (waxing)

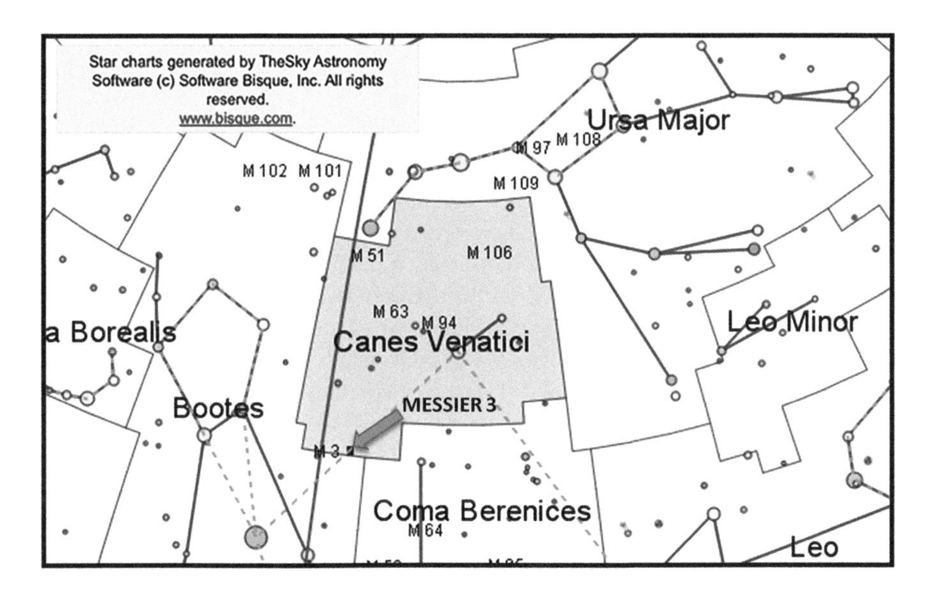

Fig. 3.9 Chart showing the location of M3

Telescope Optics
OTA: Planewave 20" CDK
Optical Design: Corrected Dall-Kirkham Astrograph
Aperture: 510 mm
Focal Length: 2280 mm (0.66 Focal Reducer Fitted)
F/Ratio: f/4.5
Guiding: Active Guiding Disabled
Mount: Planewave Ascension 200HR

Instrument Package
FLI Proline PL11002M CCD
Camera Pixel Size: 9-μm square
Resolution: 0.81 arcsec/pixel
Cooling: -30°C default
Array: 4008 x 2672 (10.7 Megapixels)

Location
Observatory: New Mexico Skies
UTC Minus 7.00 (Daylight savings time is observed)
Minimum Target Elevation: Approx 25 – 45 Degrees
(N or S) 32° 54' Decimal: 32.9 North
(W or E) 105° 31' Decimal: 105.5 West
Elevation: 2225 meters (7298 ft)

Messier 4

Messier 4, shown in Figure 3.10, is a globular cluster located in the constellation of Scorpius. Messier observed it on May 8, 1764: "I have discovered a nebula near *Antares*, & on its parallel, it is a light which has little extension, which is faint, & which is difficult to be seen: when employing a good telescope for viewing it, one can perceive very small stars. Its right ascension has been determined at 242d 16' 56", & its declination as 25d 55' 40" south" (SEDS). Famous American astronomer Robert Burnham Jr. described M4 as a "Fine star cluster, one of the largest of its type." The New General Catalog lists M4 as NGC 6121. M4 is bright enough to be seen with the naked eye under suitably dark sky conditions.

SEDS gives a distance of 7,200 ly for Messier 4. Compare this with the great globular M13, which lies at a distance of 25,100 ly. Messier 4 is 36' of arc in diameter, in comparison with M13 at only 20'. However, M4 is only 75 ly in diameter, while M13 is roughly double this. On the negative image shown in Figure 3.11, I have picked out three stars. The first is HIP 80400, a main sequence G0 star, with an apparent magnitude of 8.36. The Hipparcos Catalog puts it at a distance of

Fig. 3.10 Image of Messier 4, a globular cluster in Scorpius

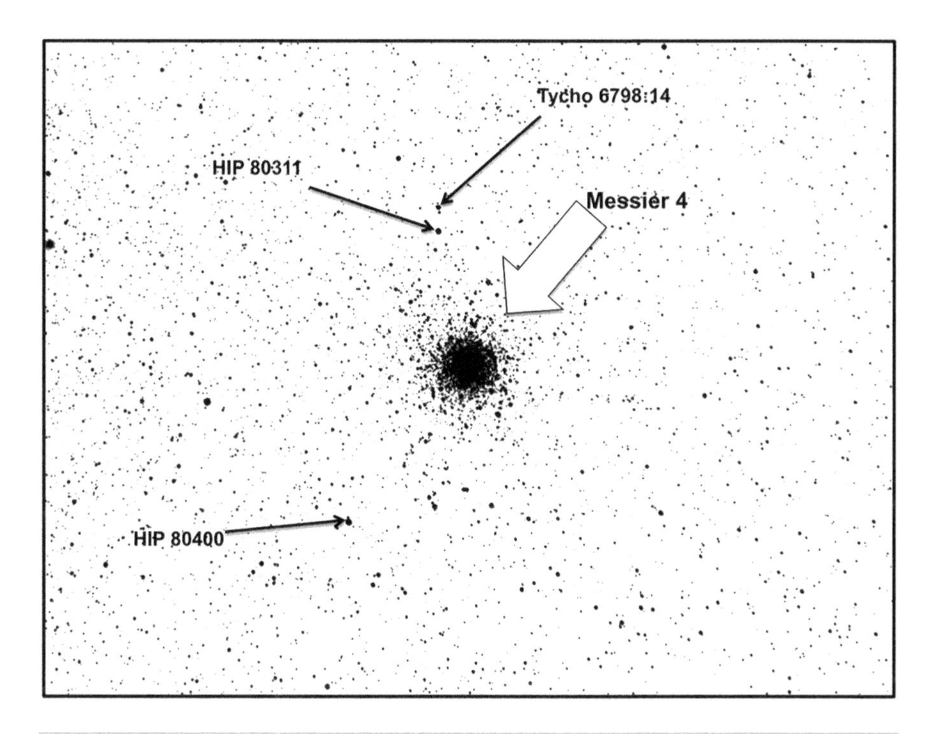

Fig. 3.11 Negative image of Messier 4

403.16 ly, making it a foreground star. The second star is HIP 80311, a hotter main sequence A0 star with an apparent magnitude of 8.92 at a distance of 419.22 ly. The third is Tycho 6798:14 with a much fainter magnitude of only 10.25. However, this star is only 37.36 ly distant. The more central stars of M4 on the solved image include many magnitude 11 stars, going down to magnitude 14.

The first thing to notice on the chart in Figure 3.12 is M4's close proximity to the star Antares in the constellation of Scorpius. Antares is a bright red star whose name translates to "Rival of Mars." The color similarity between Antares and the Red Planet is striking. There are three other Messier Objects in Scorpius: M80, a globular cluster, and M6 and M7, which are both open clusters. NGC objects in Scorpius include 12 globular clusters, two nebulae, 22 open clusters, four planetary nebulae and a galaxy. Quite a busy constellation lying on the Milky Way!

The image of Messier 4 was taken at 04h 02m 18s in Siding Spring on March 9. At that time, the Universal Time was 17h 02m 18s (March 8). The Right Ascension of M4 is 16h 24m 39s, and the Local Sidereal Time when the image was taken was 14h 04m 40s. The difference between the two is 2h 19m 59s, which represents the remaining time it would take M4 to arrive on the meridian at 06h 22m local time.

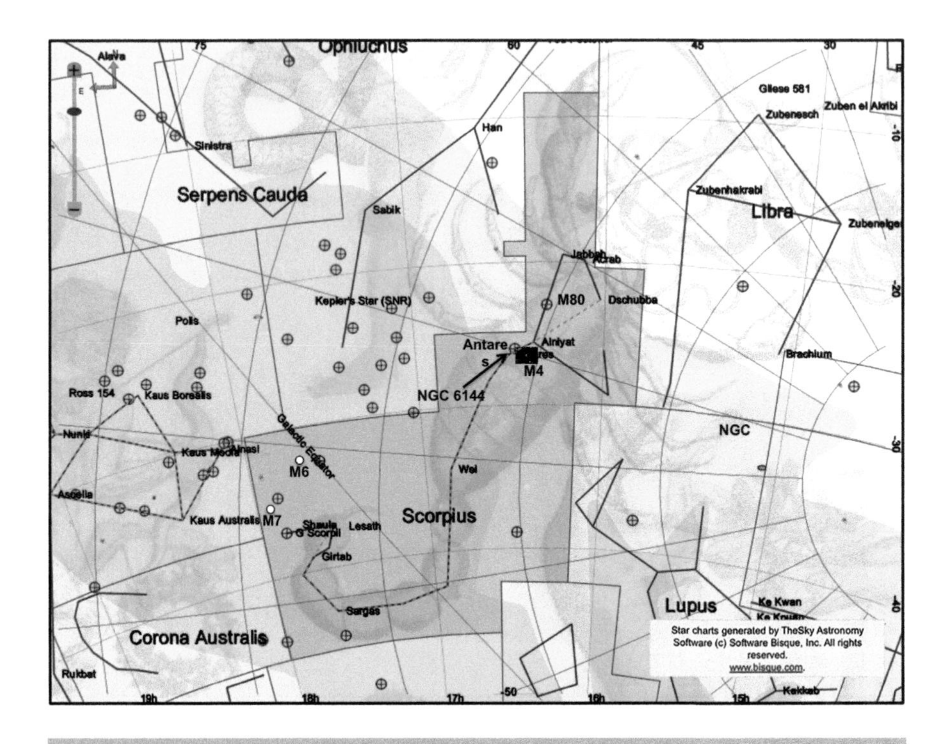

Fig. 3.12 Chart showing the location of M4

Messier 4: Data relating to the image in Figure 3.10

REMOTELY IMAGED FROM NEW SOUTH WALES, AUSTRALIA
NGC 6121
OBJECT TYPE: Globular Cluster

RA: 16h 24m 39s
DEC: -26° 33' 39"
ALTITUDE: +59° 08' 10"
AZIMUTH: 90° 15' 26"
FIELD OF VIEW: 1° 37' 44" x 1° 13' 18"
OBJECT SIZE: 36'
POSITION ANGLE: 357° 17' from North
EXPOSURE TIME: 300 s
DATE: 9th March
LOCAL TIME: 04h 02m 18s
UNIVERSAL TIME: 17h 02m 18s
SCALE: 3.67 arcsec/pixel
MOON PHASE: 83.37% (waxing)

Telescope Optics
OTA: Takahashi Sky 90
Optical Design: Apochromatic Refractor
Aperture: 90 mm
Focal Length: 417 mm
F/Ratio: f/5.6
Guiding: None
Mount: Paramount PME

Instrument Package
CCD: SBIG ST2000 XMC
Colour CMOS
Pixel Size: 7.4μm square
Sensor: Frontlit
Cooling: CNA
Array: 1600 x 1200 pixels
FOV: 60.5 x 80.7 arcmin

Location
Observatory: Siding Spring
UTC +10.00 (Australia Daylight savings time is observed)
31° 16' 24" South
149 03' 52" East
Elevation: 1122 meters (3681 ft)

Messier 5

Messier 5 (NGC 5904) is another superb globular cluster, shown in Figure 3.13. It lies in the constellation of Serpens Caput and has a rough size of 23' of arc. Charles Messier discovered it independently on May 23, 1764, although it had previously been discovered by Gottfried Kirch in 1702. The plate solution for this image confirms that the image was 0° 54' 55" x 0° 36' 36" in size and that the position angle was 90° 40' from north, meaning that north is to the right and east is to the top. Messier noted, "The night of May 23 to 24, 1764, I have discovered a beautiful nebula in the constellation of Serpens, near the star of sixth magnitude; the fifth according to the catalog of Flamsteed. That nebula doesn't contain any star; it is round, & could have a diameter of 3 arcmin; one can see it very well, under a good sky, with an ordinary refractor of one foot. I have observed that nebula in the Meridian, & I have compared it to the star Alpha Serpentis. Its position was Right Ascension 226° 39' 4" & its declination 2° 57' 16" north. On March 11, 1769, at about four o'clock in the morning, I have reviewed that nebula with a good Gregorian telescope which magnified 104 times, & I have ensured that it doesn't contain any star" (SEDS).

Fig. 3.13 Image of Messier 5, a globular cluster in Serpens Caput

M5 lies at a distance of 24,500 ly -- roughly 1/4 the size of the Galaxy. It is believed to be about 13 billion years old. The bright star is 5 Serpentis, or HIP 74975. This star has a magnitude of 5.05, so it can be very helpful in locating M5. The Hipparcos data gives it a distance of 80.61 ly – a very near neighbor in comparison to M5. It is in fact a double star and is also known as WDS STF1930. The other star labeled in the negative image in Figure 3.14 in is HIP 74907, which has a magnitude of 10.36 and is 403.66 ly away from us.

The chart in Figure 3.15 shows the location of M5 in the constellation of Serpens Caput. This constellation lies to the south of Corona Borealis and to the east of the bright star Arcturus in Bootes. There are no other Messier objects in Serpens Caput, but M16 lies in the "other" Serpens – Serpens Cauda. The actual image of M5 is shown to scale as the small vertical black rectangle arrowed on the chart. The image has been rotated 90° to put north at the top. Serpens is actually one constellation split into two parts. The part containing M5, Serpens Caput, is the head of the snake, while the other part, Serpens Cauda, is the tail of the snake.

The date this image was taken was February 1 in New Mexico at around 03h 20m. The Universal Time was 10h 20m. There was a fairly bright Moon with a Phase of about 73% that was only 38° away -- not the best time to take a deep sky image. M5 has a Right Ascension of 15h 19m 14s and the Loceal Sidereal Time was 12h 03m 53s. At the time that the image was taken, M5 was therefore 3h 15m 22s away from the meridian, due to transit at 06h 28m. The same telescope used to take the image of Messier 3 was used to take the image of Messier 5 – a 20-inch

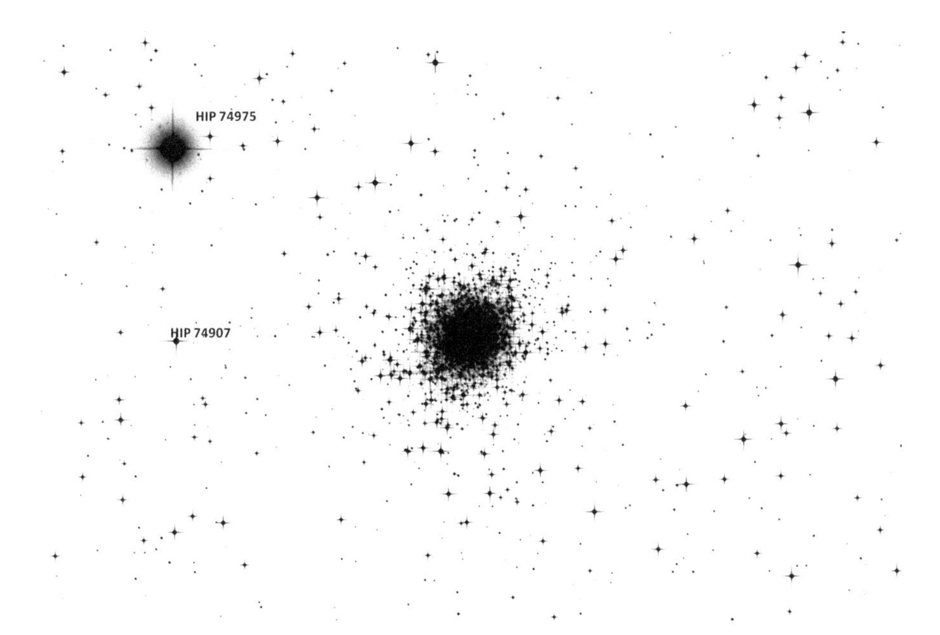

Fig. 3.14 Negative image of Messier 5

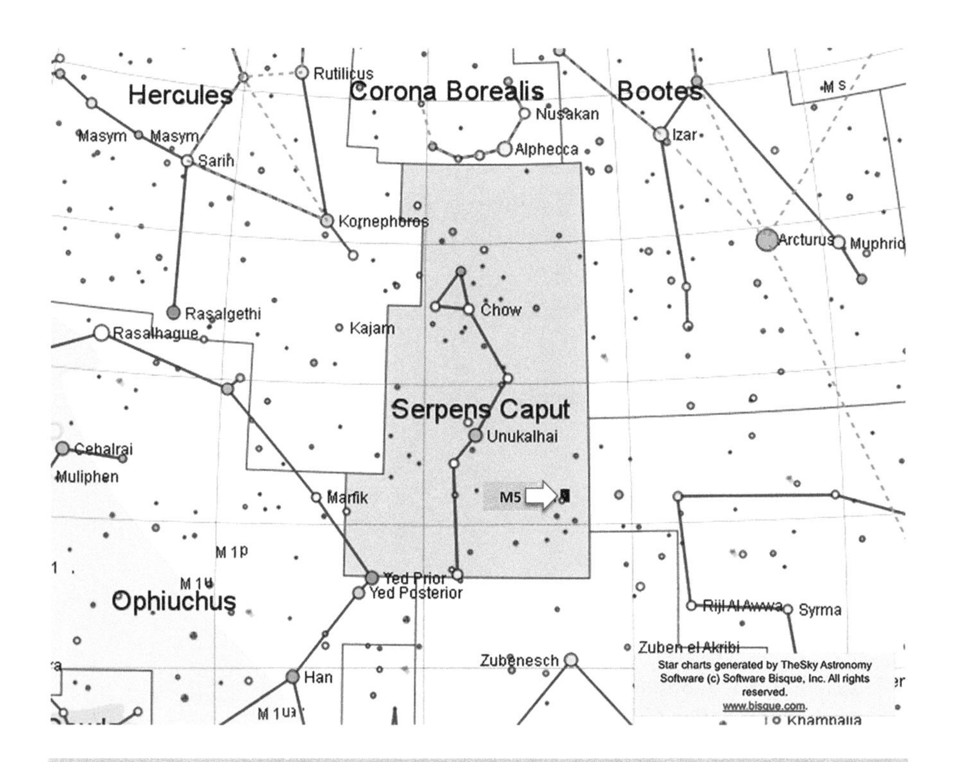

Fig. 3.15 Chart showing the location of M5

telescope with a very accurate mount that allows one to capture images of stars with magnitudes as faint as 21.5. An add-on to Adobe Photoshop was used to add some spikes to a few of the stars for effect.

Messier 5: Data relating to the image in Figure 3.13

REMOTELY IMAGED FROM MAYHILL, NEW MEXICO
NGC 5904
OBJECT TYPE: Globular Cluster

RA: 15h 19m 14s
DEC: +02° 02' 05"
ALTITUDE: +34° 54' 17"
AZIMUTH: 113° 26' 44"
FIELD OF VIEW: 0° 54' 55" x 0° 36' 36
OBJECT SIZE: 23' x 23'
POSITION ANGLE: 90° 40' from North
EXPOSURE TIME: 300 s
DATE: 1st February
LOCAL TIME: 03h 19m 59s
UNIVERSAL TIME: 10h 19m 59s
SCALE: 0.82 arcsec/pixel
MOON PHASE: 72.98% (waning)

Telescope Optics
OTA: Planewave 20" CDK
Optical Design: Corrected Dall-Kirkham Astrograph
Aperture: 510 mm
Focal Length: 2280 mm (0.66 Focal Reducer Fitted)
F/Ratio: f/4.5
Guiding: Active Guiding Disabled
Mount: Planewave Ascension 200HR

Instrument Package
FLI Proline PL11002M CCD Camera
Pixel Size: 9 µm square
Cooling: -30°C default
Array: 4008 x 2672 (10.7 Megapixels)
Location
Observatory: New Mexico Skies
UTC Minus 7.00 (Daylight savings time is observed)
Minimum Target Elevation: Approx 25 – 45 Degrees
(N or S) 32° 54' Decimal: 32.9 North
(W or E) 105° 31' Decimal: 105.5 West
Elevation: 2225 meters (7298 ft)

Messier 6

Messier 6 (NGC 6405), shown in Figure 3.16, is a galactic cluster in the constellation of Scorpius. It is known as the Butterfly Cluster (although one could argue it looks more like a dragonfly). The discovery of M6 is credited to Swiss astronomer Jean Philippe Loy de Chésaux in 1746, although it was referred to far earlier than that by Ptolemy in his catalog. Messier came across it on 23 May 1764: "On the same night of May 23 to 24, 1764, I have determined the position of a cluster of small stars between the bow of Sagittarius & the tail of Scorpius: At simple view, this cluster appears to form a nebula without stars, but the slightest instrument which one employs to examine it makes one see that it is nothing but a cluster of small stars, the diameter of which could be 15 arcmin: I have determined its position during its passage of the Meridian" (SEDS). SEDS gives the cluster a diameter of 25' of arc. The French astronomer Flammarion saw M6 as being "arranged in a remarkable pattern."

The negative image in Figure 3.17 shows a circle with a diameter of 25', derived from the scale of the 1° 37' 39" x 1° 13' 14" image. It seems quite a tight fit at 25', so M6 certainly has a diameter larger than Messier's estimate of 15'. The color image in Figure 3.16 shows most stars in the cluster as blue, with some bright

Fig. 3.16 Image of Messier 6, an open cluster in Scorpius

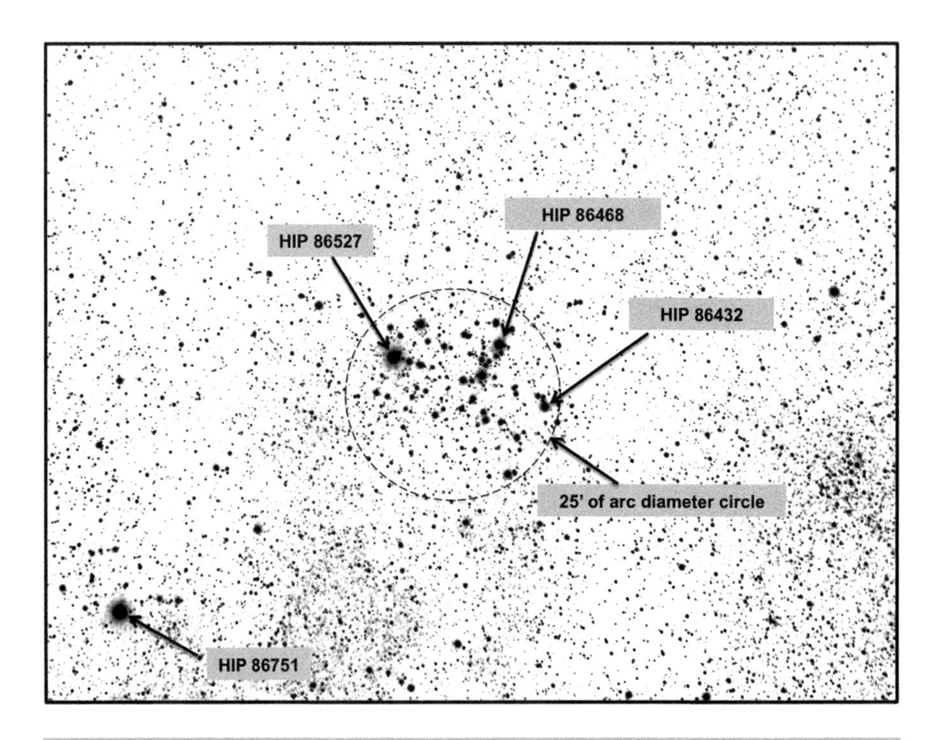

Fig. 3.17 Negative image of Messier 6

orange stars in the image frame. The star at bottom left is the magnitude 6.65 HIP 86751. The Hipparcos Catalog gives it a spectral type of M3/M4. This means it has a surface temperature of 3,400K to 3,500k – rather cool, as you might expect from its color. The other orange star that looks as though it might be in the cluster itself is the magnitude 6 HIP 86527, spectral type K3 with a surface temperature of about 4,800 K. The Hipparcos Catalog places this at 1,531 ly. The distance to M6 is in the region of 1,600 ly, so it could be a cluster member. The magnitude 6.74 main sequence star HIP 86468 is spectral type B7 with a surface temperature of 13,500K at a distance of 1,734 ly, so it is in the cluster region. The main sequence magnitude 7.16 star shown is spectral type B3 with a surface temperature of 17,600 K, but according to Hipparcos, it lies at a distance of 3,930 ly, which would put it well behind the cluster.

The chart of Scorpius in Figure 3.18 has north at the top. The image linked to the chart shows the position of M6 and its orientation. The Galactic Equator cuts just across the top right corner of the image. Scorpius has three other Messier objects within its boundaries. These are M4 (globular cluster), M7 (open cluster), M80 (globular cluster). Scorpius also contains three Caldwell objects, C69, a planetary nebula, C75 an open cluster and C76, also an open cluster. M6 in Scorpius lies just to the west of the Sagittarius teapot and looks as though it has been poured out of the spout. M7 lies just to the south of M6.

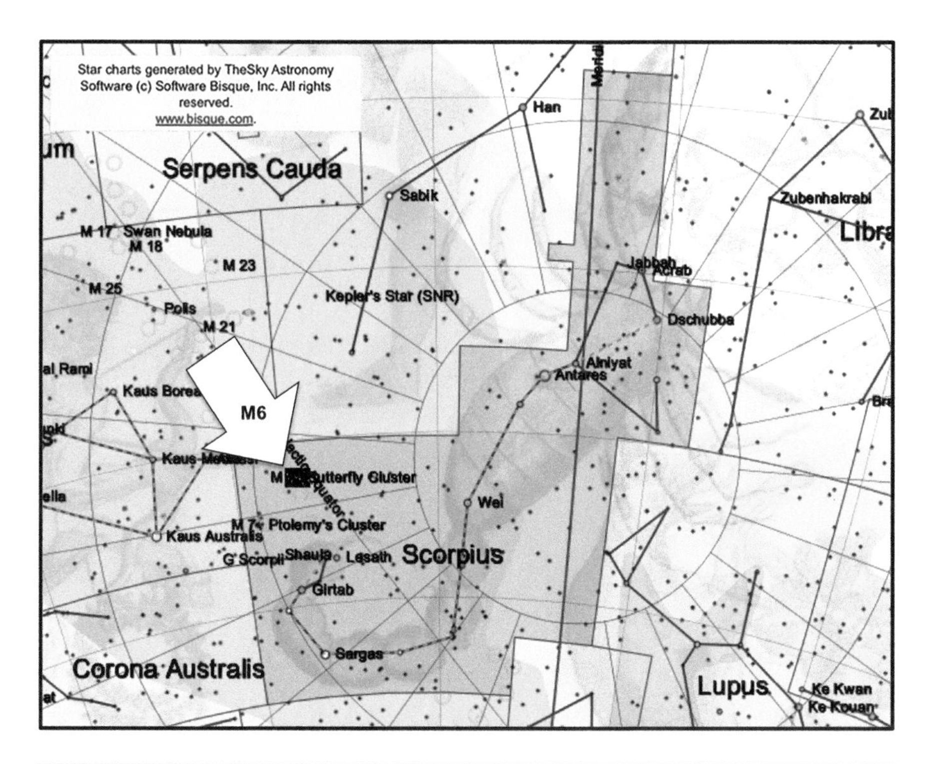

Fig. 3.18 Chart showing the location of M6

The image of M6 was taken on March 25 at 05h 14m 31s local time in Siding Spring. Universal time was 18h 14m 31s on the previous day. M6 has a Right Ascension of 17h 41m 24s, and at the time the image was taken, the LST was 16h 20m 09s, representing the right ascension on the meridian at that time. Adding the difference between these gives the approximate meridian transit time of 06h 36m.

Messier 6: Data relating to the image in Figure 3.16

REMOTELY IMAGED FROM NEW SOUTH WALES, AUSTRALIA
NGC 6405 (BUTTERFLY CLUSTER)
OBJECT TYPE: Open Cluster

RA: 17h 41m 24s
DEC: -32° 16' 18"
ALTITUDE: +72° 43' 26"
AZIMUTH: 98° 46' 07"
FIELD OF VIEW: 1° 37' 39" x 1° 13' 14"
OBJECT SIZE: 25'
POSITION ANGLE: 357° 26' from North
EXPOSURE TIME: 300 s
DATE: 25th March

LOCAL TIME: 05h 14m 31s
UNIVERSAL TIME: 18h 14m 31s (24th March)
SCALE: 3.66 arcsec/pixel
MOON PHASE: 13.54% (waning)

Telescope Optics
OTA: Takahashi Sky 90
Optical Design: Apochromatic Refractor
Aperture: 90 mm
Focal Length: 417mm
F/Ratio: f/5.6
Guiding: None
Mount: Paramount PME

Instrument Package
CCD: SBIG ST2000 XMC
Colour CMOS
Pixel Size: 7.4μm square
Sensor: Frontlit
Cooling: CNA
Array: 1600 x 1200 pixels
FOV: 60.5 x 80.7 arcmin

Location
Observatory: Siding Spring
UTC +10.00 (Australia Daylight savings time is observed)
31° 16’ 24” South
149 03’ 52” East
Elevation: 1122 meters (3681 ft)

Messier 7

Messier 7 is an open cluster known as NGC 6475, as shown in Figure 3.19. The cluster is located in the constellation of Scorpius. It is also called the Ptolemy Cluster. Ptolemy (Claudius Ptolemaeus Pelusiniensis) lived in Alexandria in Egypt c. 87 – 150 AD and listed a number of objects in his Almagest. Number 567 (M7) was described as “Following the sting of Scorpius and nebulous.” Messier observed the cluster in 1764 and noted: “I have determined, in the night of May 23 to 24, the position of another star cluster which is more considerable & of a larger extension: its diameter could occupy 30 arcminutes. This star cluster also appears at simple view like a considerable nebulosity: but when examining it with a refractor, the nebulosity disappears, & one perceives nothing but a cluster of small stars, among which there is one which has more light: this cluster is little distant from the preceding; it is between the bow of Sagittarius & the tail of Scorpius. I observed in the

Fig. 3.19 Image of Messier 7, an open cluster in Scorpius

Meridian the passage of the middle of this cluster, & compared it to the star Epsilon Sagittarii for determining its position" (SEDS). The image of M7 has north at the top and east to the left.

The negative image of M7 shown in Figure 3.20 also has north at the top. The cluster is generally estimated to be at a distance of around 800 ly. HIP 87616 is arrowed in the cluster, a magnitude 6 blue star. The Hipparcos catalog gives it a distance of 1,009.77 ly. The spectral type of this star is B9, so it is a hot young star that one would expect to be in such a cluster. I note a reference to a recent estimated distance of 1000 ly, which would make this star a probable cluster member. HIP 87460 is another hot, blue star of magnitude 5.88 and spectral type B6. The Hipparcos Catalog places this at 829.92 ly. You will notice in the color image that not all the stars are blue. The bright yellow star HIP 87569 is arrowed. This is a magnitude 5.58, cool, K3 star at a distance of 1,055.52 ly. The brightest star, the yellowish HIP 87472, has a magnitude of 5.84 and is of spectral type G8, lying at a distance of 744.65 ly.

The position of the Messier 7 cluster image in Scorpius is shown on the chart in Figure 3.21. The cluster star lies just north of the star G Scorpii, otherwise known as HIP 8726. It is a cool, spectral type K, magnitude 3.19 star. It lies at a distance of 126.86 ly. This star forms the tip of the scorpion's tail referred to by Ptolemy when he talked about the nebulous M7 near the sting. Notice on the chart that

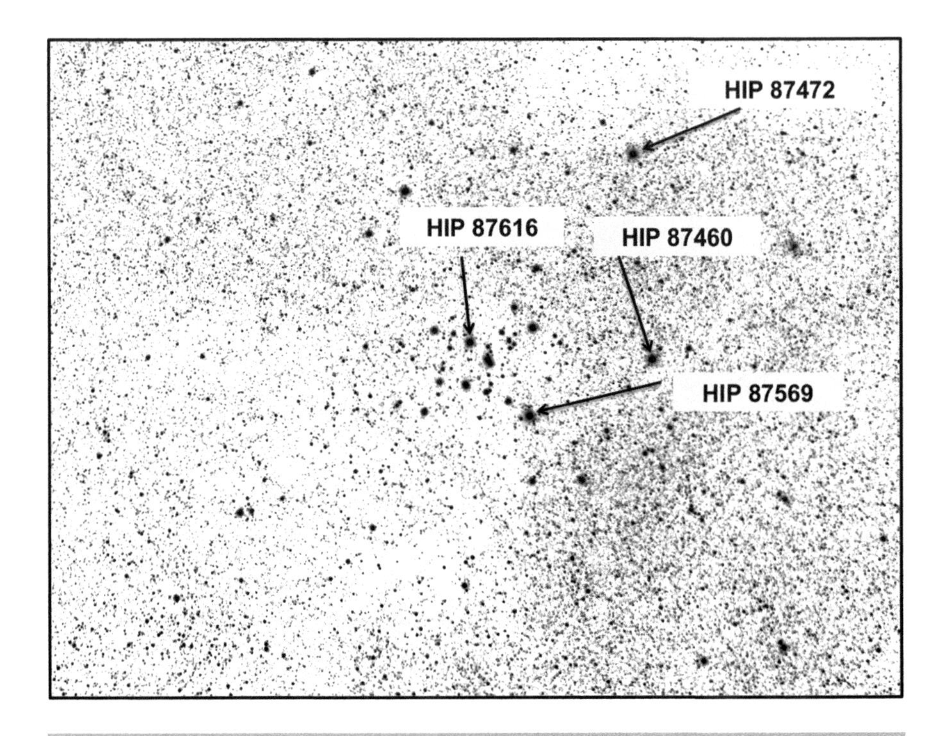

Fig. 3.20 Negative image of Messier 7

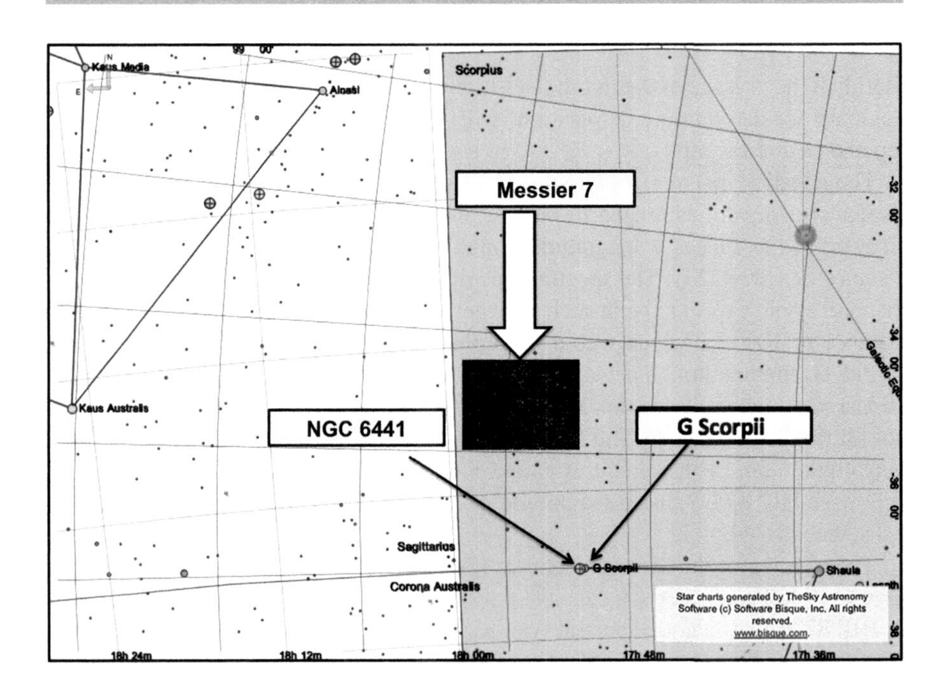

Fig. 3.21 Chart showing the location of M7

immediately to the east of G Scorpii is the globular cluster NGC 6441. Further to the east (left) of M7 is the distinctive teapot of Sagittarius, with the stars Kaus Australis (HIP 90185), Kaus Media (HIP 89931) and Alnasl (HIP 88635) forming the spout.

The 300-second exposure image of M7 was taken on March 28 from Siding Spring at 04h 06m 01s local time. The Universal Time was 17h 06m 01s. M7 has a Right Ascension of 17h 54m 43s, and the LST was 15h 23m 18s. Thus, there was roughly a 2 ½-hour difference. M7 would transit the meridian later that morning at around 06h 37m.

Messier 7: Data relating to the image in Figure 3.19

REMOTELY IMAGED FROM NEW SOUTH WALES, AUSTRALIA
NGC 6475 (PTOLEMY CLUSTER)
OBJECT TYPE: Open Cluster

RA: 17h 54m 43s
DEC: -34° 48' 31"
ALTITUDE: +58° 14' 57"
AZIMUTH: 106° 46' 18"
FIELD OF VIEW: 1° 37' 39" x 1° 13' 14"
OBJECT SIZE: 80'
POSITION ANGLE: 357° 26' from North
EXPOSURE TIME: 300 s
DATE: 28th March
LOCAL TIME: 04h 06m 01s
UNIVERSAL TIME: 17h 06m 01s
SCALE: 3.66 arcsec/pixel
MOON PHASE: 0.28% (waning)

Telescope Optics
OTA: Takahashi Sky 90
Optical Design: Apochromatic Refractor
Aperture: 90 mm
Focal Length: 417 mm
F/Ratio: f/5.6
Guiding: None
Mount: Paramount PME

Instrument Package
CCD: SBIG ST2000 XMC
Colour CMOS
Pixel Size: 7.4µm square
Sensor: Frontlit
Cooling: CNA
Array: 1600 x 1200 pixels
FOV: 60.5 x 80.7 arcmin

Location
Observatory: Siding Spring
UTC +10.00 (Australia Daylight savings time is observed)
31° 16' 24" South
149 03' 52" East
Elevation: 1122 meters (3681 ft)

Messier 8

Messier 8 (NGC 6523), shown in Figure 3.22, is a nebula located in the constellation of Sagittarius. It is difficult to determine its precise size, but SEDS gives it dimensions of 90' x 40' of arc. There is a dark rift that runs through the nebula. It is a region of star formation, and a cluster of new stars is clearly visible in the image. Messier observed M8 on the night of May 23, 1764: "I also have determined, in the same night, the position of a small star cluster which one sees in the form of a nebula, if one views it with an ordinary refractor of 3 feet but when employing a good instrument one notices a large quantity of small stars: near this cluster is a rather brilliant star which is surrounded by a very faint light: this is the

Fig. 3.22 Image of Messier 8, a nebula in Sagittarius

ninth star of Sagittarius, of seventh magnitude, according to the catalog of Flamsteed: this cluster appears in an elongated shape which extends from North-East to South-West. I observed its position during its passage of the Meridian, comparing it with the star Delta Sagittarii, & I determined its right ascension as 267d 29' 30", & its declination as 24d 21' 10" south. This star cluster could have an extension, from North-East to South-West, of about 30 minutes of arc" (SEDS).

Messier 8 is 5,200 ly away from us. It has a long side of 140 ly and a short side of 60 ly. The star cluster, known as NGC 6530, is believed to lie just in front of the Lagoon Nebula. This is a young cluster that is definitely linked to the nebulosity of M8. A single star is pointed out on the monochrome image in Figure 3.23. The star is HIP 88256, which is of magnitude 8.42 at a distance of 1,169 ly, is less than 1/5 of the distance to M8.

M8 is in the constellation of Sagittarius, as shown in the chart in Figure 3.24. The black rectangle on the chart is the actual image to scale in the correct orientation, to give an idea of its size and location. In the image, north is to the right and east is to the top, so on the chart, the image has been rotated 90° (from landscape to portrait) to put north at the top. Sagittarius is an extremely busy constellation with a number of interesting objects. For a start, the direction of the Galaxy center

Fig. 3.23 Negative image of Messier 8

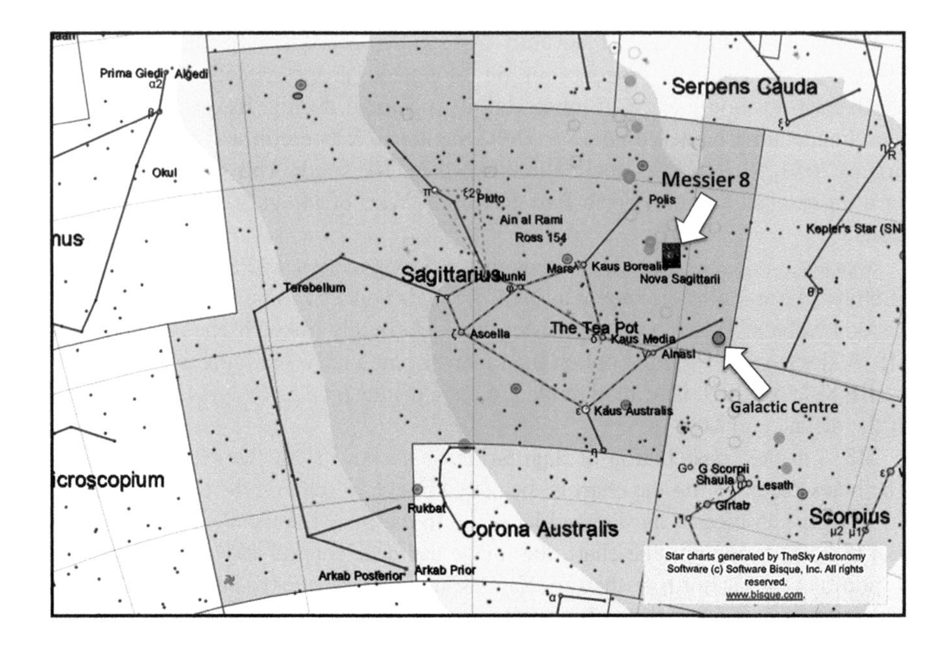

Fig. 3.24 Chart showing the location of M8

lies in this constellation and is only 6° away from M8, shown in Figure 3.24. When I say busy, consider that in addition to M8, Sagittarius contains M17, M18, M20, M21, M22, M23, M24, M25, M28, M54, M55, M69, M70 and M75 – not to mention NGC objects. Running a search in the SkyX gives a result of 75 NGC objects in Sagittarius – too much to list in the pages of this book.

The image was taken from a telescope located in Siding Spring at a local time of 20h 59m on October 7. M8 was at an altitude of just over 50°.The Lagoon Nebula has a Right Ascension of 18h 05m 22s and a Declination of -24° 18' 01. The Local Sidereal Time was 21h 01m 37s. The difference between the RA and the LST was therefore 02h 56m 15s – the time it would take for M8 to transit the meridian. The telescope used to take the image is a single-shot color camera that did not require the use of filters.

Messier 8: Data relating to the image in Figure 3.22

REMOTELY IMAGED FROM NEW SOUTH WALES, AUSTRALIA
NGC 6523 Lagoon Nebula
OBJECT TYPE: Star-forming Nebula

RA: 18h 05m 22s
DEC: -24° 18' 01"
ALTITUDE: +50° 39' 33"
AZIMUTH: 268° 53' 54"

FIELD OF VIEW: 75.3' x 100.4'
OBJECT SIZE: 90' x 40'
POSITION ANGLE: 90° 10' from Morth
EXPOSURE TIME: 600 s
DATE: 7th October
LOCAL TIME: 20h 59m 41s
UNIVERSAL TIME: 09h 59m 41s
SCALE: 3.66 arcsec/pixel
MOON PHASE: 33.12% (waxing)

Telescope Optics
OTA: Takahashi Sky 90
Optical Design: Apochromatic Refractor
Aperture: 90 mm
Focal Length: 417 mm
F/Ratio: f/5.6
Guiding: None
Mount: Paramount PME

Instrument Package
CCD: SBIG ST2000 XMC
Colour CMOS
Pixel Size: 7.4µm square
Sensor: Frontlit
Cooling: CNA
Array: 1600 x 1200 pixels
FOV: 60.5 x 80.7 arcmin

Location
Observatory: Siding Spring
UTC +10.00 (Australia Daylight savings time is observed)
31° 16' 24" South
149 03' 52" East
Elevation: 1122 meters (3681 ft)

Messier 9

Messier 9 (NGC 6333), is a globular cluster with a small, rounded appearance, certainly not displaying the same glorious appearance of M3 or M13. The image shown in Figure 3.25 was taken using the same telescope and camera setup used to image globular cluster M15, so a comparison will highlight the differences. Messier observed, "In the night of May 28 to 29, 1764, I have determined the position of a new nebula which is situated in the right leg of *Ophiuchus*, between the stars Eta

Fig. 3.25 Image of Messier 9, a globular cluster in Ophiuchus

and Rho of that constellation; that nebula doesn't contain any star; I have examined it with a Gregorian telescope which magnifies 104 times; it is round its light is faint, & its diameter is about 3 minutes of arc: its right ascension is 256d 20' 36", & its declination 18d 13' 26" south." The fact that Messier could not resolve the object into stars meant that he could only see the brighter core of the cluster, resulting in his size estimate of 3' of arc being lower than the 12' size now quoted. Messier is believed to be the original discoverer of this object. Note that north is down and east is to the right in the images.

This is quite a distant globular lying at a distance of around 25,800 ly from us. It is at a very similar distance to M13, but lying in a completely different direction in the constellation of Ophiuchus. My simple calculation of the size of M9 based on its distance and apparent size gives it a diameter of about 90 ly. Some of the cluster is obscured by intervening dust from a nearby dark nebula. On the negative image in Figure 3.26, three foreground stars have been identified. HIP 84813 is a main sequence G2 star with an apparent magnitude of 9.83 and distance of 327.47 ly, as given by the Hipparcos Catalog. It helps to visualise a foreground star's

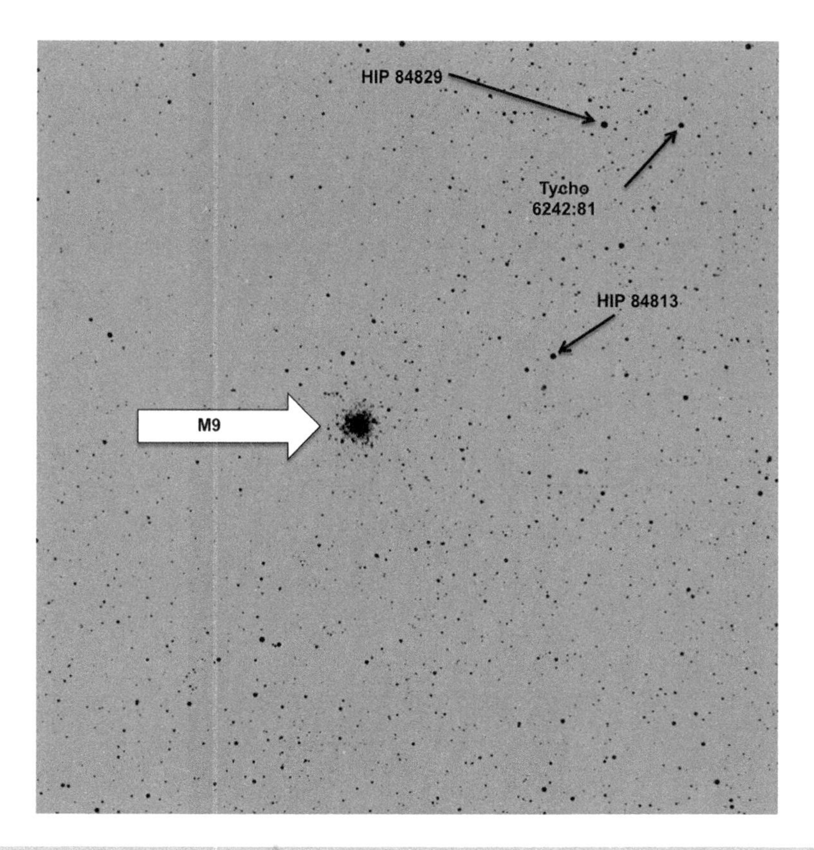

Fig. 3.26 Negative image of Messier 9

position in 3D, relative to the main object. On this basis, M9 is 79 times further away than HIP 84813, which gives some perspective to the image. HIP 84829 is a magnitude 9.27 star of spectral type B2. This hot star is 1,583.29 ly away from us -- so is 1/16th of the way to M9. The final foreground star identified is Tycho 6242:81, a magnitude 10.45 star 90.6 ly away from us, meaning that M9 is 285 times further away. Thinking of it another way, the distance to the latter star is roughly the same as the diameter of Messier 9.

The chart in Figure 3.27 shows the location of globular cluster M9 in the constellation of Ophiuchus. There are six other Messier objects in Ophiuchus and, surprisingly, they are all globular clusters, making a total of 7 Messier globulars in the constellation. The others are M10, M12, M14, M19, M62 and M107. Clearly, Messier enjoyed exploring this constellation. To find M9 visually, start from the star Sabik just to the Northwest of the tiny dark square of the solved M9 image overlaid on the chart. The distance between Sabik and M9 is less than 4°. The constellation of Ophuchus is sandwiched between Hercules and Scorpius, with Serpens Cauda to the east and Serpens Caput to the west.

The image was taken at local time 4h 54m 06s on March 9 in New Mexico. The Universal Time was 11h 54 06s. Messier 9 has a Right Ascension of 17h 20m 12s, and the Local Sidereal Time was 16h 00m 17s. Thus, M9 was 1h 19m 55s away from the meridian. The grid lines shown on the chart are lines of azimuth. The azimuth recorded at the time the image was taken is 156° 37' 54". The azimuth of the meridian is 180°, so the difference of over 23° can be observed on the calibrated chart, which has 15° separation on the azimuth grid lines. The telescope used was a Takahashi TOA – a 150 mm achromatic refractor.

Messier 9: Data relating to the image in Figure 3.25

REMOTELY IMAGED FROM MAYHILL, NEW MEXICO
NGC 6333
OBJECT TYPE: Globular Cluster

RA: 17h 20m 12s
DEC: -18° 31' 51"
ALTITUDE: +35° 14' 45"
AZIMUTH: 156° 37' 54" "
FIELD OF VIEW: 0° 47' 08" x 0° 47' 08"
OBJECT SIZE: 12' X 12'
POSITION ANGLE: 180° 23' from North
EXPOSURE TIME: 300 s
DATE: 9th March
LOCAL TIME: 04h 54m 06s
UNIVERSAL TIME: 11h 54m 06s
SCALE: 1.38 arcsec/pixel
MOON PHASE: 89.29% (waxing)

Telescope Optics
OTA: Takahashi TOA-150
Optical Design: Apochromatic Refractor
Aperture: 150 mm
Focal Length: 1095 mm
F/Ratio: f/7.3
Guiding: Internal
Mount: Paramount GTS

Instrument Package
SBIG ST-4000M One-Shot Color CCD
A/D Gain: 0.6e-/ADU
Pixel Size: 7.4um square
Resolution: 1.45 arcsec/pixel
Sensor: Frontlit
Cooling: -20°C Winter (-10°C Summer)
Array: 2048 x 2048 (8.3 Megapixels)
FOV: 49.6 x49.6 arcmin

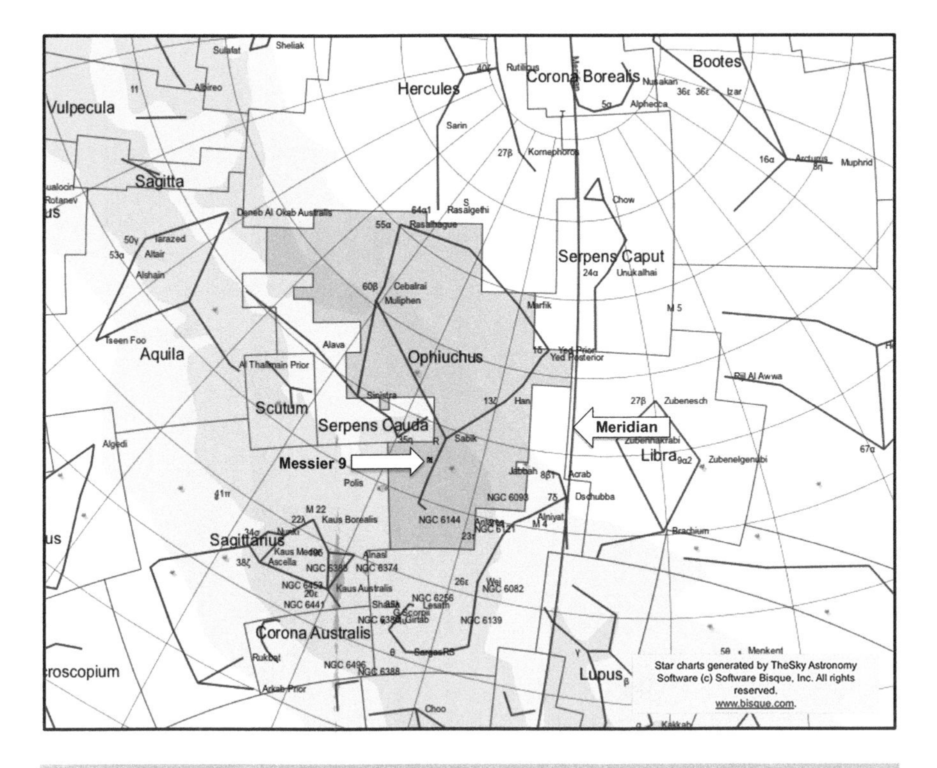

Fig. 3.27 Chart showing the location of M9

Location
Observatory: New Mexico Skies
UTC Minus 7.00 (Daylight savings time is observed)
Minimum Target Elevation: Approx. 25 – 45 Degrees
(N or S) 32° 54' Decimal: 32.9 North
(W or E) 105° 31' Decimal: 105.5 West
Elevation: 2225 meters (7298 ft)

Messier 10

Messier 10 (NGC 6254), shown in Figure 3.28, is one of the brightest globular clusters in Ophiuchus. It has a diameter of about 20' of arc, which will appear less in smaller telescopes. Charles Messier discovered the object on May 29, 1764. He noted, "I have determined the position of a nebula which I have discovered in the girdle of *Ophiuchus*, near the 30th star of that constellation, of sixth magnitude, according to the catalog of Flamsteed. When having examined that nebula with a Gregorian telescope which magnified 104 times, I have not seen

Fig. 3.28 Image of Messier 10, a globular cluster in Ophiuchus

any star there: it is round & beautiful, its diameter is about 4 minutes of arc; one sees it with difficulty with an ordinary [non-achromatic] refractor of one foot [FL]. Near that nebula one perceives a small telescopic star. I have determined the right ascension of that nebula as 251d 12' 6", & its declination as 3d 42' 18" south." The New General Catalog lists M10 as NGC 6254. North is up and east is to the left in Figure 3.28.

M10 lies at a distance of 14,300 ly (SEDS), so it is more than half the distance of the globular M9, which is at a distance of 25,800 ly and is about 15.5 ° away from M10 in the same constellation. Using this distance and the 20' diameter, you can calculate that the actual size of globular M10 is roughly 83 ly across. The negative image in Figure 3.29 shows three fairly bright foreground stars forming a triangle around the globular cluster, although it is not centralized. The first star is HIP 82905, which is of magnitude 9.94. It lies at a distance of 886.30 ly from us – only 1/16th the distance to M10. The second is the magnitude 9.73 G5 star HIP 82945, and the third is Tycho 5059:814, a magnitude 10.78 star a mere 45 ly from us.

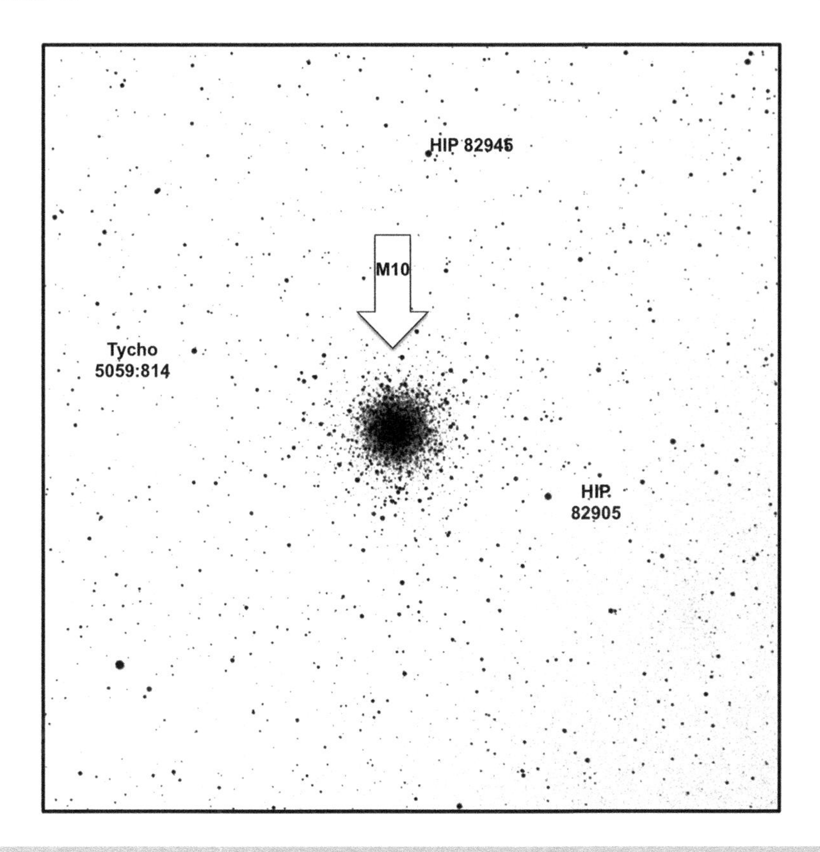

Fig. 3.29 Negative Image of Messier 10

The chart in Figure 3.30 shows the position of the M10 globular cluster within the constellation of Ophiuchus. In his notes, Messier referred to the proximity of "the 30th star of the constellation." The inset in the chart shows a magnified view of the area surrounding M10, which identifies this star. It is star 30 from Flamsteed's catalog, with the alternate designation of HIP 83262 from the Hipparcos catalog. HIP 83262 has a magnitude of 4.82 and is a cool K4 type star. Messier referred to this as a magnitude 6 star, and indeed, on the Flamsteed catalog 30 Ophiuchus was indeed given a magnitude 6 rating. The star lies at a distance of 402.17 ly from us.

The 300-second exposure image of M10 began at 04h 03m 19s on March 14 in New Mexico. The Universal Time was 10h 03m 19s. The Right Ascension of M10 is 16h 58m 03s, and at the time the image was taken, the Local Sidereal Time was 14h 28m 55s. The difference of about 2 ½ hours indicates the time before M10 would cross the meridian. The azimuth of M10 was 129° 03' 01", meaning that there was roughly a 50° azimuthal angle between the object and meridian, corresponding to the 2 ½ hours mentioned. The same telescope used for M9 was used to image M10 – a New Mexico-based 150-mm refractor.

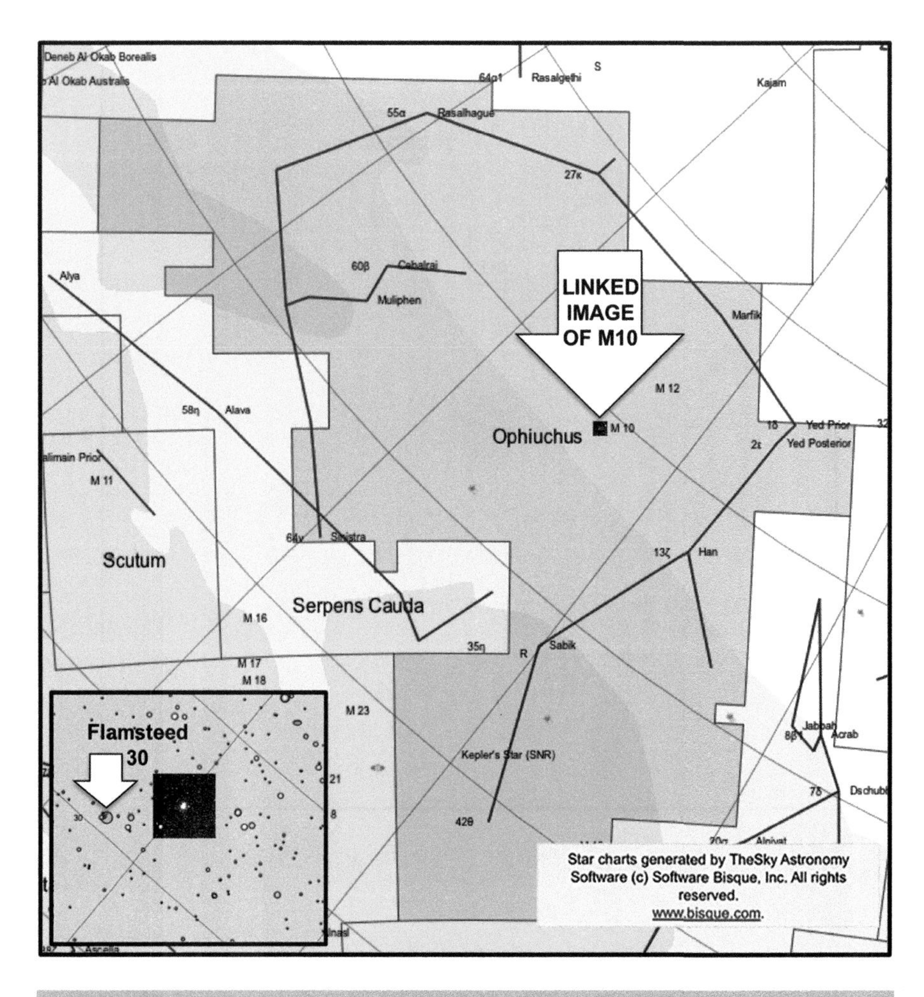

Fig. 3.30 Chart showing the location of M10

Messier 10: Data relating to the image in Figure 3.28

REMOTELY IMAGED FROM MAYHILL, NEW MEXICO
NGC 6254
OBJECT TYPE: Globular Cluster

RA: 16h 58m 03s
DEC: -04° 07' 27"
ALTITUDE: +38° 55' 12"
AZIMUTH: 129° 03' 01"
FIELD OF VIEW: 0° 47' 06" x 0° 47' 06"
OBJECT SIZE: 20’ X 20’
POSITION ANGLE: 180° 25' from North

EXPOSURE TIME: 300 s
DATE: 14th March
LOCAL TIME: 04h 03m 19s
UNIVERSAL TIME: 10h 03m 19s
SCALE: 1.38 arcsec/pixel
MOON PHASE: 96.61%

Telescope Optics
OTA: Takahashi TOA-150
Optical Design: Apochromatic Refractor
Aperture: 150 mm
Focal Length: 1095 mm
F/Ratio: f/7.3
Guiding: Internal
Mount: Paramount GTS

Instrument Package
SBIG ST-4000M One-Shot Color CCD
A/D Gain: 0.6e-/ADU
Pixel Size: 7.4um square
Resolution: 1.45 arcsec/pixel
Sensor: Frontlit
Cooling: -20°C Winter (-10°C Summer)
Array: 2048 x 2048 (8.3 Megapixels)
FOV: 49.6 x49.6 arcmin

Location
Observatory: New Mexico Skies
UTC Minus 7.00 (Daylight savings time is observed)
Minimum Target Elevation: Approx. 25 – 45 Degrees (N or S)
32° 54' Decimal: 32.9 North
(W or E) 105° 31' Decimal: 105.5 West
Elevation: 2225 meters (7298 ft)

Messier 11

Messier 11 (NGC 6705) is an open cluster in the constellation of Scutum. It is known as the Wild Duck Cluster and is shown in Figure 3.31. The cluster does not resemble an individual wild duck (although you could make most any shape by joining carefully selected dots/stars) – it looks like a flock of ducks flying by. It was so named by the astronomer Admiral William Henry Smyth, author of the well-known *Cycle of Celestial Objects*. Messier reported: "In the night of May 30 to 31, 1764, I have discovered, near the star Kappa of *Antinous*, a cluster of a large number of small stars which one perceives with good instruments; I have employed for

Fig. 3.31 Image of Messier 11, an open cluster in Scutum

this a Gregorian telescope which magnifies 104 times. When one examines it with an ordinary refractor of 3 & a half feet this star cluster resembles a comet; the center is brilliant, there is among the small stars one star of eighth magnitude; two other, one of the ninth & one of the tenth: this cluster is intermixed with a faint light, & its diameter is about 4 minutes of arc" (SEDS). Messier referred to the constellation of Antinous, which has become obsolete.

As always, there are various versions of the diameter of the object quoted in the literature. SEDS quoted 14' of arc, so I measured the size on the solved plate using the software tool incorporated into the SkyX professional software. A diameter of 14' does seem to include any stars that could be in the cluster -- based on appearance only. A circle is drawn of this approximate diameter on the negative image shown in Figure 3.32. Bear in mind that the size of the image is 47'. The distance to M11 is 6,000 ly. Using the small angle formula with that distance and a 14' diameter, we can determine the actual diameter to be roughly 25 ly across. The bright orange foreground star at the top right is HIP 92391, a cool K-type star 1,489.30 ly from us, only ¼ of the distance to M11.

The position of M11 in the constellation of Scutum is shown in the chart in Figure 3.33. In the chart, north is up and east is to the left. When the image was

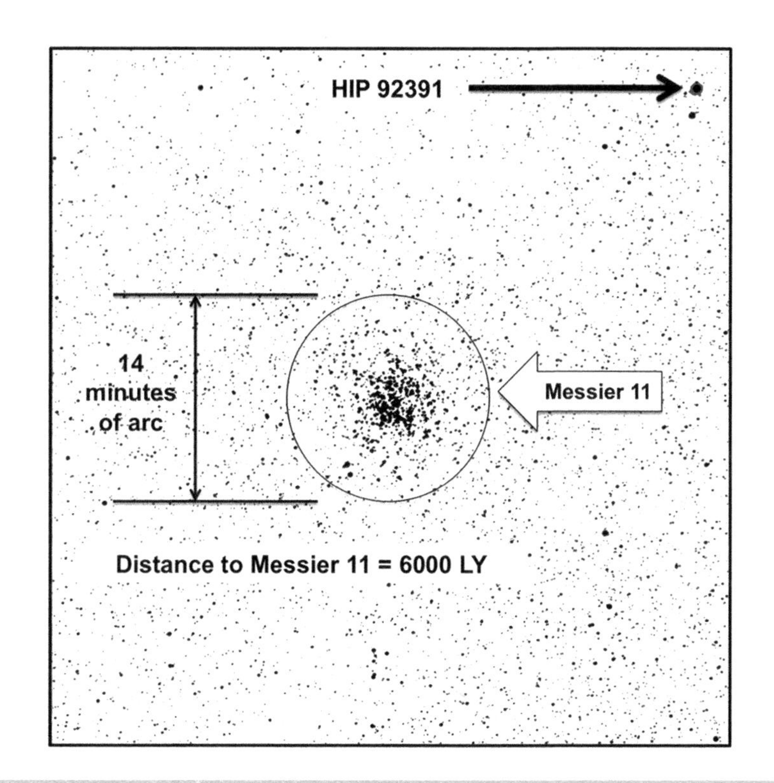

Fig. 3.32 Negative image of Messier 11

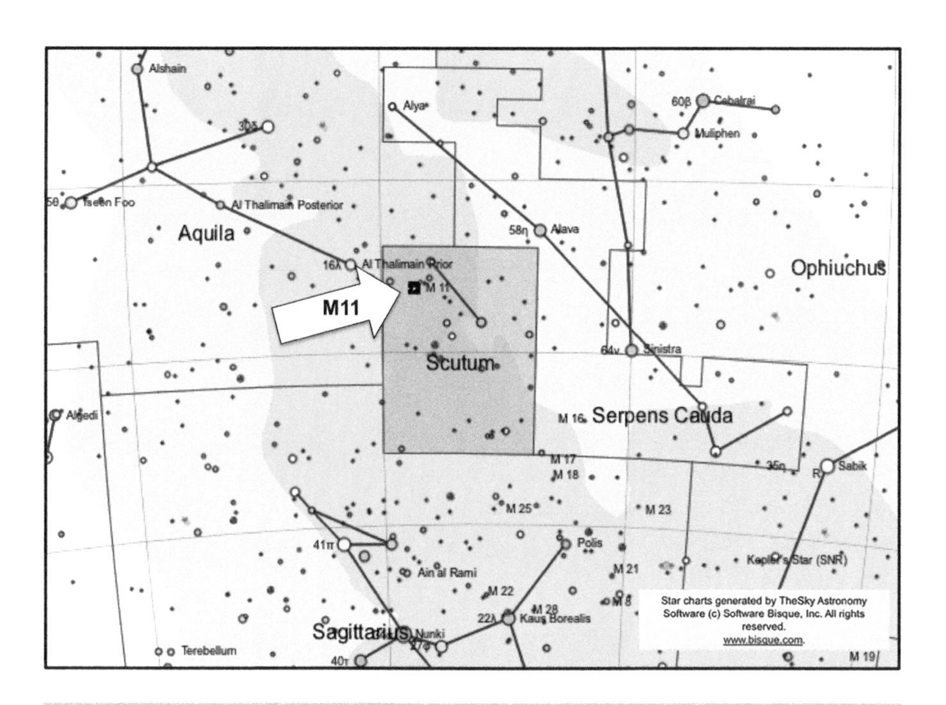

Fig. 3.33 Chart showing the location of M11

taken, the position angle of the camera was 180°, that is to say, the camera was "upside down" with respect to north. Figures 3.32 and 3.33 have both been roated to have north at the top. With my own equipment, I always try to align the camera as closely as possible with north so the position angle is 0° (or 360°). With remote telescopes, you have to accept the position of the camera, although if you could remotely control an electric rotator (a feature which is now sometimes available), you could adjust the position angle as you wish. Solving the plate allows the software to work out what the actual position angle of the camera is with respect to the sky. Scutum lies between the constellations of Aquila and Serpens Cauda above Sagittarius.

The image was taken at local time 04h 27m 10s in New Mexico on March 17. The Universal time was 10h 27m 10s. The Right Ascension of Messier 11 is 18h 52m 00s, while the Local Sidereal Time was 15h 04m 39s. Thus, there was about 3h 48m to go before M11 crossed the meridian just before 08h 15m. When the image was taken, the altitude of M11 was only 23° 28' 08" and the azimuth was 114° 52' 39". The image was taken with a with a 150-mm refractor.

Messier 11: Data relating to the image in Figure 3.31

REMOTELY IMAGED FROM MAYHILL, NEW MEXICO
NGC 6705 (WILD DUCK CLUSTER)
OBJECT TYPE: Open Cluster

RA: 18h 52m 00s
DEC: -06° 14' 43"
ALTITUDE: +23° 28' 08"
AZIMUTH: 114° 52' 39"
FIELD OF VIEW:47' 6''' X 47' 6"
OBJECT SIZE: 14' X 14'
POSITION ANGLE: 180° 25' from North
EXPOSURE TIME: 300 s
DATE: 17th March
LOCAL TIME: 04h 27m 10s
UNIVERSAL TIME: 10h 27m 10s
SCALE: 1.38 arcsec/pixel
MOON PHASE: 78.83% (waning)

Telescope Optics
OTA: Takahashi TOA-150
Optical Design: Apochromatic Refractor
Aperture: 150 mm
Focal Length: 1095 mm
F/Ratio: f/7.3
Guiding: Internal
Mount: Paramount GTS

Instrument Package
SBIG ST-4000M One-Shot Color CCD
A/D Gain: 0.6e-/ADU
Pixel Size: 7.4um square
Resolution: 1.45 arcsec/pixel
Sensor: Frontlit
Cooling: -20°C Winter (-10°C Summer)
Array: 2048 x 2048 (8.3 Megapixels)
FOV: 49.6 x49.6 arcmin

Location
Observatory: New Mexico Skies
UTC Minus 7.00 (Daylight savings time is observed)
Minimum Target Elevation: Approx 25 – 45 Degrees
(N or S) 32° 54' Decimal: 32.9 North
(W or E) 105° 31' Decimal: 105.5 West
Elevation: 2225 meters (7298 feet)

Messier 12

Messier 12 (NGC 6218) is yet another globular cluster in Messier's list, shown in Figure 3.34. It is located in the constellation of Ophiuchus, not far from a similar globular, Messier 10. M12 has a diameter of 16' of arc, whereas M10 has a diameter of 20' (SEDS). The image has north at the top and east to the left. M12 lies at a distance of 16,000 ly, while M10 lies at a distance of 14,300 ly. M12 was discovered by Messier, who noted, "In the night of May 30 to 31, 1764, I have discovered a nebula in Serpens, between the arm & left side of *Ophiuchus*, according to the charts of Flamsteed: That nebula doesn't contain any star; it is round, its diameter can be 3 minutes of arc, its light is faint; on sees it very well with an ordinary refractor of 3 feet I have determined its position, by comparing with the star Delta *Ophiuchi*." The 5-minute monochrome image shown in Figure 3.34 is a cropped section of the original image. Comparing this M12 image with that of M10, you can see that the latter seems more concentrated, with a brighter appearance towards the center. Bear in mind that the M12 image is cropped and enlarged, whereas the M10 image is not.

The negative image shown in Figure 3.35 also has north at the top and east to the left. Some foreground stars of similar apparent magnitude are indicated on the image to put things into perspective. The star Tycho 5054:4 has an apparent magnitude of 10.10 and is at a distance of 189.63 ly, about 84 times nearer to us then M12. From the apparent magnitude and distance, the star has a calculated absolute magnitude of 6.28. A typical star selected in the cluster itself was in the region of magnitude 15, so at a distance of 16000 ly, its absolute magnitude would be 1.55. This shows just how bright a star has to be to appear to be magnitude 15 to us from

Fig. 3.34 Image of Messier 12, a globular cluster in Ophiuchus

a distance of 16,000 ly. Tycho 5054:85 has a magnitude of 10.66 and lies at 46.40 ly. This is a lot nearer to us than Tycho 5054:4, so it must be a lot fainter in absolute terms. Tycho 5054:85 has a calculated absolute magnitude of 9.89. Tycho 5050:86 has an apparent magnitude of 10.52 and lies at a distance of 37.71 ly, while Tycho 5054:17 has an apparent magnitude of 10.80 and lies at a distance of 258.85 ly. The apparent magnitudes and distances were taken from the Tycho Catalog.

The chart in Figure 3.36 shows the location of Messier 12 in the constellation of Ophiuchus. Notice that M10 is not too far away from M12 in the same constellation. There is a distance of about 3° 15' separation on the chart. Ophiuchus has a total of 7 Messier globular clusters. In addition to this, there are roughly 14 additional globulars clusters, such as NGC 6633. There are no Caldwell objects in this constellation.

The image was taken at 04h 02m 27s on March 7 from New Mexico. The Universal Time was 11h 02m 27s. The Right Ascension of M12 is 16h 48m 08s, and the Local Sidereal Time when the image run was started was 15h 00m 37s. The object would therefore transit the meridian at around 05h 50m local time. The azimuth at the start of the exposure was 138° 33' 10", and the altitude of M12 was +46° 57' 03". The 300-second image was taken with a corrected Dall-Kirkham Astrograph.

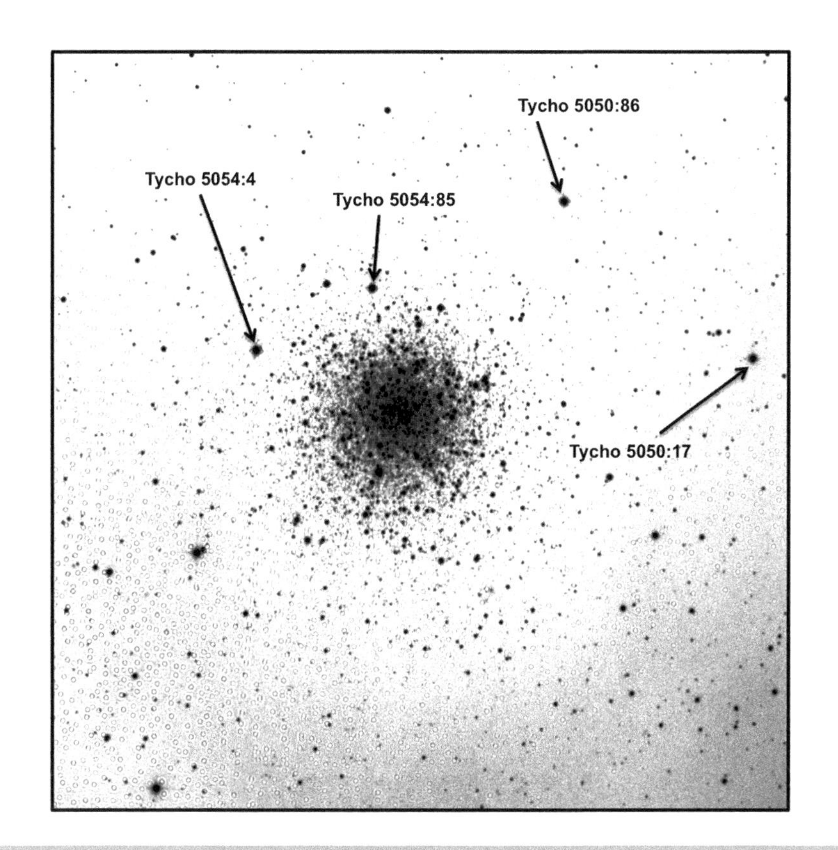

Fig. 3.35 Negative image of Messier 12

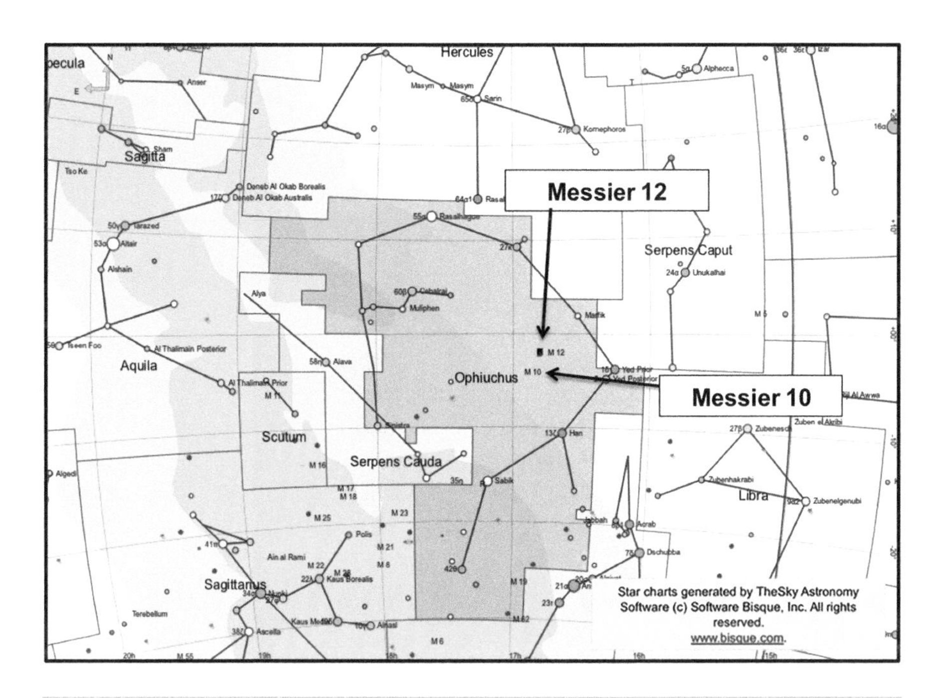

Fig. 3.36 Chart showing the location of M12

Messier 12: Data relating to the image in Figure 3.34

REMOTELY IMAGED FROM MAYHILL, NEW MEXICO
NGC 6218
OBJECT TYPE: Globular Cluster

RA: 16h 48m 08s
DEC: -01° 58' 36"
ALTITUDE: +46° 57' 03"
AZIMUTH: 138° 33' 10"
FIELD OF VIEW: 0° 54' 47" x 0° 36' 31"
OBJECT SIZE: 16' x 16'
POSITION ANGLE: 271° 11' from North
EXPOSURE TIME: 300 s
DATE: 7th March
LOCAL TIME: 04h 02m 27s
UNIVERSAL TIME: 11h 02m 27s
MOON PHASE: 71.82% (waxing)

Telescope Optics
OTA: Planewave 20" CDK
Optical Design: Corrected Dall-Kirkham Astrograph
Aperture: 510 mm
Focal Length: 2280 mm (0.66 Focal Reducer Fitted)
F/Ratio: f/4.5
Guiding: Active Guiding Disabled
Mount: Planewave Ascension 200HR

Instrument Package
FLI Proline PL11002M CCD
Camera Pixel Size: 9 μm square
Resolution: 0.81 arcsec/pixel
Cooling: -30°C default
Array: 4008 x 2672 (10.7 Megapixels)

Location
Observatory: New Mexico Skies
UTC Minus 7.00 (Daylight savings time is observed)
Minimum Target Elevation: Approx 25 – 45 Degrees
(N or S) 32° 54' Decimal: 32.9 North
(W or E) 105° 31' Decimal: 105.5 West
Elevation: 2225 meters (7298 feet)

Messier 13

Messier 13 is probably the most spectacular of the globular clusters on Messier's List, as shown in Figure 3.37. With an approximate diameter of 20' of arc, it appears as a massive ball of stars. M13 was originally discovered in 1714 by Edmund Halley of Halley's Comet fame but was catalogued by Messier 50 years later on June 1, 1764. Of M13, Messier commented, "Nebula without star, discovered in the belt of Hercules; it is round & brilliant, the center [is] more brilliant than the edges, one perceives it with a telescope of one foot; it is near two stars, the one & the other of 8th magnitude, the one above and the other below it: the [position of the] nebula was determined by comparing it with Epsilon Herculis" (SEDS). M13 can be visible to the naked eye in dark skies, as pointed out by Halley himself. British astronomer Reverend T.W. Webb regarded it as "*spangled with glittering points* in a 5 ½ foot achromat, becoming a superb object in large telescopes."

Messier 13 lies at a distance of 25,100 ly from us (SEDS). This is roughly ¼ of the distance of the major axis of our Galaxy. The negative image in Figure 3.38 shows foreground star Tycho 2588:1662, which has a magnitude of 10.99 and lies at a distance of only 72 ly. The cluster members appear fainter than this star in the region of magnitude 12. There have been many calculations over the years as to the number of stars in this globular cluster -- perhaps you could try your own estimate from the photograph! The cluster subtends an angle of 20 arcmin at a distance of 25,100 ly. My rough calculations based on these figures give an approximate diameter of about 150 ly.

Fig. 3.37 Image of Messier 13, a globular cluster in Hercules

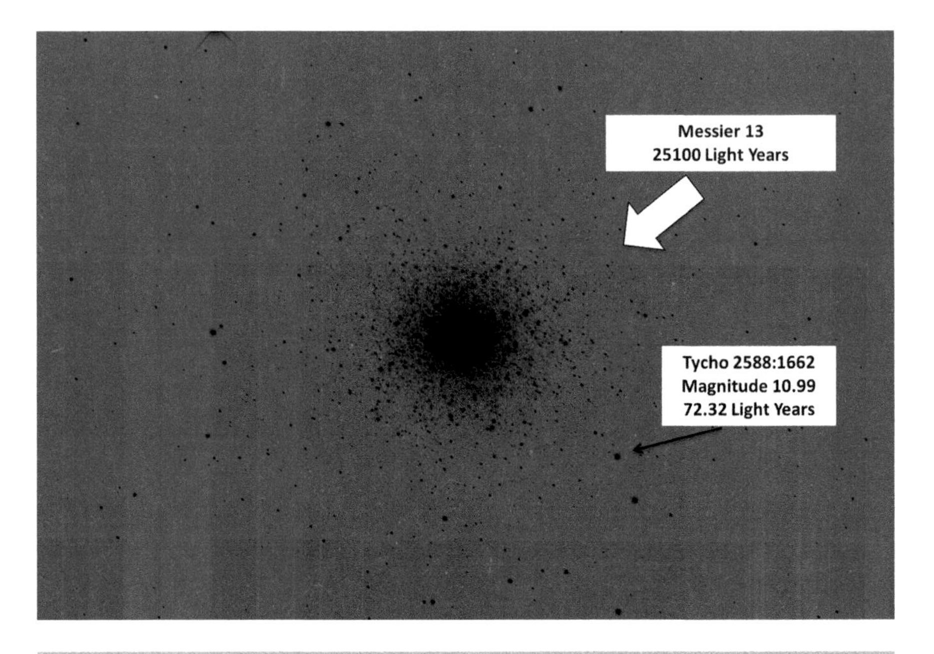

Fig. 3.38 Negative image of Messier 13

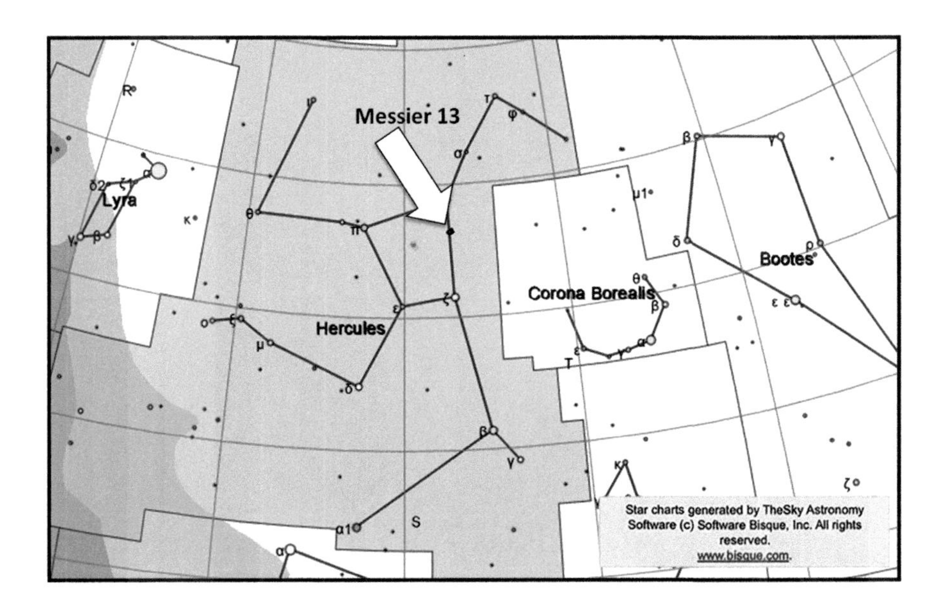

Fig. 3.39 Chart showing the location of M13

M13 lies in the constellation of Hercules, as shown on the chart in Figure 3.39. Again, the black rectangle is the actual image superimposed on the chart. The plate solution gives a position angle of 220° 26' from North (position angle is measured clockwise from the north – so the image has to be rotated 220 degrees counter-clockwise to put north at the top). The image has an angular size of 0° 37' 16" x 0° 24' 51". M13 lies between the stars ζ and η, towards the western side of the constellation. Of course, there is another globular cluster in Hercules listed by Messier – the globular M92 -- that we will see later in the book. M92 and M13 are the only Messier objects in Hercules. Note that there are no Caldwell objects in this constellation.

The image was taken on January 8 at 07h 13m 54s local time in Spain. At the time, the Moon was almost 60° away at an altitude of 22° and a phase of 17.6% (waning), causing no interference. This corresponded to a Universal Time of 06h 13m 54s. The Right Ascension of M13 is 16h 42m 09s, and the Local Sidereal Time was 13h18m 08s. Subtracting one from the other gives a difference of 3h 24m 01s -- the time still to go before M13 crosses the meridian.

The Air Mass, which indicates how much atmosphere the telescope is looking through, was 1.30. The declination of M13 is just less than 37°, by coincidence very close to the Latitude of the Spanish site. When an object that has a declination equal to the Latitude is targeted, the object will pass directly overhead. When that happens, the Air Mass is 1, which corresponds to the minimum thickness of atmosphere – the best condition for imaging. Thus, M13 would have been at its best when it crossed the meridian – if it had not been broad daylight! The telescope used to take the image of M13 was a Corrected Dall-Kirkham Astrograph with an aperture of 318 mm mounted on a Paramount ME.

Messier 13: Data relating to the image in Figure 3.37

REMOTELY IMAGED FROM NERPIO, SPAIN
NGC 6205
OBJECT TYPE: Globular Cluster

RA: 16h 42m 09s
DEC: +36° 26' 06"
ALTITUDE: +49° 39' 04"
AZIMUTH: 74° 57' 13"
FIELD OF VIEW: 0° 37' 16" x 0° 24' 51"
OBJECT SIZE: 20’
POSITION ANGLE: 220° 26' from North
EXPOSURE TIME: 300 s
DATE: 8th January
LOCAL TIME: 07h 13m 54s
UNIVERSAL TIME: 06h 13m 54s
SCALE: 0.73 arcsec/pixel
MOON PHASE: 17.6% (waning)

Telescope Optics
OTA: Planewave CDK
Optical Design: Corrected Dall-Kirkham Astrograph
Aperture: 318 mm
Focal Length: 2541 mm
F/Ratio: f/7.9
Guiding: External
Mount: Paramount ME

Instrument Package
CCD: SBIG-STXL-6303E
A/D Gain: 1.47 e-/ADU
Pixel Size: 9μm square
Resolution: 0.73 arcsec/pixel
Sensor: Front Illuminated
Cooling: -20°C default
Array: 3072 x 2048 (6.3 Megapixels)
FOV: 37.41 x 24.94 arcmin

Location
Observatory: AstroCamp – MPC Code -189
UTC +1.00 (Daylight savings time is observed)
Minimum Target Elevation: Approx 30 – 40 Degrees
N 38° 09'
W 002° 19'
Elevation: 1650 meters (5413 ft)

Messier 14

Messier 14 (NGC 6402) is a globular cluster located in the constellation of Ophiuchus. Quite a number of publications about Messier objects fail to mention M14, or merely gloss over it. It is much fainter than the magnificent globular M13, but with a 5-minute exposure on a good telescope (Image shown in Figure 3.40), it is spectacular in its own right. This is what Messier had to say when he discovered it on the night of June 1, 1764: "I have discovered a new nebula in the garb which dresses the right arm of *Ophiuchus;* on the charts of Flamsteed it is situated on the parallel of the star Zeta Serpentis: that nebula is not considerable, its light is faint, yet it is seen well with an ordinary refractor of 3 feet & a half it is round, & its diameter can be 2 minutes of arc; above it & very close to it is a small star of the ninth magnitude. I have employed for seeing this nebula nothing but the ordinary refractor of 3 feet & a half with which I have not noticed any star; maybe with a larger instrument one could perceive one. I have determined the position of that nebula by its passage of the Meridian, comparing it with Gamma *Ophiuchi*" (SEDS). M14 lies at a distance of 30,300 ly from us, in comparison with M13's distance of 25,100 ly.

Fig. 3.40 Image of Messier 14, a globular cluster in Ophiuchus

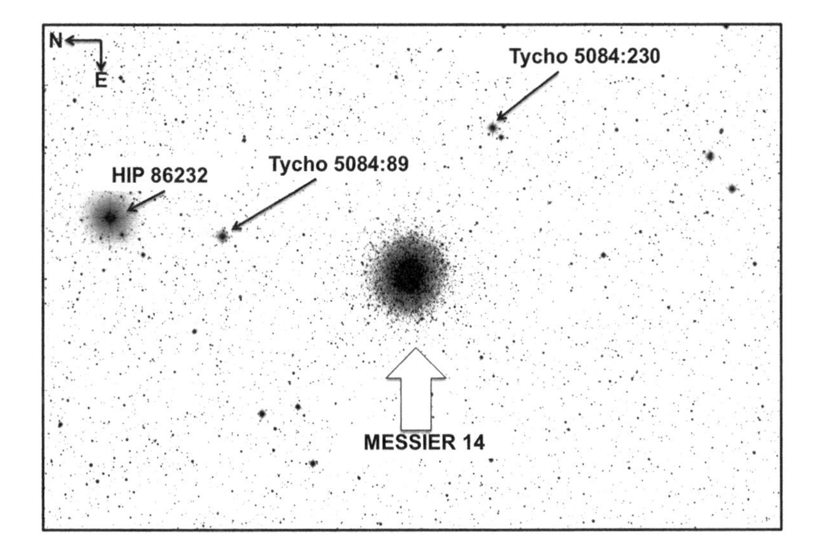

Fig. 3.41 Negative image of Messier 14

The negative image in Figure 3.41 shows M14 and a nearby bright star. This star is HIP 86232, which is of magnitude 7.43 and spectral type G5. The Hipparcos Catalog places this at a distance of 243.40 ly, so, although they may look close on the image, M14 is actually 125 times further away than HIP 86232. Messier made refer-

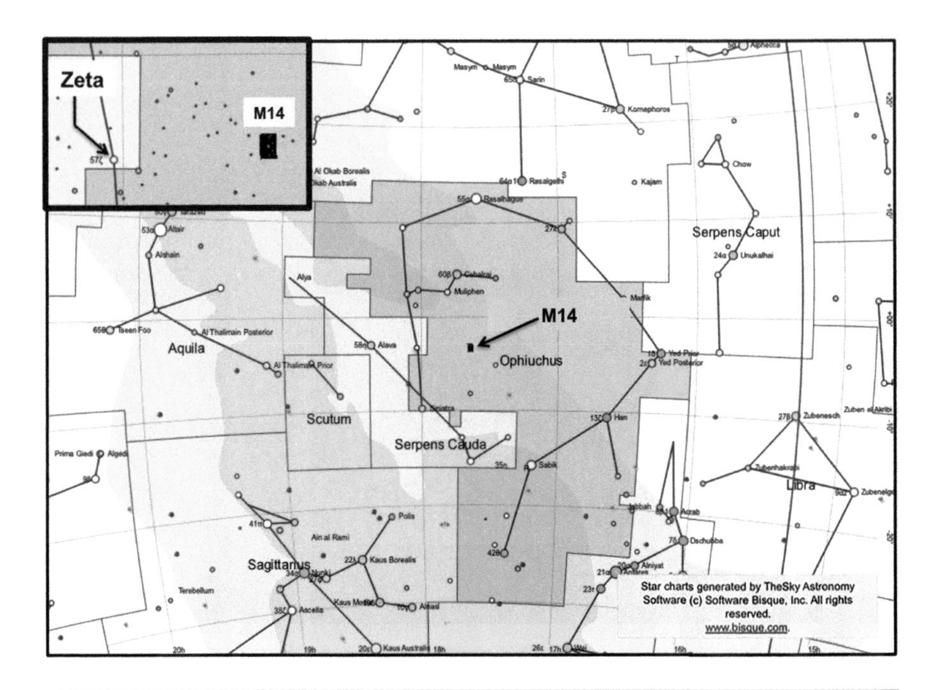

Fig. 3.42 Chart showing the location of M14

ence to a close star of magnitude 9. Could this be the star Tycho 5084:89 arrowed on the negative image? This star has a magnitude of 9.27. Note that north is to the left in Figure 2.54. Another candidate for Messier'a 9th magnitude star is the star Tycho 5084:230 which has an apparent magnitude of 9.58 and is at a distance of 1,019.24 ly. Stars in the outer edge of the cluster are in the region of magnitude 14 or 15. The diameter of Messier 14 appears in the image to be only a few arcmin, but in fact does extend out to around 11 arcmin. At a distance of 30,300 ly, this would give it an actual diameter of 97 ly, using the small angle formula.

The chart in Figure 3.42 shows the position of M14 in Ophiuchus. Note that unlike the images that have north to the left, this shows north at the top. This corresponds to the image having a position angle of 271° 11' from North. Messier mentioned that M14 was on the same parallel as Zeta Serpentis. The inset showing the enlarged area confirms that this is the case. M14 has a declination of -03° 15' 15", and Zeta Serpentis has a declination of -03° 41' 22" -- fairly close declination values. However, Zeta Serpentis is only 76 ly away from us, which makes the star 400 times nearer than M14.

The image was taken at the local time 03h 32m 24s on April 3 from New Mexico. The corresponding Universal Time was 09h 32m 24s. Messier 14 has a Right Ascension of 17h 38m 30s, and the Local Sidereal Time was 15h 16m 46s, with a difference of 2h 21m 44s. Adding this to the local time means that the local meridian transit of M14 would be at 05h 54m 07s later that morning. At the time the image was taken, the altitude of M14 was +40° 46' 46" and the azimuth was

130° 08' 33". At meridian transit, the altitude of M14 would be +53° 57' 45" and the azimuth would be 180°. The image was taking using a 510-mm telescope and a monochrome camera.

Messier 14: Data relating to the image in Figure 3.40

REMOTELY IMAGED FROM MAYHILL, NEW MEXICO
NGC 6402
OBJECT TYPE: Globular Cluster

RA: 17h 38m 30s
DEC: -03° 15' 15"
ALTITUDE: +40° 46' 46"
AZIMUTH: 130° 08' 33"
FIELD OF VIEW: 0° 54' 47" x 0° 36' 31"
OBJECT SIZE: 11' x 11'
POSITION ANGLE: 271° 11' from North
EXPOSURE TIME: 300 s
DATE: 3rd April
LOCAL TIME: 03h 32m 24s
UNIVERSAL TIME: 09h 32m 24s
MOON PHASE: 45.82% (waxing)

Telescope Optics
OTA: Planewave 20" CDK
Optical Design: Corrected Dall-Kirkham Astrograph
Aperture: 510 mm
Focal Length: 2280 mm (0.66 Focal Reducer Fitted)
F/Ratio: f/4.5
Guiding: Active Guiding Disabled
Mount: Planewave Ascension 200HR

Instrument Package
FLI Proline PL11002M CCD
Camera Pixel Size: 9-µm square
Resolution: 0.81 arcsec/pixel
Sensor:
Cooling: -30°C default
Array: 4008 x 2672 (10.7 Megapixels)

Location
Observatory: New Mexico Skies
UTC Minus 7.00 (Daylight savings time is observed)
Minimum Target Elevation: Approx 25 – 45 Degrees
(N or S) 32° 54' Decimal: 32.9 North
(W or E) 105° 31' Decimal: 105.5
West Elevation: 2225 meters (7298 ft)

Messier 15

Messier 15 (NGC 7078) is another of the Galaxy's globular clusters and is located in the constellation of Pegasus. It is about 18' of arc in diameter and is shown in Figure 3.43. On June 3, 1764, Messier wrote that M15 was "[A] Nebula without a star, between the head of Pegasus and that of Equuleus; it is round, in the center it is brilliant, its position was determined by comparison with Delta Equulei. M. Maraldi, in the Memoirs of the Academy of 1746, reports of this nebula: 'I have found, he says, between the stars Epsilon Pegasi and Beta Equulei, a fairly bright nebulous star, which is composed of many stars; its right ascension is 319d 27' 6", and its northern declination is 11d 2' 22" (Diam. 3')" (SEDS). As Messier pointed out, M15 was originally discovered in 1746 by Jean-Dominique Maraldi, some 18 years before his note on the object. You may recall that Maraldi had also discovered M2 when observing Comet de Chésaux. Maraldi came across M15 on September 7, 1746, and M2 just a few days later on September 11, 1746.

Fig. 3.43 Image of Messier15, a globular cluster in Pegasus

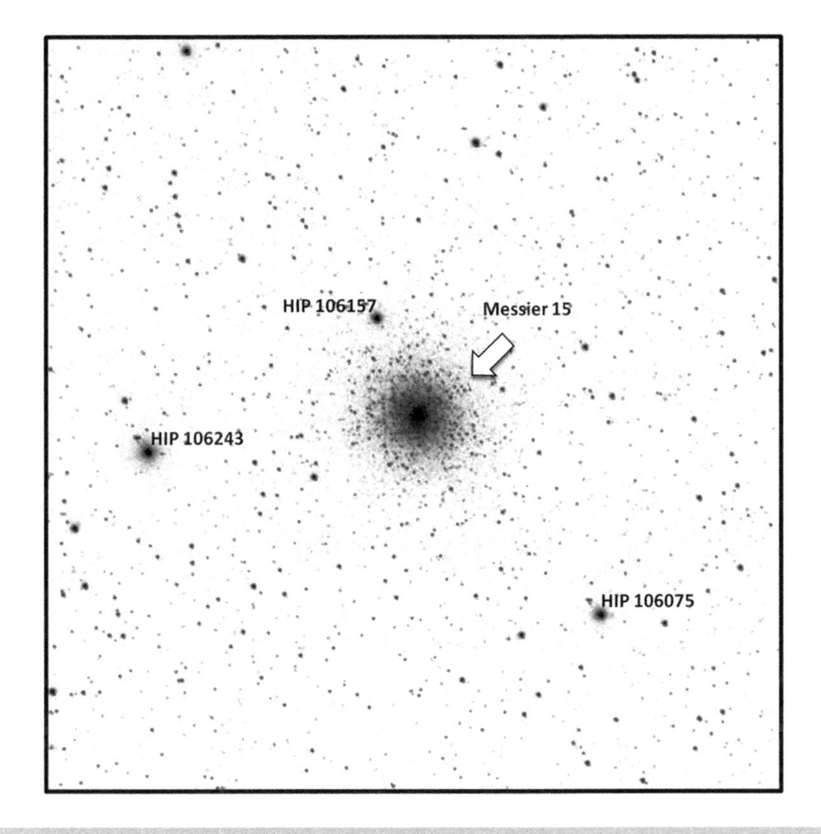

Fig. 3.44 Negative image of Messier 15

There are three bright line of sight stars surrounding M15, indicated on the negative image in Figure 3.44. The brightest of these, HIP 106243, is magnitude 6.1 and lies at a distance of 417.0 ly. The second brightest of the three stars is HIP 106075, which is magnitude 7.31 and is 718.4 ly distant. The faintest of the three is HIP 106157 with magnitude 7.62, lying at a distance of a mere 197.67 ly. Clearly, the first and second stars must be intrinsically brighter than the third. HIP 106243 appears bluish in color and in fact is a hot blue star of spectral type B9.5, indicating a surface temperature of around 11,000 K. HIP 106075 appears reddish and is a cool star of spectral type K2 – a temperature of approximately 446 0K. The third star HIP 106157 has a whitish appearance and is spectral type G0 with a temperature of around 6,000 K. Messier 15 itself is at a distance of about 33,600 ly (SEDS).

M15 is located in the constellation of Pegasus, adjacent to the constellation of Andromeda, as shown in the chart in Figure 3.45. The black square is the actual image to scale. M15 is the only Messier object in Pegasus, but there are a number of interesting other objects, including the spiral galaxy NGC 7331 (also known as Caldwell 30, or C30). English amateur astronomer Patrick Moore generated a list of objects that Messier had not included but are worth highlighting. Pegasus

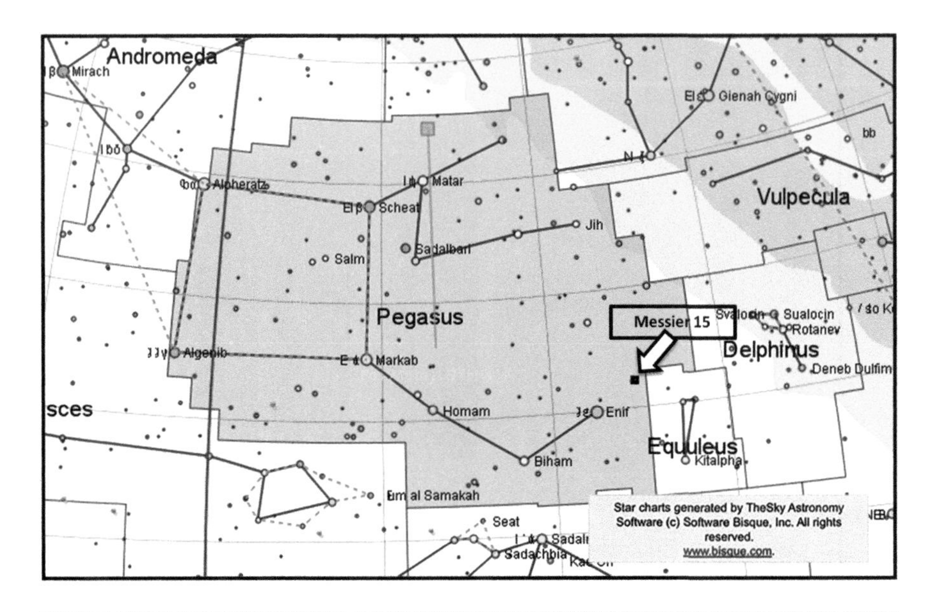

Fig. 3.45 Chart showing the location of M15

contains three other Caldwell Objects -- C43 (NGC 7814), C44 (NGC 7479) and C57 (NGC 6822). The plate solution gives a position angle of 358° 17', so the image has north towards the top but rotated by slightly less than 2° from the north-south line.

The image was taken in New Mexico at local time 22h 55m 39s, when M15 was at an altitude of just over 50° and at an azimuth of 247° 51' 16". The lunar phase at the time was 45.82% (waning). M15 has a Right Ascension of 21h 30m 47s and, at the time that the image was taken, the Local Sidereal Time was 00h 00m 33s. Thus, the First Point of Aries (now in Scorpius, not Aries) was coincidentally within 33 seconds of the meridian. This means that M15 was roughly two and a half hours to the west of the meridian at that time. The transit time would have been roughly 08h 26m local time. It is worth becoming familiar with the link between RA and LST to help in your imaging planning. Remember: the best time to image an object is when it is on the meridian. However, factors such as lunar interference and telescope access do not always make that possible.

Messier 15: Data relating to the image in Figure 3.43

REMOTELY IMAGED FROM MAYHILL, NEW MEXICO
NGC 7078
OBJECT TYPE: Globular Cluster

RA: 21h 30m 47s
DEC: +12° 14' 44"
ALTITUDE: +50° 05' 59"

AZIMUTH: 247° 51' 16"
FIELD OF VIEW: 0° 47' 06" x 0° 47' 06"
OBJECT SIZE: 18' x 18'
POSITION ANGLE: 358° 17' from North
EXPOSURE TIME: 300 s
DATE: 22nd October
LOCAL TIME: 22h 55m 39s
UNIVERSAL TIME: 04h 55m 39s
SCALE: 1.38 arcsec/pixel
MOON PHASE: 45.82% (waning)

Telescope Optics
OTA: Takahashi TOA-150
Optical Design: Apochromatic Refractor
Aperture: 150 mm
Focal Length: 1095 mm
F/Ratio: f/7.3
Guiding: Internal
Mount: Paramount GTS

Instrument Package
SBIG ST-4000M One-Shot Color CCD
A/D Gain: 0.6e-/ADU
Pixel Size: 7.4um square
Resolution: 1.45 arcsec/pixel
Sensor: Frontlit
Cooling: -20°C Winter (-10°C Summer)
Array: 2048 x 2048 (8.3 Megapixels)
FOV: 49.6 x49.6 arcmin

Location
Observatory: New Mexico Skies
UTC Minus 7.00 (Daylight savings time is observed)
Minimum Target Elevation: Approx. 25 – 45 Degrees
(N or S) 32° 54' Decimal: 32.9 North
(W or E) 105° 31' Decimal: 105.5 West
Elevation: 2225 meters (7298 ft)

Messier 16

Messier 16 (NGC 6611) is a rather spectacular Messier object in that it consists of a star cluster embedded in a glowing nebula (shown in Figure 3.46). It is located in the constellation of Serpens Cauda. Messier noted, "In the same night of June 3 to

Fig. 3.46 Image of Messier 16, an open cluster and emission nebula in Serpens Cauda

4, 1764, I have discovered a cluster of small stars, mixed with a faint light, near the tail of Serpens, at little distance from the parallel of the star Zeta of that constellation: this cluster may have 8 minutes of arc in extension: with a weak refractor, these stars appear in the form of a nebula; but when employing a good instrument one distinguishes these stars, & one remarks in addition a nebulosity which contains three of these stars. I have determined the position of the middle of this cluster; its right ascension was 271d 15' 3", & its declination 13d 51' 44" south" (SEDS). With his telescope, Messier was able to observe the cluster, but not the extended nebula that modern telescopes equipped with cameras can now bring out. He was aware of the nebulosity, however, as his notes above show. The cluster itself is given the designation NGC 6611 and the nebula IC 4703, but the "object" is generally referred to as M16, or the Eagle Nebula.

M16 is listed by SEDS as lying at a distance of 7,000 ly and having a diameter of 7' of arc for the cluster. The Messier object shot to fame when images taken by the Hubble Space Telescope in 1995 illustrated star formation in progress in the spectacular "Pillars of Creation." The location of these are identified on the negative image in Figure 3.47. Another area that Hubble imaged was the "Stellar Spire,"

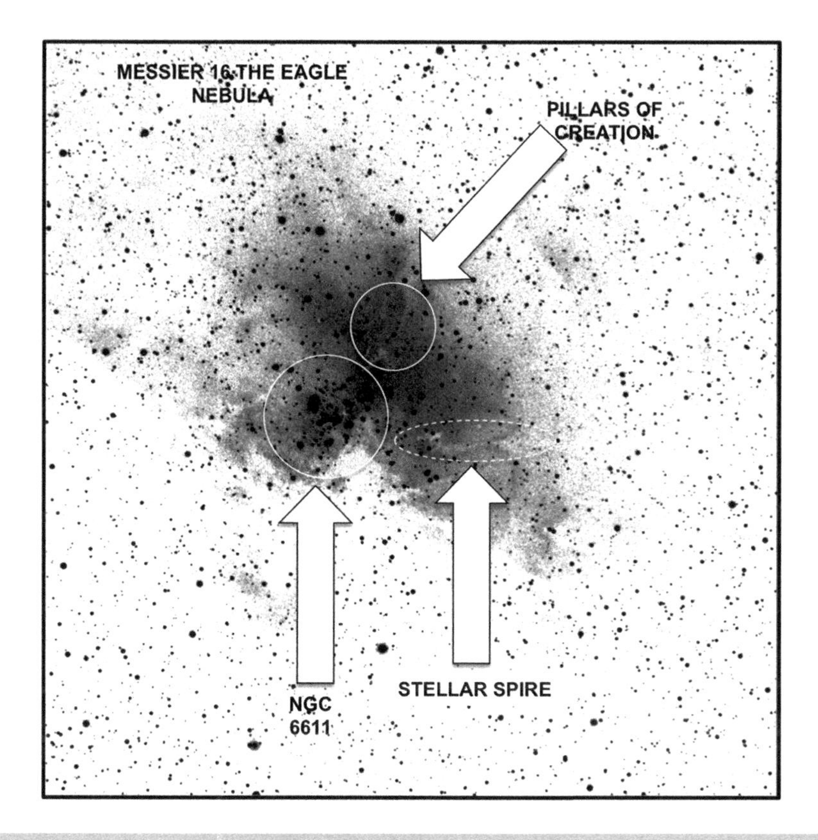

Fig. 3.47 Negative image of Messier 16

described as "a creature from a fairy tale atop a pedestal." The general area is identified in the negative image. You should be able to identify the features from the color image, but the 6-inch Earth-based refractor used to take the image is not quite a match for the Hubble Space Telescope. Perhaps one day, it will be a routine matter for amateur astronomers to take control of a space-based telescope for imaging! At present, it is a far too expensive technique to implement. It is the star cluster shown in the negative image that is actually powering the nebula.

The chart in Figure 3.48 shows the location of M16 towards the southern end of the constellation of Serpens Cauda. The constellation of Serpens has two parts. Serpens Cauda is the tail of the serpent, while Serpens Caput is the head of the serpent. The serpent constellations straddle the constellation of Ophiuchus. Messier mentioned the proximity of the star Zeta (ζ) Serpentis. This is shown on the chart. It is now known as HIP 88175. This star has a magnitude of 4.62 and is a main sequence F3 star. The Hipparcos Catalog gives it a distance of 75.66 ly. Contrast this with the vast distance to Messier 16 at 7,000 ly, and it means that in effect, the star is in our backyard.

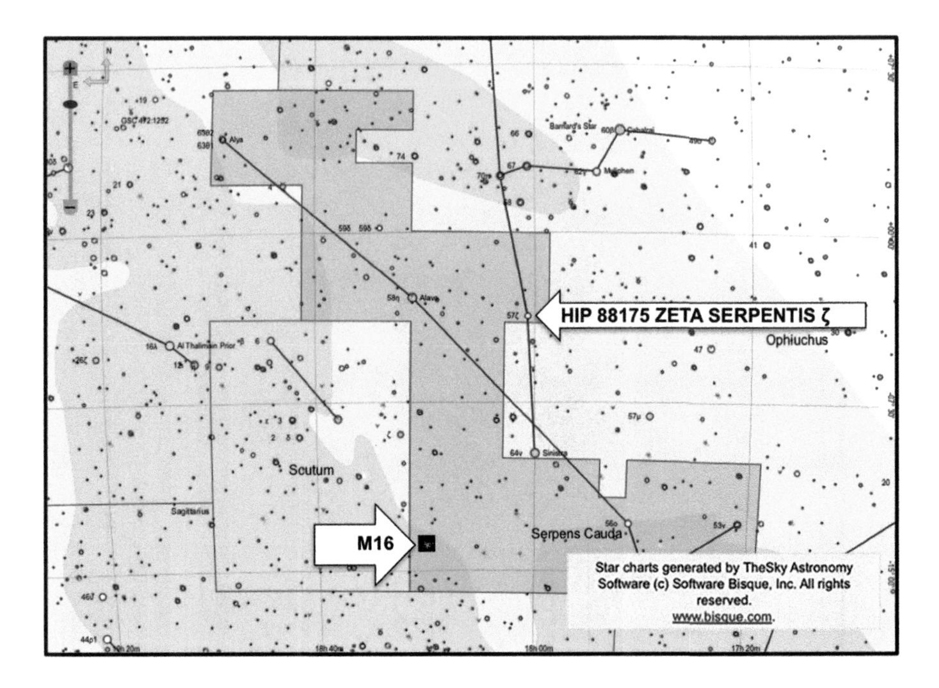

Fig. 3.48 Chart showing the location of M16

The image of M16 was taken in New Mexico at 05h 04m 54s on March 9. The Universal Time was 12h 04m 54s. The Right Ascension of M16 is 18h 19m 52s, and the LST was 16h 11m 07s. The difference is 2h 08m 45s. M16 was therefore heading towards the meridian and would reach it at roughly 07h 13m. The azimuth angle was 141° 18' 55", so it would move through about 39° of azimuth to reach the meridian. The telescope used was a 6-inch refractor by Takahashi.

Messier 16: Data relating to the image in Figure 3.46

REMOTELY IMAGED FROM MAYHILL, NEW MEXICO
NGC 6611
OBJECT TYPE: OPEN CLUSTER AND NEBULA

RA: 18h 19m 52s
DEC: -13° 50' 08"
ALTITUDE: +34° 09' 04"
AZIMUTH: 141° 18' 55"
FIELD OF VIEW: 0° 47' 08" x 0° 47' 08"
OBJECT SIZE: 7' X 7' (Cluster)
POSITION ANGLE: 180° 24' from North
EXPOSURE TIME: 300 s
DATE: 9th March

LOCAL TIME: 05h 04m 54s
UNIVERSAL TIME: 12h 04m 54s
SCALE: 1.38 arcsec/pixel
MOON PHASE: 89.34% (waxing)

Telescope Optics
OTA: Takahashi TOA-150
Optical Design: Apochromatic Refractor
Aperture: 150 mm
Focal Length: 1095 mm
F/Ratio: f/7.3
Guiding: Internal
Mount: Paramount GTS

Instrument Package
SBIG ST-4000M One-Shot Color CCD
A/D Gain: 0.6e-/ADU
Pixel Size: 7.4um square
Resolution: 1.45 arcsec/pixel
Sensor: Frontlit
Cooling: -20°C Winter (-10°C Summer)
Array: 2048 x 2048 (8.3 Megapixels)
FOV: 49.6 x49.6 arcmin

Location
Observatory: New Mexico Skies
UTC Minus 7.00 (Daylight savings time is observed)
Minimum Target Elevation: Approx. 25 – 45 Degrees
(N or S) 32° 54' Decimal: 32.9 North
(W or E) 105° 31' Decimal: 105.5
West Elevation: 2225 meters (7298 ft)

Messier 17

Messier 17 (NGC 6618) is another spectacular star-powered nebula of similar apparent size to M16. It also goes by the name of the Swan Nebula, or sometimes the Omega Nebula. Figure 3.49 clearly shows the swan shape, but note that south must be oriented at the top to give this appearance. Other patterns have been found in this object, such as the Greek capital letter Omega and either a horseshoe or lobster (as it is occasionally called). The appearance of the object varies considerably with the equipment used. The 5-minute exposure shown would have seemed miraculous to Messier observing it in 1764. He wrote, "In the same night [June 3 to 4, 1764], I have discovered at little distance of the cluster of stars of which I just

Fig. 3.49 Image of Messier 17, an emission nebula in Sagittarius

have told, a train of light of five or six minutes of arc in extension, in the shape of a spindle, & in almost the same as that in the girdle of Andromeda [M31]; but of a very faint light, not containing any star; one can see two of them nearby which are telescopic & placed parallel to the Equator: in a good sky one perceives very well that nebula with an ordinary refractor of 3 feet & a half" (SEDS).

In his notes, Messier referred to two stars that are telescopic and "placed parallel to the equator," I believe he was referring to the two magnitude 8 stars, HIP 89956 and HIP 89963, which are shown on the negative image in Figure 2.66. These stars are certainly telescopic in that they are too faint to be seen with the naked eye. HIP 89956 has a magnitude of 8.08 and a Declination of -16° 21' 54.449", while HIP 89963 has a magnitude of 8.36 and a Declination of -16° 21' 48.630". There is only a 6-second difference in their Declination, making them effectively parallel to the celestial equator. These stars lie at 1,614.64 ly and 1,342.21 ly respectively, so they are roughly 270 ly apart and are not physically associated. M17 lies at a distance of about 5,000 ly, so the two stars are in the foreground. For reference in the negative image, the bright star HIP 89851 is arrowed. This star has a magnitude of 5.39 and thus should definitely be visible with the naked eye under good viewing conditions.

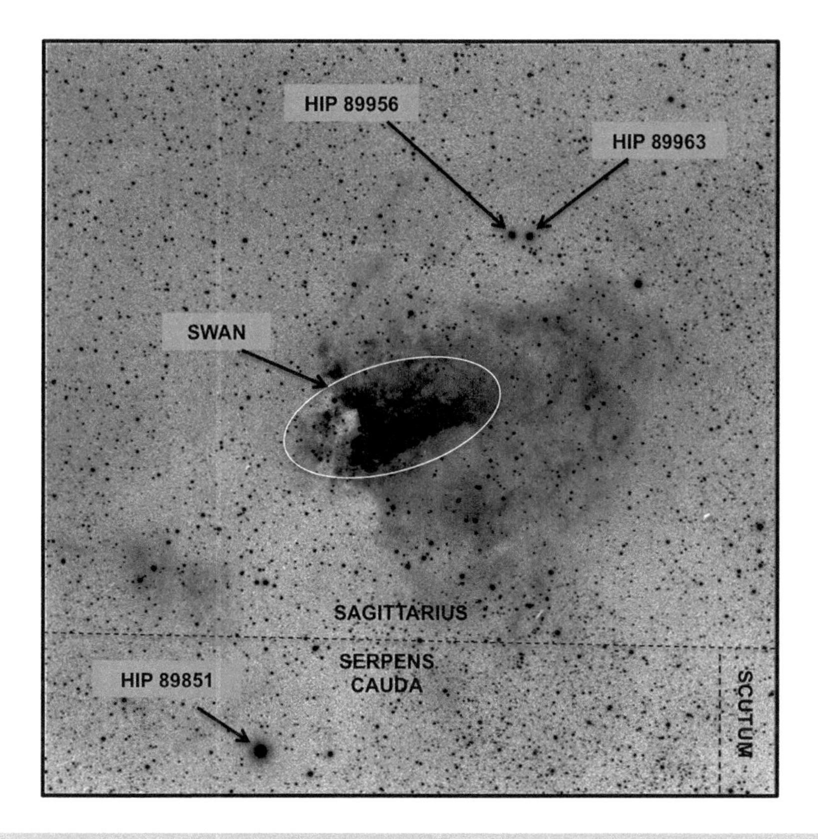

Fig. 3.50 Negative image of Messier 17

Note however that this star actually lies in a different constellation. Figure 3.50 captures parts of 3 constellations: Sagittarius, Serpens Cauda and Scutum. Note that north is down.

The chart of Sagittarius in Figure 3.51 has north at the top and illustrates just how close to the northern border of the constellation M17 lies. Sagittarius has 15 Messier objects as a result of its position in a rich part of the Milky Way. These are listed in the description of Messier 8, as is the number of NGC objects in this constellation. Sagittarius could keep an observer busy for years! The center of the Galaxy lies in this constellation. The star below M17 on the chart is called Polis, or alternatively, 13-Mu Sagittarii. This is a hot, magnitude 3.84, type B2 star that is only 30 ly distant.

The image of the Omega Nebula was taken at 04h 57m 32s on March 11 from New Mexico. The Universal Time was 11h 57m 32s. The Right Ascension of M17 is 18h 21m 47s, and the charting software gave a Local Sidereal Time of 16h 11m 37s. The difference is 2hr 10m 10s. Thus, M17 would cross the meridian about 7 minutes past 7 am local time. The telescope used was the 6-inch New Mexico refractor, which gives excellent single-shot color.

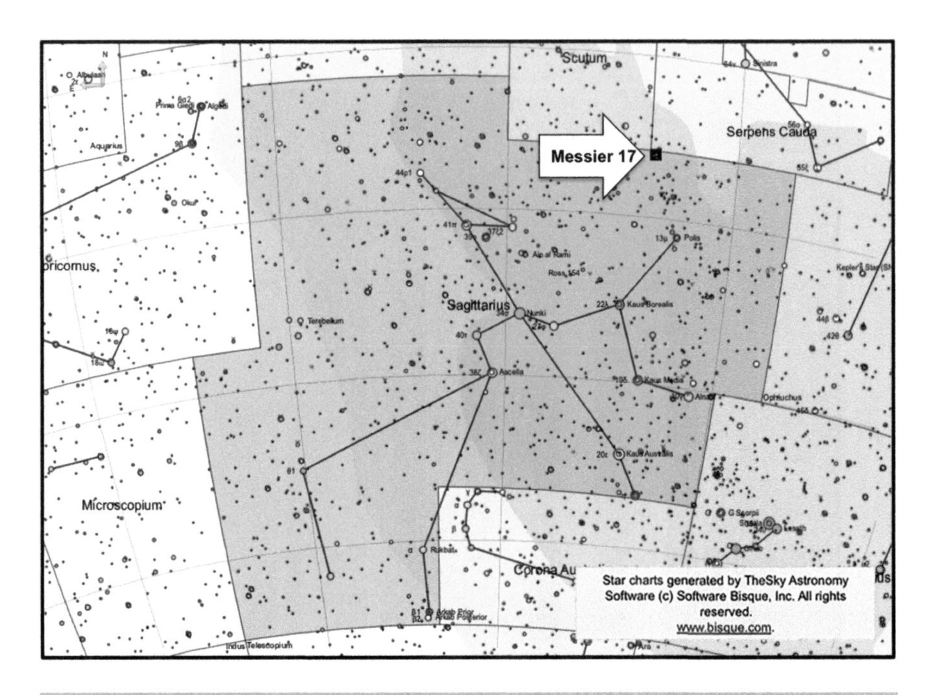

Fig. 3.51 Chart showing the location of M17

Messier 17: Data relating to the image in Figure 3.49

REMOTELY IMAGED FROM MAYHILL, NEW MEXICO
NGC 6618 (SWAN OR OMEGA NEBULA)
OBJECT TYPE: OPEN CLUSTER AND NEBULA

RA: 18h 21m 47s
DEC: -16° 10' 23"
ALTITUDE: +31° 59' 48"
AZIMUTH: 142° 28' 18"
FIELD OF VIEW: 0° 47' 06" x 0° 47' 06"
OBJECT SIZE: 11’ X 11’
POSITION ANGLE: 180° 25' from North
EXPOSURE TIME: 300 s
DATE: 11th March
LOCAL TIME: 04h 57m 32s
UNIVERSAL TIME: 11h 57m 32s
SCALE: 1.38 arcsec/pixel
MOON PHASE: 98.62% (waxing)

Telescope Optics
OTA: Takahashi TOA-150
Optical Design: Apochromatic Refractor
Aperture: 150 mm
Focal Length: 1095 mm
F/Ratio: f/7.3
Guiding: Internal
Mount: Paramount GTS

Instrument Package
SBIG ST-4000M One-Shot Color CCD
A/D Gain: 0.6e-/ADU
Pixel Size: 7.4um square
Resolution: 1.45 arcsec/pixel
Sensor: Frontlit
Cooling: -20°C Winter (-10°C Summer)
Array: 2048 x 2048 (8.3 Megapixels)
FOV: 49.6 x49.6 arcmin

Location
Observatory: New Mexico Skies
UTC Minus 7.00 (Daylight savings time is observed)
Minimum Target Elevation: Approx. 25 – 45
Degrees (N or S) 32° 54' Decimal: 32.9 North
(W or E) 105° 31' Decimal: 105.5 West
Elevation: 2225 meters (7298 ft)

Messier 18

Messier 18 (NGC 6613) is an open cluster found in the constellation of Sagittarius. Messier wrote, "On the night of June 3 to 4, 1764, I have discovered a bit below the nebula reported here above, a cluster of small stars, environed in a thin nebulosity; its extension may be 5 minutes of arc: its appearances are less sensible in an ordinary refractor of 3 feet & a half than that of the two preceding [M16 and M17] with a modest refractor, this star cluster appears in the form of a nebula; but when employing a good instrument, as I have done, one sees well many of the small stars: after my observations I have determined its position: Cluster of small stars, which contains a slight nebulosity, a little below the train of light which has been mentioned" (SEDS). The image in Figure 3.52 shows dark lanes caused by interstellar dust that blocks the light from background stars. Messier referred to a "train of light," which is of course Messier 17. If you look back to the M17 color image, you will see why Messier referred to it in this way. M17 and M18 are in very close proximity in the sky.

Fig. 3.52 Image of Messier 18, an open cluster in Sagittarius

The negative image of M18 in Figure 3.53 highlights six stars in its cluster. Star A is HIP 89831 and is a hot B star of magnitude 8.67. Star B is magnitude 12.79, star C is magnitude 9.25, star D is magnitude 9.36, star E is magnitude 12.78 and star F is magnitude 10.38. American astronomer Robert Burnham Jr. referred to this small galactic cluster as a minor object in the constellation and regarded it as one of the most neglected Messier objects. He also pointed out that, even though large modern telescopes can detect some nebulosity, the instruments used by Messier could not have possibly observed this, and that it must have been the effect of "the unresolved background of faint stars." Burnham also noted that "the brighter members were arranged in several coarse pairs." Stars A and B as well as stars D and E are no doubt examples of what he was referring to. A 2016 European Southern Observatory press release featured Messier 18, together with a high-resolution image taken with the Chile-based VLT Survey Telescope (VST). The very faint nebulosity caused by ionized hydrogen gas is visible in the VST image. ESO gave a distance of 4,600 ly to Messier 18 and an age of about 30 million years for the young cluster stars.

The chart in Figure 3.54 points out the location of M18 in Sagittarius. The constellation is packed with Messier objects – 15 in all. M18 is one of a number of open clusters in the constellation. There are 14 NGC objects, which are open clusters and not Messier objects, so if you like to examine open clusters, Sagittarius is a good place to start. The nearest star to M18 that helps form the constellation outline is named Polis. This is labeled HIP 89341 and is a magnitude 3.84 star to the southwest

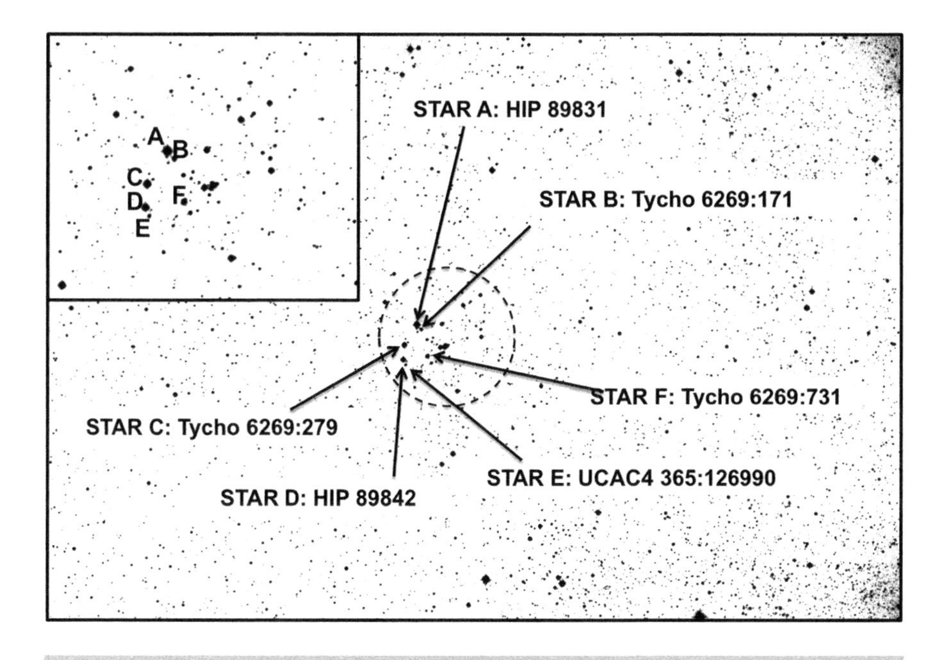

Fig. 3.53 Negative image of Messier 18

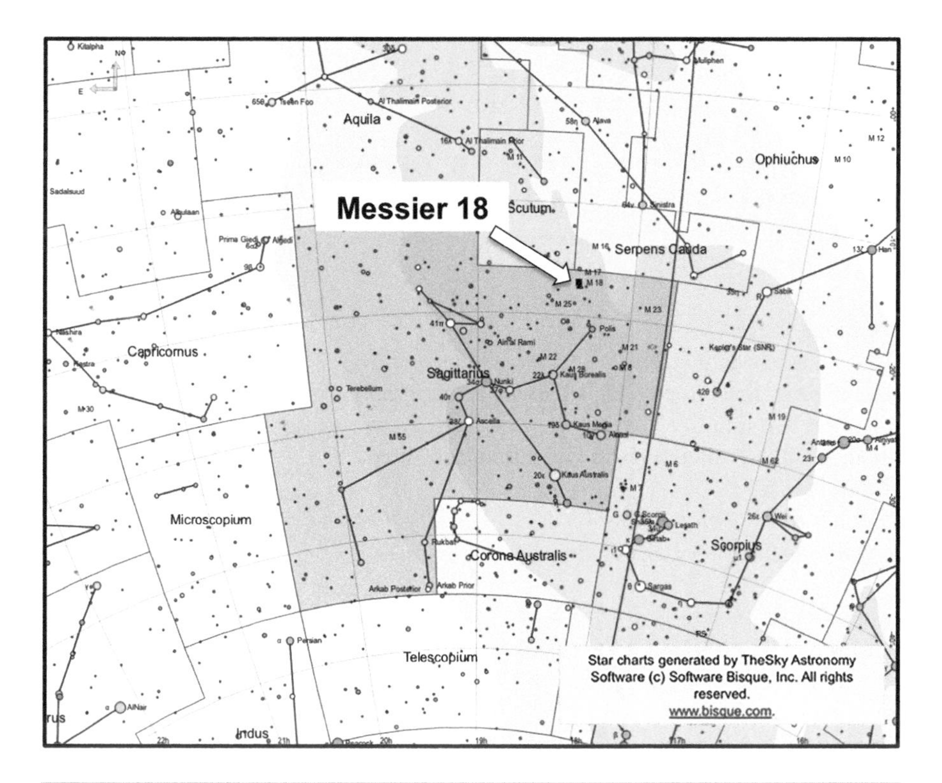

Fig. 3.54 Chart showing the location of M18

of M18. It is a remote, hot, giant B-type star with a radius of over 100 times that of the Sun. The star is roughly 4° away from M18. Note that the chart in Figure 3.54 has north at the top, whereas Figures 3.52 and 3.53 have north to the left.

The image of Messier 18 was taken at 5h 49m 38s on April 6 in New Mexico. The Universal Time was 11h 49m 38s. The Right Ascension of M18 is 18h 20m 58s, and the Local Sidereal Time was 17h 46m 12s. The difference between the two gives the time to go to the meridian transit of M18. The difference of 0h 34m 46s added to the local time provides a meridian transit time of 6h 24m 24s. Thus, M18 would be due south on the meridian at this time. The image was taken with M18 at an altitude of +39° 26' 13" and an azimuth of 169° 13' 17". The telescope used was a 20-inch corrected Dall Kirkham on a mount that is accurate enough to take unguided exposures of up to ten minutes and reach magnitudes as faint as 21.5.

Messier 18: Data relating to the image in Figure 3.52

REMOTELY IMAGED FROM MAYHILL, NEW MEXICO
NGC 6613
OBJECT TYPE: Open Cluster

RA: 18h 20m 58s
DEC: -17° 05' 30"
ALTITUDE: +39° 26' 13"
AZIMUTH: 169° 13' 17"
FIELD OF VIEW: 0° 54' 47" x 0° 36' 31"
OBJECT SIZE: 9' x 9'
POSITION ANGLE: 271° 10' from North
EXPOSURE TIME: 300 s
DATE: 6th April
LOCAL TIME: 05h 49m 38s
UNIVERSAL TIME: 11h 49m 38s
MOON PHASE: 78.11% (waxing)

Telescope Optics
OTA: Planewave 20" CDK
Optical Design: Corrected Dall-Kirkham Astrograph
Aperture: 510 mm
Focal Length: 2280 mm (0.66 Focal Reducer Fitted)
F/Ratio: f/4.5
Guiding: Active Guiding Disabled
Mount: Planewave Ascension 200HR

Instrument Package
FLI Proline PL11002M CCD
Camera Pixel Size: 9 μm square
Resolution: 0.81 arcsec/pixel
Cooling: -30°C default
Array: 4008 x 2672 (10.7 Megapixels)

Location
Observatory: New Mexico Skies
UTC Minus 7.00 (Daylight savings time is observed)
Minimum Target Elevation: Approx 25 – 45 Degrees
(N or S) 32° 54’ Decimal: 32.9 North
(W or E) 105° 31’ Decimal: 105.5 West
Elevation: 2225 meters (7298 feet)

Messier 19

Messier 19 is a globular cluster in the constellation of Ophiuchus. The first thing to note in Figure 3.55 is that M19 is not of the conventional circular shape, but is rather elongated roughly in the north-south direction to form an oblate spheroid. Messier wrote, “In the night of June 5 to 6, 1764, I have discovered a nebula, situated on the parallel of *Antares*, between Scorpius and the right foot of *Ophiuchus:* that nebula is round & doesn't contain any star; I have examined it with a Gregorian

Fig. 3.55 Image of Messier 19, a globular cluster in Ophiuchus

telescope which magnified 104 times, it is about 3 minutes of arc in diameter: one sees it very well with an ordinary refractor of 3 feet & a half. I have observed its passage of the Meridian, & compared it with that of the star *Antares;* I have determined the right ascension of that nebula of 252d 1' 45", & its declination of 25d 54' 46" south. The known star closest to that nebula is the 28th of the constellation *Ophiuchus*, after the catalog of Flamsteed, of sixth magnitude" (SEDS). The diameter of M19 is given as 17' of arc by SEDS and other sources, which seems high in comparison to the Messier estimate of 3'. However, measuring the bright core on the image shown, which is probably similar to Messier's view through his telescope, gives a value of 4'.

The negative image of M19 in Figure 3.56 shows the long axis of the cluster in relation to north. I have assumed north is directly to the left, which is correct to within 1°. As shown, there is a position angle of about 20° on my charting software, but defining the long axis required a bit of hit and miss. The cluster has an actual size of around 140 ly along its long axis. Inspecting the stars on the periphery of the cluster on my solved image gave many magnitudes in the region of 13 or 14. The bright star at the top right in Figure 3.56 is HIP 83296, which has a magnitude of 8.11. The Hipparcos Catalog places it at a distance of 448.63 ly. M19 itself is

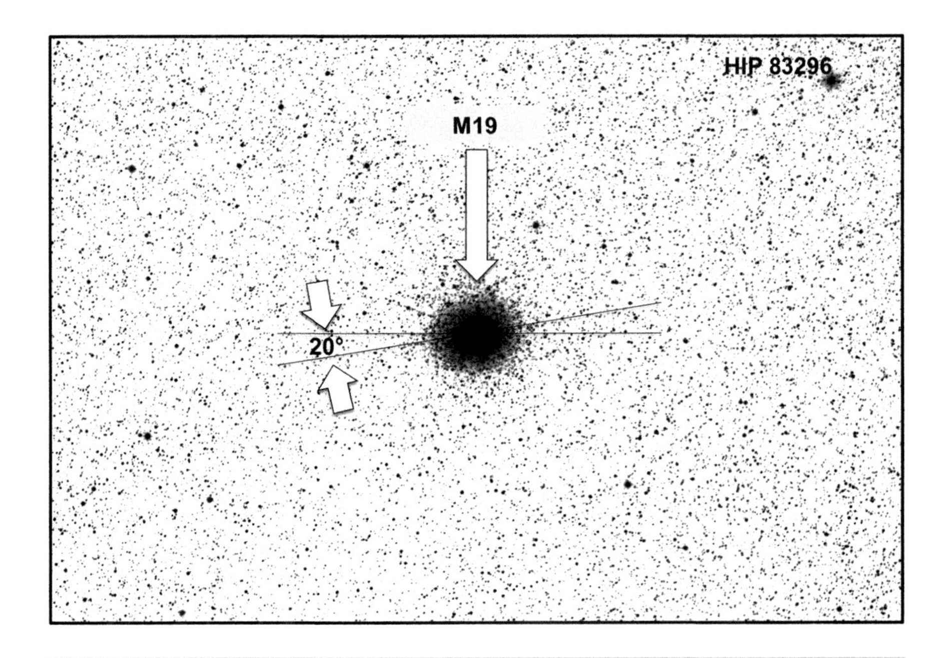

Fig. 3.56 Negative image of Messier 19

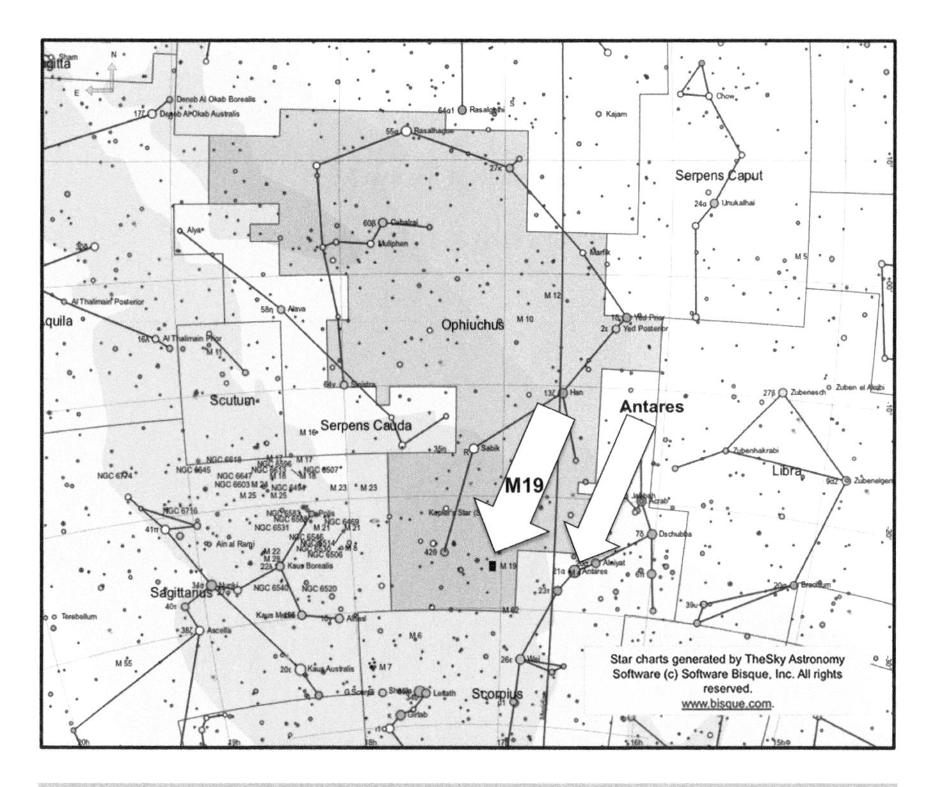

Fig. 3.57 Chart showing the location of M19

said to be at a distance of about 28,000 ly and is believed to be just over 5,000 ly from the Galactic center.

The chart in Figure 3.57 shows the position of M19 in Ophiuchus. Note that the image has been rotated to put north at the top. Messier referred to the red magnitude 1 star Antares ("the rival of Mars"), which is situated less than 8° away from M19 to the west in the adjacent constellation of Scorpius. Messier also referred to the 28th star of Ophiuchus in the Flamsteed Catalog, which is HIP 83508 in the Hipparcos Catalog. It has a magnitude of 6.66 and lies at a distance of 613.08 ly. It lies just 40' away from M19.

The M19 image was taken on the morning of March 27 at the local time in New Mexico of 05h 21m 53s. The Universal Time was 11h 21m 53s. Messier 19 has a Right Ascension of 17h 03m 42s, and the Local Sidereal Time was 16h 38m 57s. Thus, it would cross the meridian at 05h 47m, some 25 minutes later. The altitude of M19 was only +30° 38' 07" when the image was taken and would only achieve an altitude of +30° 55' 41" at meridian transit. The azimuth of M19 was 173° 33' 08". The same telescope used to image M18 was used to image M19: a 20-inch Corrected Dall Kirkham, or CDK.

Messier 19: Data relating to the image in Figure 3.55

REMOTELY IMAGED FROM MAYHILL, NEW MEXICO
NGC 6273
OBJECT TYPE: Globular Cluster

RA: 17h 03m 42s
DEC: -26° 17' 19"
ALTITUDE: +30° 38' 07"
AZIMUTH: 173° 33' 08"
FIELD OF VIEW: 0° 54' 47" x 0° 36' 31"
OBJECT SIZE: 17' x 17'
POSITION ANGLE: 271° 09' from North
EXPOSURE TIME: 300 s
DATE: 27th March
LOCAL TIME: 05h 21m 53s
UNIVERSAL TIME: 11h 21m 53s
MOON PHASE: 0.61% (waning)

Telescope Optics
OTA: Planewave 20" CDK
Optical Design: Corrected Dall-Kirkham Astrograph
Aperture: 510 mm
Focal Length: 2280 mm (0.66 Focal Reducer Fitted)
F/Ratio: f/4.5
Guiding: Active Guiding Disabled
Mount: Planewave Ascension 200HR

Instrument Package
FLI Proline PL11002M CCD
Camera Pixel Size: 9 μm square
Resolution: 0.81 arcsec/pixel
Cooling: -30°C default
Array: 4008 x 2672 (10.7 Megapixels)

Location
Observatory: New Mexico Skies
UTC Minus 7.00 (Daylight savings time is observed)
Minimum Target Elevation: Approx 25 – 45 Degrees
(N or S) 32° 54' Decimal: 32.9 North
(W or E) 105° 31' Decimal: 105.5 West
Elevation: 2225 meters (7298 feet)

Messier 20

Messier 20 (NGC 6514) is a bright nebula and star cluster in the constellation of Sagittarius. It is commonly known as the Trifid Nebula. The color image taken from Siding Spring (Figure 3.58) shows M20 and also includes at upper left M21. Messier's observations from June 5 to 6, 1764 read: "I have determined the position of two clusters of stars [M20 and M21] which are close to each other, a bit above the Ecliptic, between the bow of Sagittarius & the right foot of Ophiuchus: the known star closest to these two clusters is the 11th of the constellation Sagittarius, of seventh magnitude, after the catalog of Flamsteed: the stars of these clusters are, from the eighth to the ninth magnitude, environed with nebulosity" (SEDS). British astronomer Sir William Herschel observed M20 and noted the pronounced pattern of dark lanes. His son John is believed to have given M20 the name of the Trifid Nebula.

The negative image shown in Figure 3.59 identifies M20 and M21 and indicates three bright stars in the field. The central nebula, which is red in the color image, is an emission nebula, while the upper nebula, which is blue in the color image, is a reflection nebula. The emission nebula glows red because energy from nearby energetic (young) stars ionizes hydrogen gas, which then emit light in the red end of the spectrum. The reflection nebula is blue as a result of interstellar dust that

Fig. 3.58 Image of Messier 20, a bright nebula and cluster in Sagittarius

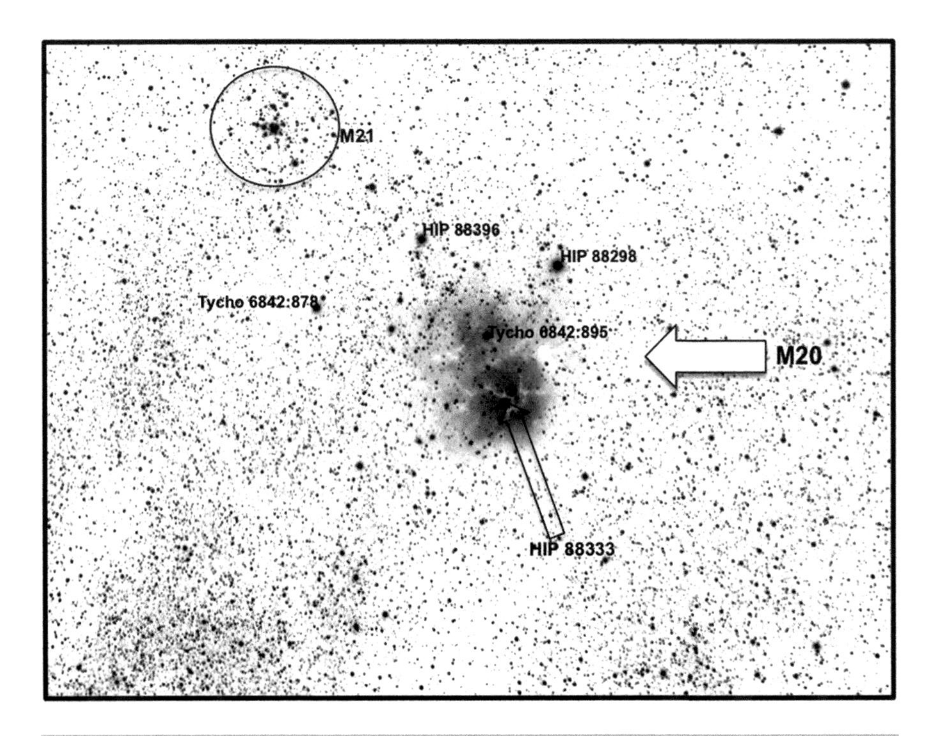

Fig. 3.59 Negative image of Messier 20

scatters lights. The dust particles are more efficient at scattering blue light (hence why Earth's sky appears blue). Nearby stars are putting out insufficient energy to ionize the surrounding hydrogen gas. The star HIP 88333 (HD 164492) is the main energetic source powering the emission nebula. This type-O6 star is also the brightest X-ray source in the nebula. M20 lies at a distance of 5,200 ly. The three bright stars shown are HIP 88298 with magnitude 5.72, HIP 88396 with magnitude 6.73 and Tycho 6842:878 with magnitude 7.14.

The chart in Figure 3.60 shows the location of M20 in Sagittarius, which is crowded with interesting objects. M20 is extremely close to M21, about 41' of arc center to center. M8 is 83' away from M20, M23 is just over 4°, M28 is about 5 ½°, M24 is 6°, M18 is just over 7°, M22 is just less than 8°, M25 is also just less than 8°, M17 is 8°, M69 is just over 11°, M70 just less than 13°, M54 is 14°, M55 is 23°, and M75 less than 29°, and yes -- all in the same constellation!

The image was taken at the local time of 04h 04m 44s on March 29 from Siding Spring. Universal Time was 17h 04m 44s. The Right Ascension of M20 is 18h 03m 31s, and the Local Sidereal Time was 15h 25m 57s. The difference is 2h 37m 34s, so the object would transit at 06h 42m 18s. The single-shot color image camera was used with a 90-mm refractor.

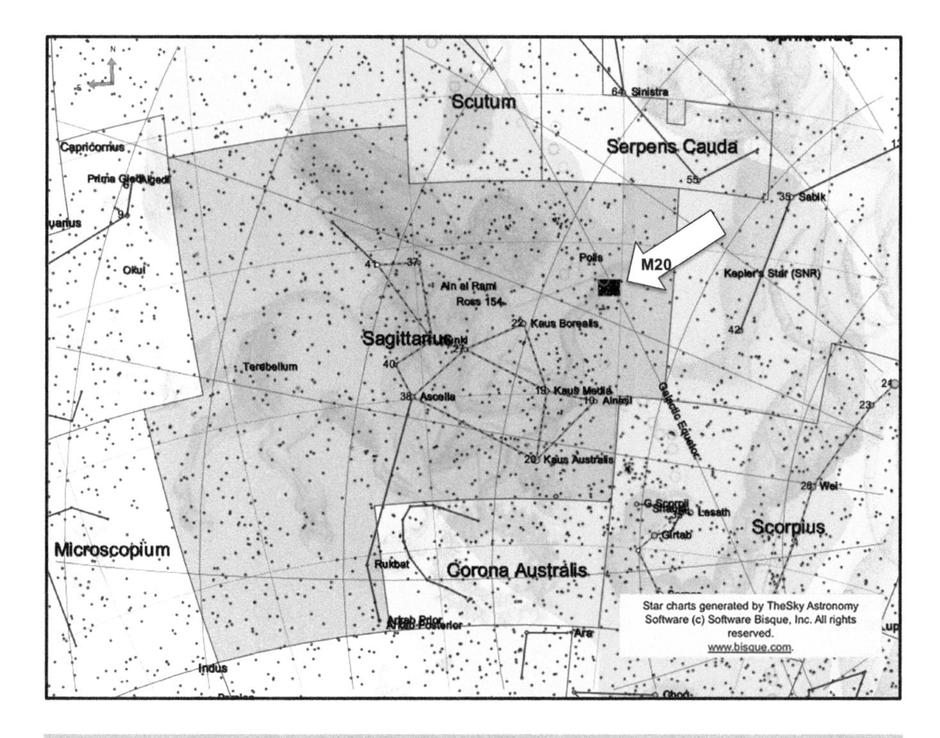

Fig. 3.60 Chart showing the location of M20

Messier 20: Data relating to the image in Figure 3.58

REMOTELY IMAGED FROM NEW SOUTH WALES, AUSTRALIA
NGC 6514 (TRIFID NEBULA)
OBJECT TYPE: Bright Nebula

RA: 18h 03m 31s
DEC: -22° 59' 08"
ALTITUDE: +54° 10' 40"
AZIMUTH: 86° 33' 37"
FIELD OF VIEW: 1° 37' 39" x 1° 13' 14"
OBJECT SIZE: 28’
POSITION ANGLE: 357° 19' from North
EXPOSURE TIME: 300 s
DATE: 29th March
LOCAL TIME: 04h 04m 44s
UNIVERSAL TIME: 17h 04m 44s
SCALE: 3.66 arcsec/pixel
MOON PHASE: 0.57% (waxing)

Telescope Optics
OTA: Takahashi Sky 90
Optical Design: Apochromatic Refractor
Aperture: 90 mm
Focal Length: 417 mm
F/Ratio: f/5.6
Guiding: None
Mount: Paramount PME

Instrument Package
CCD: SBIG ST2000 XMC
Colour CMOS
Pixel Size: 7.4μm square
Sensor: Frontlit
Cooling: CNA
Array: 1600 x 1200 pixels
FOV: 60.5 x 80.7 arcmin

Location
Observatory: Siding Spring
UTC +10.00 (Australia Daylight savings time is observed)
31° 16' 24" South
149 03' 52" East
Elevation: 1122 meters (3681 ft)

Messier 21

Messier 21 is an open cluster located in the constellation of Sagittarius. It is situated next to the Trifid Nebula, as can be seen on the color image in Figure 3.61. Messier reported that on the night of June 5/6, 1764: "I have determined the position of two clusters of stars which are close to each other, a bit above the Ecliptic, between the bow of Sagittarius & the right foot of *Ophiuchus:* the known star closest to these two clusters is the 11th of the constellation Sagittarius, of seventh magnitude, after the catalog of Flamsteed: the stars of these clusters are, from the eighth to the ninth magnitude, environed with nebulosity. I have determined their positions" (SEDS). Of course, the other cluster is the Trifid Nebula M20. M21 is located at a distance of 4250 ly and is said to have an age of 4.6 million years (SEDS), in which case, this is a very young cluster.

The negative image of M21 shown in Figure 3.62 incorporates an expanded view of the cluster. Three cluster stars have been identified with arrows. The stars are: Tycho 6842:246, which has a magnitude of 7.21, Tycho 6263:63, which has a magnitude of 9.13, and UCAC4 338:122274, which has a magnitude of 12.53. To give an idea of size, the circle of stars at the top of the cluster that contains Tycho 6263:63 and UCAC4 338:122274 is 2' of arc across. At a distance of 4,250 ly, this little circle would have an actual size of just less than 2.5 ly. The distance of 4,250 ly to M21 is less than the distance to the Trifid Nebula M20, which is 5,200 ly.

Fig. 3.61 Image of Messier 21, an open cluster in Sagittarius

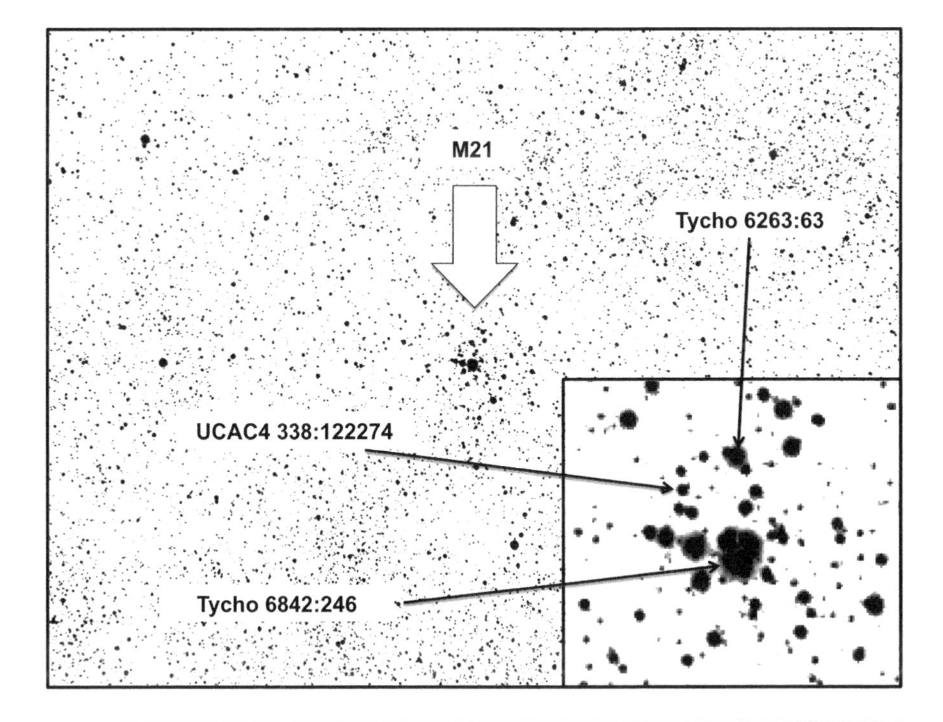

Fig. 3.62 Negative image of M21

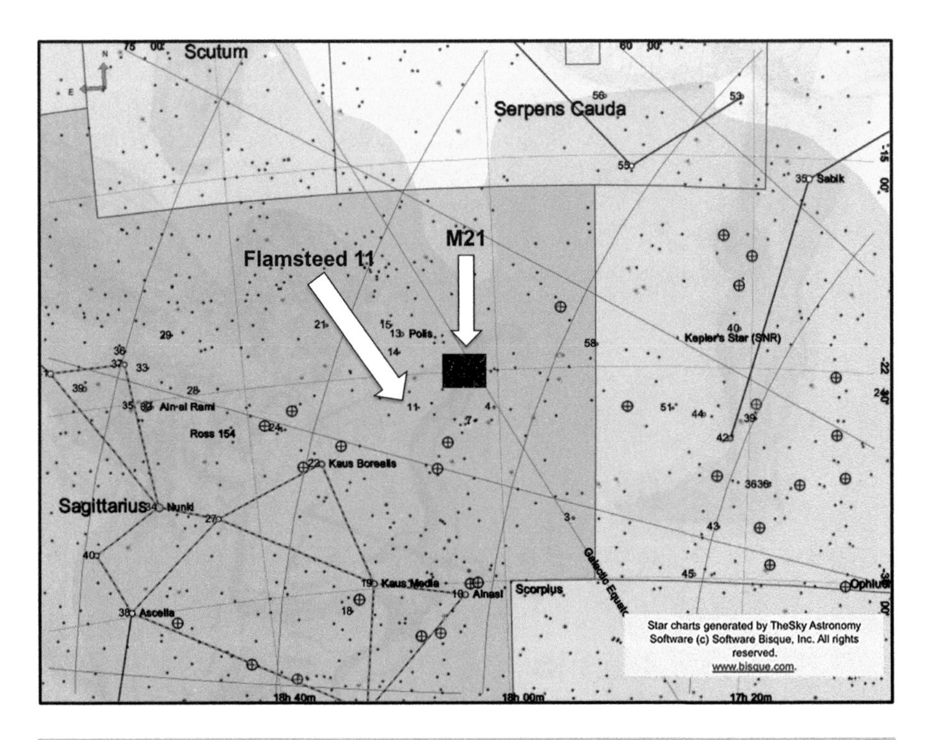

Fig. 3.63 Chart showing the location of M21

The chart in Figure 3.63 shows M21 in the constellation of Sagittarius. Messier's observations of M20 and M21 referred to the nearby star Flamsteed 11. Astronomer Robert Burnham Jr. commented that "Messier identified the brightest star of the cluster as 11 Sagittarii, which is something of a puzzle, as the star which now bears that designation is more than 2° distant, towards the SE." I measured the distance of Flamsteed 11 as being 2° 10' to the SE,which agrees with Burnham, but that star (HIP 89153 in the Hipparcos Catalog) has a magnitude 4.96 rather than the magnitude 7 mentioned by Messier. It does not seem like Messier was saying that this star was a cluster member or the brightest star in that cluster, but rather that it was "the nearest neighboring known star" to M20 and M21. What is puzzling is that the star HIP 88298, which has a magnitude of 5.72, is right next to M20 (c. 16' of arc) and close to M21(c. 42'), which Messier must have noticed in his observations of both objects. So why did he refer to Flamsteed 11? One can assume that he was indeed only referring to stars in Flamsteed's Catalog.

The image of M21 was taken at 04h 24m 58s on the morning of March 28 in Siding Spring. The Universal Time was 17h 24m 58s (March 27). The Right Ascension of M21 was18h 05m 15s, and the Local Sidereal Time was 15h 42m 18s. The difference is 2h 22m 57s, so M21 would transit the meridian at 06h 47m 55s. The azimuth of M21 at the time the image was taken was 83° 27' 08", and the altitude was +57° 05' 43". The same 90-mm refractor telescope that was used to image M20 was used to image M21.

Messier 21: Data relating to the image in Figure 3.61

REMOTELY IMAGED FROM NEW SOUTH WALES, AUSTRALIA
NGC 6531
OBJECT TYPE: Open Cluster

RA: 18h 05m 15s
DEC: -22° 29' 10"
ALTITUDE: +57° 05' 43"
AZIMUTH: 83° 27' 08"
FIELD OF VIEW: 1° 37' 39" x 1° 13' 14"
OBJECT SIZE: 13’
POSITION ANGLE: 357° 18' from North
EXPOSURE TIME: 300 s
DATE: 28th March
LOCAL TIME: 04h 24m 58s
UNIVERSAL TIME: 17h 24m 58s (27th March)
SCALE: 3.66 arcsec/pixel
MOON PHASE: 0.26% (waning)

Telescope Optics
OTA: Takahashi Sky 90
Optical Design: Apochromatic Refractor
Aperture: 90 mm
Focal Length: 417 mm
F/Ratio: f/5.6
Guiding: None
Mount: Paramount PME

Instrument Package
CCD: SBIG ST2000 XMC
Colour CMOS
Pixel Size: 7.4μm square
Sensor: Frontlit
Cooling: CNA
Array: 1600 x 1200 pixels
FOV: 60.5 x 80.7 arcmin

Location
Observatory: Siding Spring
UTC +10.00 (Australia Daylight savings time is observed)
31° 16’ 24” South
149 03’ 52” East
Elevation: 1122 meters (3681 ft)

Messier 22

Messier 22, shown in Figure 3.64, is an impressive globular cluster located in the constellation of Sagittarius. One should realize that *it exceeds the size of the full Moon in the sky* at 32' of arc. It was the first globular cluster to be discovered, and credit is normally given to Johann Abraham Ihle, an amateur astronomer born in Germany in 1627. Of his observations taken on the night of June 5 to 6, 1764, Messier wrote, "I have observed a nebula situated a bit below the ecliptic, between the head & the bow of Sagittarius, near the star of seventh magnitude, the twenty-fifth of that constellation, according to the catalog of Flamsteed. That nebula didn't appear to me to contain any star, although I have examined it with a good Gregorian telescope which magnified 104 times: it is round, & one sees it very well with an ordinary refractor of 3 feet & a half; its diameter is about 6 minutes of arc. I have determined its position by comparing with the star Lambda Sagittarii." John Herschel (son of William Herschel) said on July 1, 1826 that M22 was a "magnificent globular cluster….the stars from 12 to 20 m. Those 12 m are equally scattered over it, but those of 20m form the central mass."

In the negative image shown in Figure 3.65, a circle has been drawn to illustrate the current accepted diameter of M22. Two stars are arrowed to give an idea of the magnitudes. Star Tycho 6858:955 is of magnitude 9.61, and star Tycho 6858:579

Fig. 3.64 Image of Messier 22, a globular cluster in Sagittarius

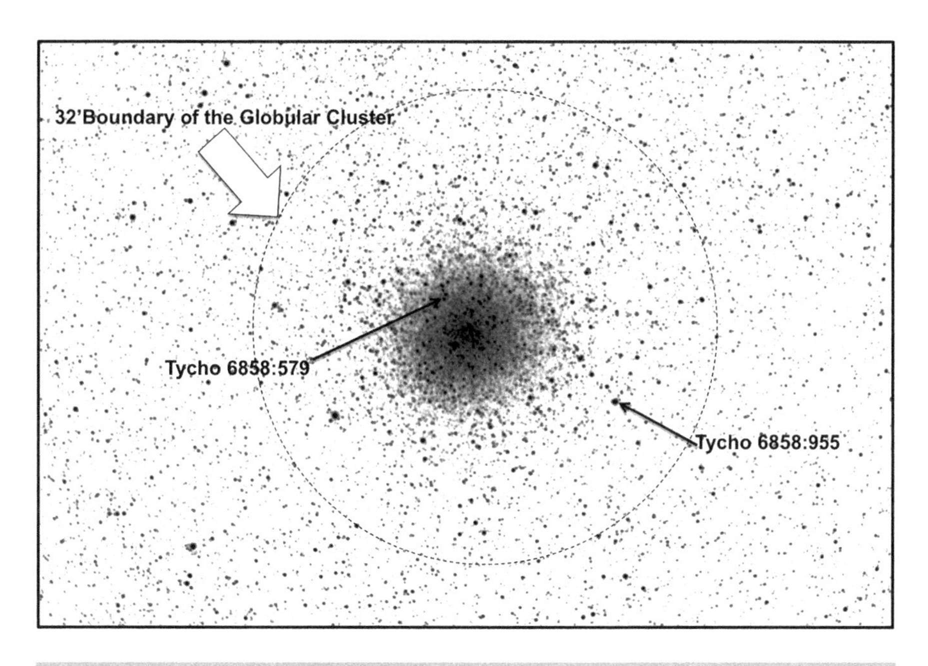

Fig. 3.65 Negative Image of Messier 22

has a magnitude of 12.03. M22 contains hundreds of thousands of stars and is relatively close to us at a distance of 10,400 ly. At that distance, the 32' apparent diameter would give the cluster an actual diameter of almost 100 ly. The diameter of the great globular cluster M13 is only 20'. It is, however, over twice the distance of M22. M13 is about one and a half times the actual size of M22.

The chart in Figure 3.66 shows the location of M22 in Sagittarius. Messier referred to its proximity to Flamsteed 25, a magnitude 7 star. The star is marked 25 and can be easily identified in the inset. The star is HIP 91066 and has a magnitude of 6.53, not far off Messier's magnitude 7 estimation. The bright star to the south west of M22 is Kaus Borealis, which forms part of the Sagittarius figure. It is a magnitude 2.82 star otherwise known as 22-Lambda Sagittarii. The star to the south east of M22 in the Sagittarius figure is the fainter HIP 92041, a magnitude 3.17 star. This is 27-Phi Sagittarii. Further to the east is the bright star Nunki, which is magnitude 2.04 and is 34-Sigma Sagittarii.

The image was taken at 04h 42m 20s on April 7 from New Mexico. The Universal Time was 10h 42m 20s. The Right Ascension of M22 is 18h 37m 27s, and the Local Sidereal Time was 16h 42m 39s. Subtracting the LST from the RA gives a value of 1h 54m 48s, which means M22 would reach the meridian at a transit time of 06h 37m 8s. The telescope used to take the image was 500 mm in aperture, based in New Mexico.

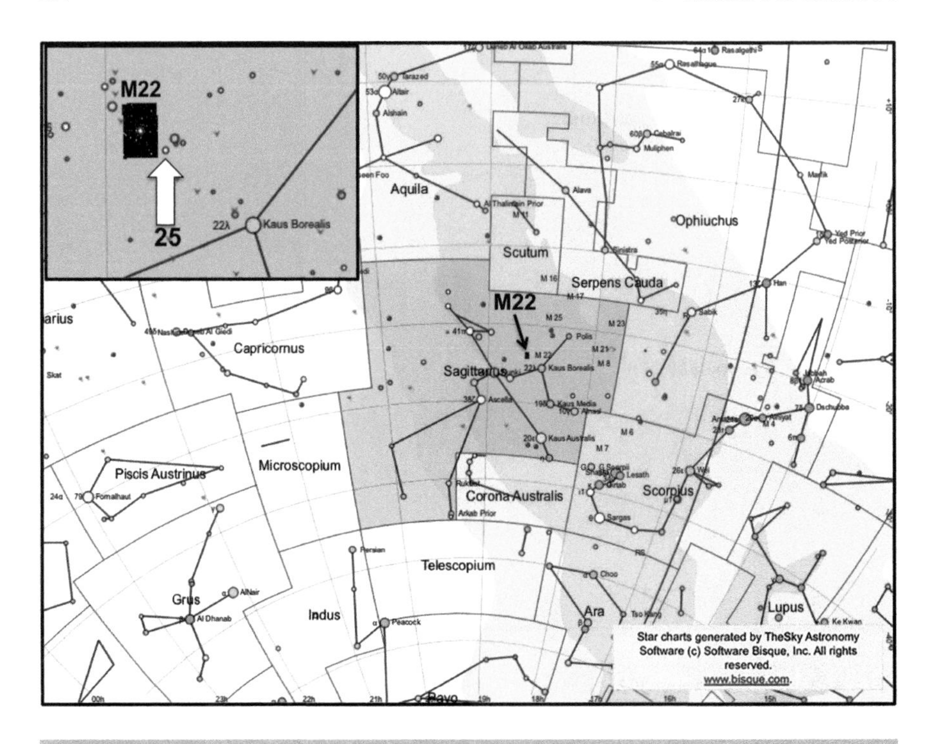

Fig. 3.66 Chart showing the location of M22

Messier 22: Data relating to the image in Figure 3.64

REMOTELY IMAGED FROM MAYHILL, NEW MEXICO
NGC 6656
OBJECT TYPE: Globular Cluster

RA: 18h 37m 27s
DEC: -23° 53' 07"
ALTITUDE: +27° 04' 06"
AZIMUTH: 150° 27' 25"
FIELD OF VIEW: 0° 54' 43" x 0° 36' 28"
OBJECT SIZE: 32'
POSITION ANGLE: 271° 09' from North
EXPOSURE TIME: 300 s
DATE: 7th April
LOCAL TIME: 04h 42m 20s
UNIVERSAL TIME: 10h 42m 20s
MOON PHASE: 85.81% (waxing)

Telescope Optics
OTA: Planewave 20" CDK
Optical Design: Corrected Dall-Kirkham Astrograph
Aperture: 510 mm
Focal Length: 2280 mm (0.66 Focal Reducer Fitted)
F/Ratio: f/4.5
Guiding: Active Guiding Disabled
Mount: Planewave Ascension 200HR

Instrument Package
FLI Proline PL11002M CCD
Camera Pixel Size: 9-μm square
Resolution: 0.81 arcsec/pixel
Sensor:
Cooling: -30°C default
Array: 4008 x 2672 (10.7 Megapixels)

Location
Observatory: New Mexico Skies
UTC Minus 7.00 (Daylight savings time is observed)
Minimum Target Elevation: Approx 25 – 45 Degrees
(N or S) 32° 54' Decimal: 32.9 North
(W or E) 105° 31' Decimal: 105.5 West
Elevation: 2225 meters (7298 ft)

Messier 23

Messier 23 (NGC 6494) is an open cluster located in a very dense part of the Milky Way, as can be seen from Figure 3.67. Messier wrote, "In the night of June 20 to 21, 1764, I determined the position of a cluster of small stars which is situated between the northern extremity of the bow of Sagittarius & the right foot of *Ophiuchus*, very close to the star of sixth magnitude, the sixty-fifth of the latter constellation, after the catalog of Flamsteed: These stars are very close to each other; there is none which one can see easily with an ordinary refractor of 3 feet & a half, & which was taken for these small stars. The diameter of all is about 15 minutes of arc. I have determined its position by comparing the middle with the star Mu Sagittarii." Although Messier estimated its size at 15' of arc, it is now given a diameter of 27'. Of course, it is difficult to decide where a cluster ends. The telescope and camera combination used gives a field of view of 3° 53' 32" x 2° 35' 41". This is the largest field of view of any of the telescopes used in this project.

The negative image in Figure 3.68 illustrates the large field of view with the full Moon-size central cluster M23 and a nearby 6 magnitude star, HIP 87782 of magnitude 6.48. It is a hot main sequence star of type A1, giving it a surface temperature of around 9330 K. The M23 cluster is at a distance of 2150 ly. The inset at bottom

Fig. 3.67 Image of Messier 23, an open cluster in Sagittarius

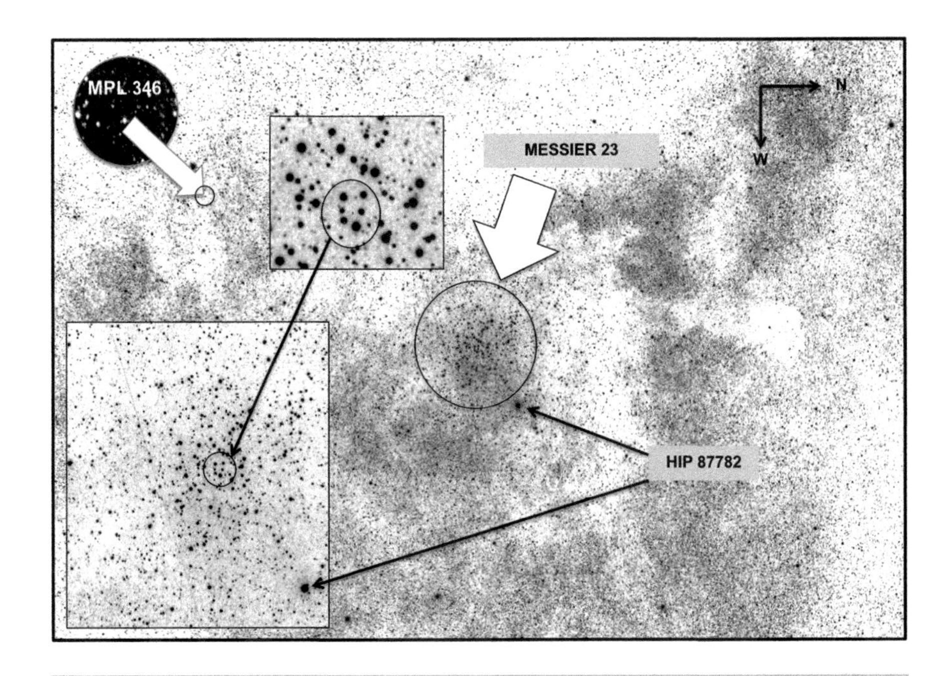

Fig. 3.68 Negative image of Messier 23

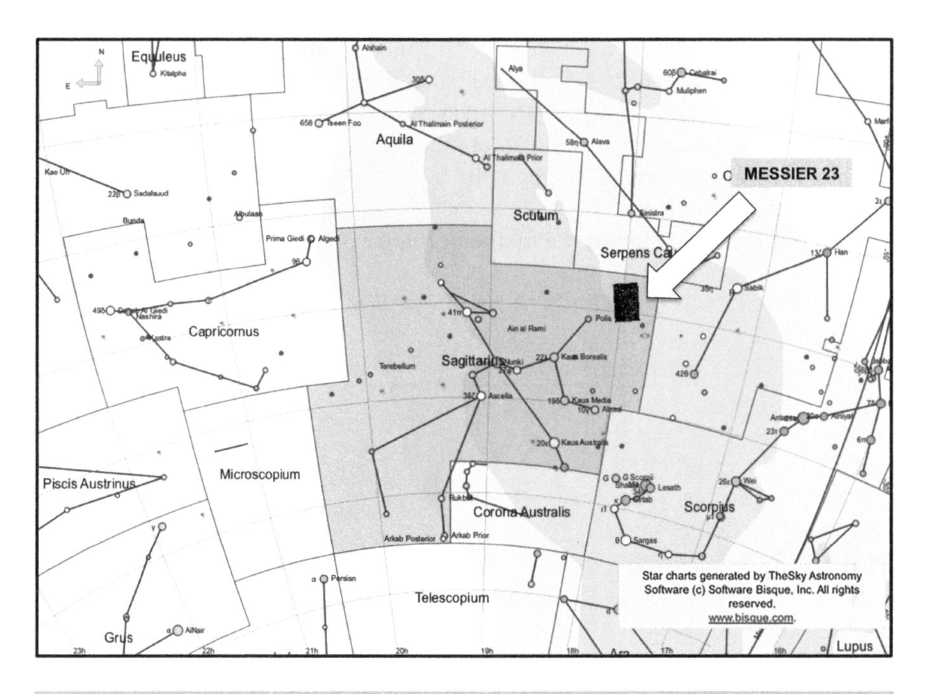

Fig. 3.69 Chart showing the location of M23

left gives a closer view of the object. Astronomer Robert Burnham Jr. noted, "From color and magnitude studies, the majority of the members of M23 are somewhat reddened main sequence stars; the most luminous members are of type B9. Several G-type stars exist in this group, however; the brightest example in the main mass of the cluster will be found in the small elliptical formation of 6 stars at the cluster center; it is the westernmost star of this flattened ring." On the negative image, I have circled and enlarged what could be this ring, but there are other possibilities.

Before looking at M23 in Figure 3.69, you may have noticed that at the top left of the negative image (Figure 3.68) there is a circle entitled MPL 346, which is pointing at an impossibly small dot that you probably cannot see! Whenever you take an image of an object, it is always worth looking at the likelihood of something being recorded in the frame that you were not expecting. Within the circle is an enlarged view of the "dot," which turned out to be the asteroid 346 Hermentaria that just happened to be passing through Sagittarius at the time. It was rather faint at magnitude 12.86 when the image was taken. The image was of sufficient resolution for considerable enlargement, which allowed me to spot this intruder. If you find a bright spot on a galaxy image not recorded in other images – check carefully – you may have discovered a supernova! The chart shows the position of M23 in Sagittarius. Note that just below and to the right of the solved image is the symbol for Saturn, which was there at the time the image was taken.

The 5-minute exposure was taken at 05h 03m 45s on March 9 in New Mexico. The Universal Time was 12h 03m 45s. M23 has a Right Ascension of 17h 58m 00s, and the Local Sidereal Time was 16h 09m 58s. The difference is 1h 48m 2s, so the cluster would transit the meridian at 06h 51m 47s. The telescope used was a 106-mm Petzval Apochromat Astrograph.

Messier 23: Data relating to the image in Figure 3.67

REMOTELY IMAGED FROM MAYHILL, NEW MEXICO
NGC 6494
OBJECT TYPE: OPEN CLUSTER

RA: 17h 58m 00s
DEC: -19° 00' 58"
ALTITUDE: +32° 07' 20"
AZIMUTH: 149° 32' 19"
FIELD OF VIEW: 3° 53' 32" x 2° 35' 41"
OBJECT SIZE: 27' X 27'
POSITION ANGLE: 93° 05' from North
EXPOSURE TIME: 300 s
DATE: 9th March
LOCAL TIME: 05h 03m 45s
UNIVERSAL TIME: 12h 03m 45s
SCALE: 3.5 arcsec/pixel
MOON PHASE: 89.34% (waxing)

Telescope Optics
OTA: Takahashi FSQ Fluorite
Optical Design: Petzval Apochromat Astrograph
Aperture: 106 mm
Focal Length: 530 mm
F/Ratio: f/5.0
Guiding: External
Mount: Paramount GTS-1100S

Instrument Package
SBIG STL-11000M
A/D Gain: 2.2e-/ADU
Pixel Size: 9um square
Resolution: 3.5 arcsec/pixel
Sensor: Frontlit
Cooling: -15°C default
Array: 4008 x 2672 (10.7 Megapixels)
FOV: 155.8 x 233.7 arcmin

Location
Observatory: New Mexico Skies
UTC Minus 7.00 (Daylight savings time is observed)
Minimum Target Elevation: Approx. 25 – 45 Degrees
(N or S) 32° 54' Decimal: 32.9 North
(W or E) 105° 31' Decimal: 105.5 West
Elevation: 2225 meters (7298 ft)

Messier 24

Messier 24 is an unusual object for Messier's list, as it is a star cloud in Sagittarius. It is referred to as the Small Sagittarius Star Cloud. A wide-angle camera/telescope arrangement was used to capture the image, which has been centered on the open cluster NGC 6603. You can just see the cluster in the center of Figure 3.70. NGC 6603 sits within the roughly rectangular-shaped M24 star field. On June 20 to 21,

Fig. 3.70 Image of Messier 24, a star cloud in Sagittarius

1764, Messier wrote, "I have discovered on the same parallel as the star cluster I have just been talking about & near the extremity of the bow of Sagittarius, in the milky way, a considerable nebulosity, of about one degree & a half extension: in that nebulosity there are several stars of different magnitudes; the light which is between these stars is divided in several parts." British astronomer W.H. Smyth commented, "This object was discovered by Messier in 1764, and described as a mass of stars…. a great nebulosity of which the light is divided in several parts. This was probably owing to want of power in the instruments used, as the whole is fairly resolvable, though there is a gathering spot with much stardust." A "gathering spot" is a good description of the way that NGC 6603 appears in the wide-angle image. A little task for you: can you spot Messier 18 in Figure 3.70?

The telescope in the photograph used a camera with an equal number of pixels (4,096) in the X and Y directions to give a square image. The image gives a field of view of 238.8' (14328"), so with each side measuring 4,096 pixels, each pixel covers 3.5". The four-part image shows objects in the field as enlargements. These were all taken from the original image, which is capable of considerable enlargement. At top left of Figure 3.71 is Barnard 92 (B92), a dark nebula that contains a

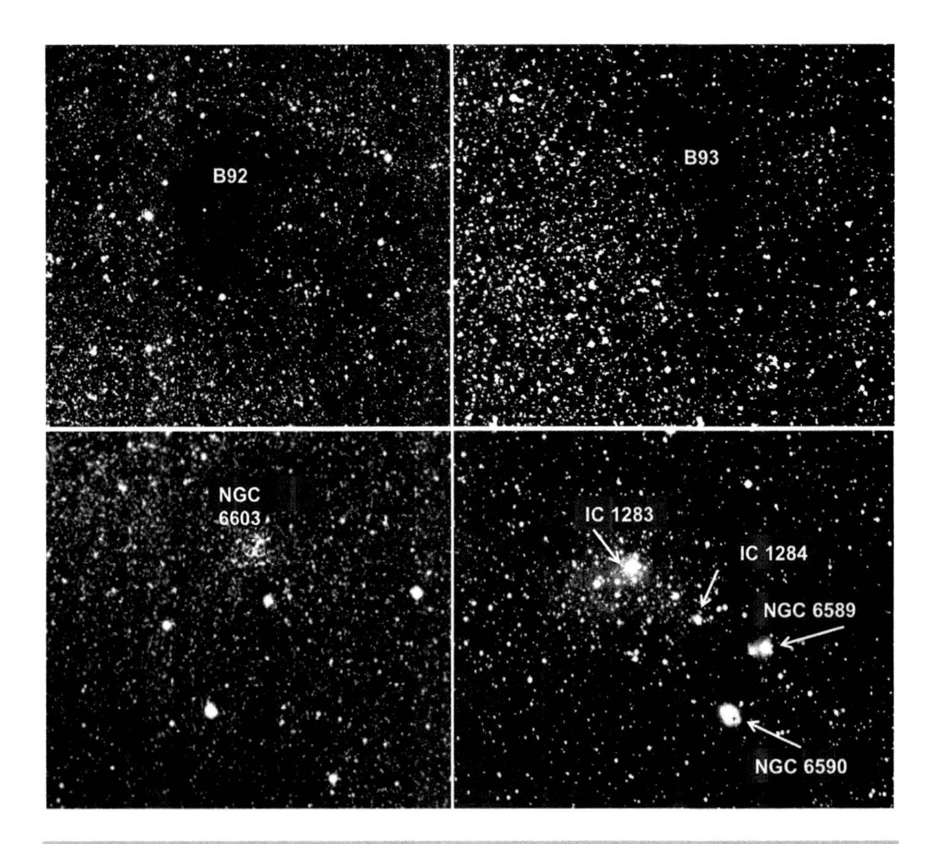

Fig. 3.71 Magnified sections of Messier 24

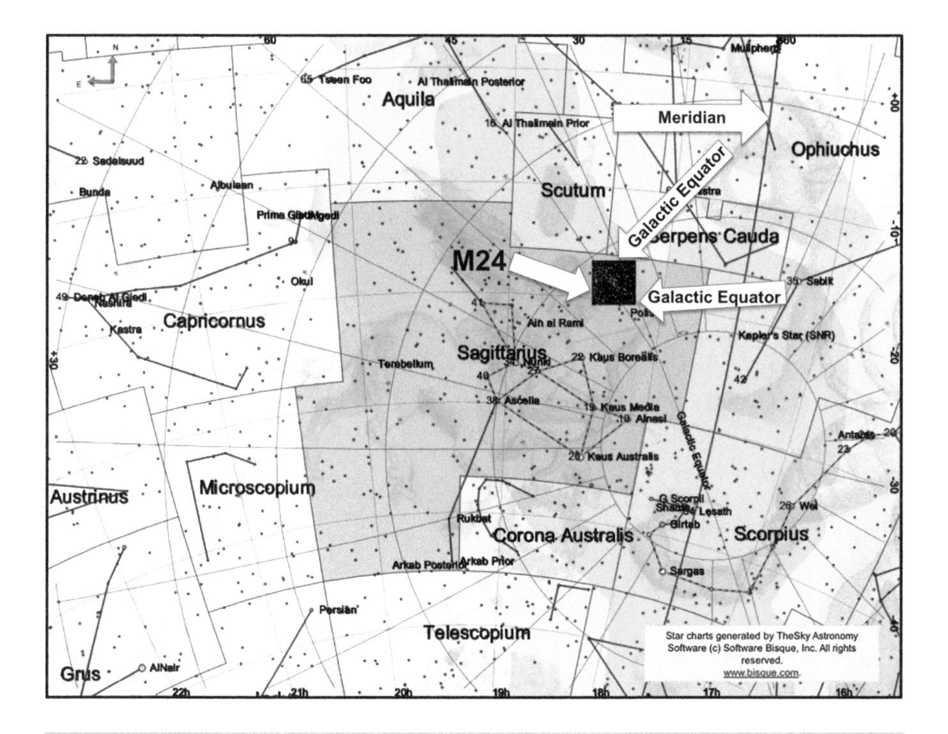

Fig. 3.72 Chart showing the location of M24

single "bright" foreground star: UCAC4 359:131948, which is magnitude 11.17. You may be able to spot this dark "hole" on the original image slightly above and to the right of NGC 6603. At top right is the dark nebula Barnard 93 (B93), which can be found on the original image between NGC 6603 and B92. At lower left is a closer view of the open cluster NGC 6603. The Revised NGC Catalog gives this a diameter of 4' of arc. At bottom right, four objects are identified. These are the reflection nebulae NGC 6589 and NGC 6590 and the emission nebulae IC1283 and IC1284. All of these can be found in the dark area towards the bottom of the image.

The chart in Figure 3.72 shows the position of the solved square image of M24 in Sagittarius with north at the top. The chart represents the situation when the image was taken. The Galactic Equator runs directly through the image, so we are looking at a great depth of stars. Arrows indicate where the Galactic Equator enters and leaves the image. Following the Galactic Equator downwards to be on a level with the spout of the teapot will bring you very close to the direction of the Galactic Center. The meridian to the right (west) of M24 at the time the image was taken is arrowed.

The image was taken at 04h 02m 07s on the morning of April 15 at Siding Spring. The corresponding Universal Time was 18h 02m 07s (April 14). The Right Ascension of M24 is 18h 19m 28s, and the Local Sidereal Time was 17h 30m 31s. The difference (RA – LST) is 48 minutes 57 second, so adding this to the current local time gives M24 as 04h 51m 04s.

Messier 24: Data relating to the image in Figure 3.69

REMOTELY IMAGED FROM NEW SOUTH WALES, AUSTRALIA
NGC 6603 (Small Sagittarius Star Cloud)
OBJECT TYPE: Star Cloud

Centered on RA: 18h 19m 28s
DEC: -18° 23' 47"
ALTITUDE: +73° 02' 53"
AZIMUTH: 43° 36' 40"
FIELD OF VIEW: 3° 59' 12" x 3° 59' 12"
OBJECT SIZE: 90’ (M24)
POSITION ANGLE: 0° 41' from North
EXPOSURE TIME: 300 s
DATE: 15th April
LOCAL TIME: 04h 02m 07s
UNIVERSAL TIME: 18h 02m 07s (14th April)
SCALE: 3.5 arcsec/pixel
MOON PHASE: 88.88% (waning)

Telescope Optics
OTA: Takahashi FSQ ED
Optical Design: Petzval Apochromatic Astrograph
Aperture: 106 mm
Focal Length: 106 mm
F/Ratio: f/5.0
Guiding: External
Mount: Paramount PME

Instrument Package
CCD: FLI Microline 16803
Pixel Size: 9μm square
Sensor: KAF – 16803 Frontlit
Cooling: -25°C Summer (-30°C Winter)
Array: 4096 x 4096 pixels
FOV: 238.8 x 238.8 arcmin

Location
Observatory: Siding Spring
UTC +10.00 (Australia Daylight savings time is observed)
31° 16’ 24” South
149 03’ 52” East
Elevation: 1122 meters (3681 ft)

Messier 25

Messier 25 (IC 4725) shown in Figure 3.73 is an open cluster in the constellation of Sagittarius. You may have noticed that the cluster has an IC (Index Catalog) number rather than an NGC number, which is the catalog containing most Messier objects. For some reason, John Herschel did not include it in his General Catalog (GC) of 5079 objects, the predecessor of the New General Catalog (NGC). Danish-Irish astronomer J.L.E. Dreyer recompiled the GC to produce the NGC with 7840 objects. Dreyer later added two additions called the Index Catalogs, which included M25 and added another 5386 objects. (IC1 –1529 objects, IC2 – 3857 objects). On the night of June 20 to June 21 1764, Messier reported, "I have determined the position of another star cluster in the vicinity of the two preceding, between the head & the extremity of the bow of Sagittarius, & almost on the same parallel as the two others: the closest known star is that of the sixth magnitude, the twenty-first of Sagittarius, in the catalog of Flamsteed: this cluster is composed of small stars which one sees with difficulty with an ordinary refractor of 3 feet: it doesn't contain any nebulosity, & its extension may be 10 minutes of arc."

Fig. 3.73 Image of Messier 25, an open cluster in Sagittarius

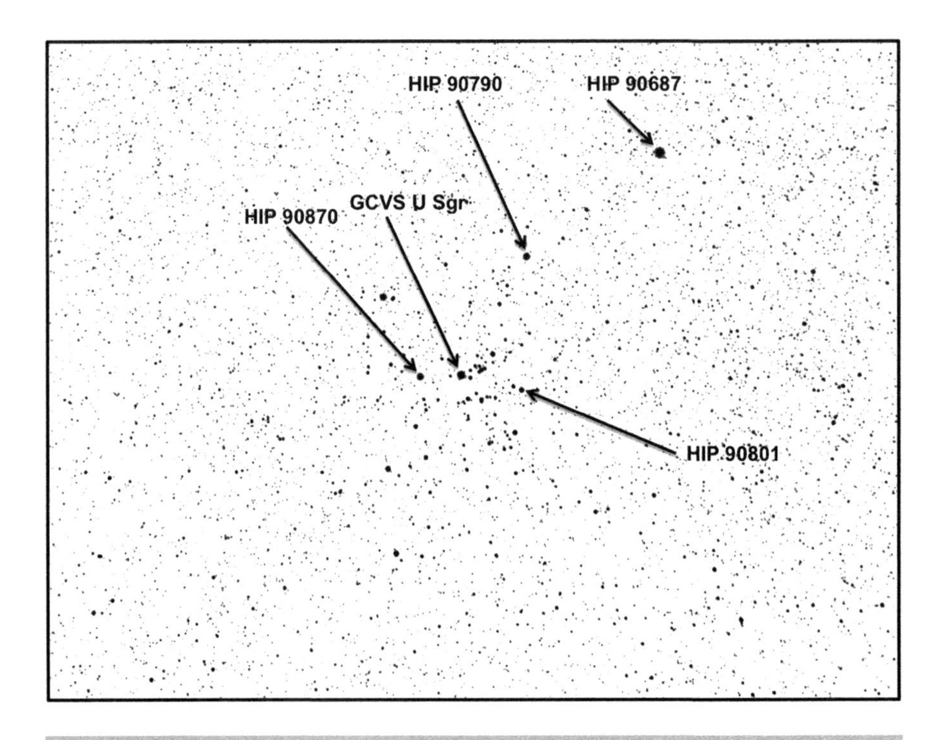

Fig. 3.74 Negative image of Messier 25

M25 lies at a distance of about 2,000 ly and has an apparent diameter of about 32' of arc, which is equivalent to 19 ly at that distance. Although there are a few orange stars, there are many bluish stars. Some of the stars are identified on the negative image in Figure 3.74. Comparing this with the color image, you can see which stars are cool (orange) and which ones are hot (blue). In particular, look at the variable star GCVS U Sgr. This is a Cepheid Variable, which is part of the actual cluster, not a line of sight member. This is pretty rare in a galactic cluster. Cepheids have a known relationship between their actual luminosity and period, so they can be used as distance indicators by comparing their absolute magnitude with their apparent magnitude. U Sgr has a period of 6.75 days and varies between magnitude 6.28 and 7.15. Star HIP 90870 is magnitude 7.30 and is spectral type K0 but lies at 1,393.83 ly (Hipparcos Catalog). Star HIP 90790 has a magnitude of 6.81 and is of type K1, only 43 ly distant. Star HIP 90687 has a magnitude of 5.63 and is type K0 at 293.04 ly distant.

The chart in Figure 3.75 shows the location of M25 in the constellation of Sagittarius. Messier referred to the "closest known star" as Flamsteed 21. You can spot this on the chart fairly easily, as its position is highlighted. Messier referred to it as a magnitude 6 star. The Hipparcos Catalog has it numbered as HIP 90289 but gives it a magnitude of 4.8. The same catalog gives main sequence 21 Sagittarii a spectral type of A1/A2 and a distance of 597.36 ly. M25 sits immediately to the north of the lid of the Sagittarius teapot.

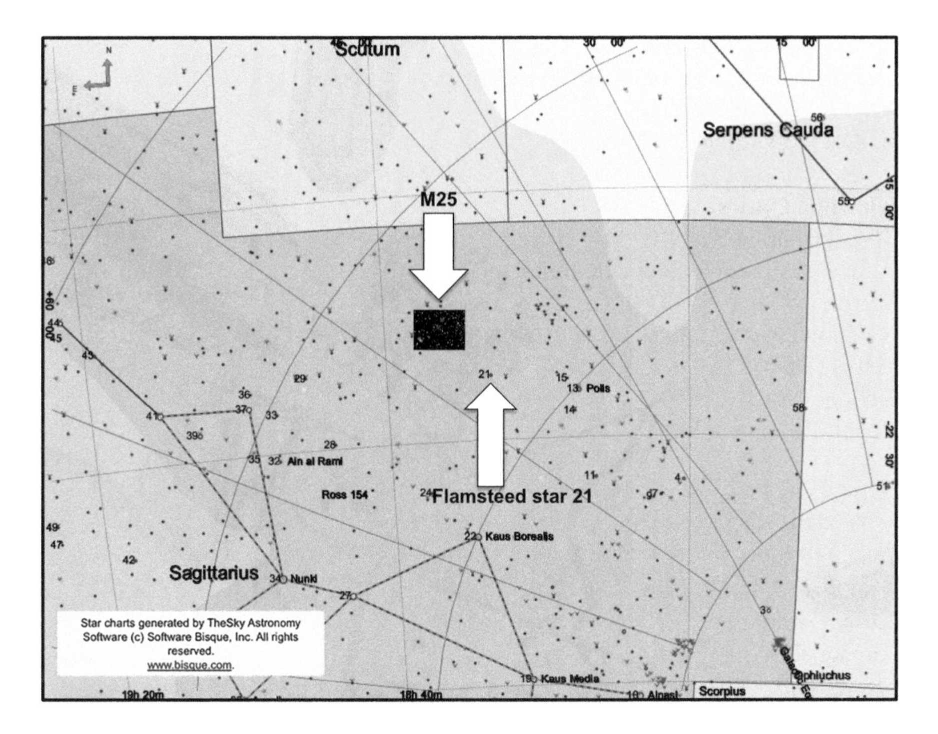

Fig. 3.75 Chart showing the location of M25

The image was taken at 04h 01m 51s on the morning of April 14 in Siding Spring. The Universal Time was 18h 01m 51s. The Right Ascension of M25 is 18h 32m 47s, and the Local Sidereal Time was 17h 26m 18s. Thus, M25 was approaching the meridian with 1h 6m 29s to go. This means that the cluster's transit time would be at 05h 08m 20s. The image of M25 was taken with the diminutive 90-mm telescope with a single-shot color camera.

Messier 25: Data relating to the image in Figure 3.73

REMOTELY IMAGED FROM NEW SOUTH WALES, AUSTRALIA
IC 4725
OBJECT TYPE: Open Cluster

RA: 18h 32m 47s
DEC: -19° 05' 59"
ALTITUDE: +70° 42' 46"
AZIMUTH: 54° 55' 17"
FIELD OF VIEW: 1° 37' 44" x 1° 13' 18"
OBJECT SIZE: 32'
POSITION ANGLE: 357° 17' from North
EXPOSURE TIME: 300 s
DATE: 14th April

LOCAL TIME: 04h 01m 51s
UNIVERSAL TIME: 18h 01m 51s (13th April)
SCALE: 3.67 arcsec/pixel
MOON PHASE: 94.08% (waning)

Telescope Optics
OTA: Takahashi Sky 90
Optical Design: Apochromatic Refractor
Aperture: 90 mm
Focal Length: 417 mm
F/Ratio: f/5.6
Guiding: None
Mount: Paramount PME

Instrument Package
CCD: SBIG ST2000 XMC
Colour CMOS
Pixel Size: 7.4μm square
Sensor: Frontlit
Cooling: CNA
Array: 1600 x 1200 pixels
FOV: 60.5 x 80.7 arcmin

Location
Observatory: Siding Spring
UTC +10.00 (Australia Daylight savings time is observed)
31° 16' 24" South
149 03' 52" East
Elevation: 1122 meters (3681 ft)

Messier 26

Messier 26 (NGC 6694) is an open cluster in the constellation of Scutum. As you can see in Figure 3.76, it is difficult to distinguish it from the background Milky Way stars. M26 cannot compare with its neighbor M11, which is a spectacular galactic open cluster, also in Scutum, that could easily be mistaken for a globular cluster. There is no way that M26 could be mistaken for a globular! The cluster was discovered by Messier himself. On the night of June 20, he wrote, "I discovered another cluster of stars near *n* & *o* of *Antinous*, among which there is one which is brighter than the others: with a refractor of three feet, it is not possible to distinguish them, it requires to employ a strong instrument: I saw them very well with a Gregorian telescope which magnified 104 times: among them one doesn't see any nebulosity, but with a refractor of 3 feet & a half, these stars don't appear individually, but in the

Fig. 3.76 Image of Messier 26, an open cluster in Scutum

form of a nebula; the diameter of that cluster may be 2 minutes of arc. I have determined its position with regard to the star *o* of *Antinous.*"Antinous is an obsolete constellation. Can you spot an oval-shaped dark ring around M26? This mystery ring was investigated by James Cuffey in 1940, and the issue remains open as to whether it is caused by obscuring material or simply a lack of stars.

M26 lies at a distance of 5,000 ly from us. It is difficult to define the borders of the cluster, so on the negative image in Figure 3.77 vertical lines are drawn roughly 11 arcmin apart. Assuming the 5,000 ly distance of M26, that would make the distance between these lines about 16 ly. I leave it to the reader to decide the actual size of the cluster. The bright hot A0 star HIP 92009 is in the foreground and is magnitude 9.13. One can assume that this is the star that Messier referred to when he said that one star is brighter than the others. The Hipparcos Catalog puts this at a distance of 1,689.93 ly, so it is about 1/3 of the way to M26. I suppose you could head straight for that star in your spaceship at the speed of light, confident that when you reach it, you would only have to do the same distance again twice over to arrive at M26. Clearly, some form of stasis is required! The maximum brightness of the other stars that appear to be in the cluster is around 10. The magnitude of Tycho 5696:282 (arrowed) is 10.21. The estimated age of the cluster is about 90 million years.

The chart in Figure 3.78 shows the position of the open cluster M26 in the constellation of Scutum. The image has been rotated so that north is at the top of the chart. Scutum has the constellation of Aquila above and to the east, and the constellation of

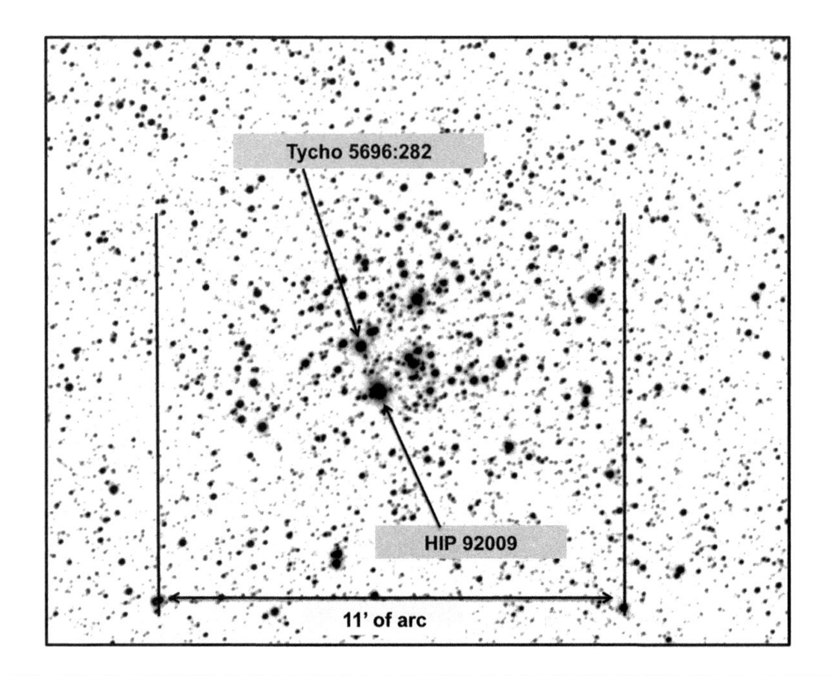

Fig. 3.77 Negative image of Messier 26

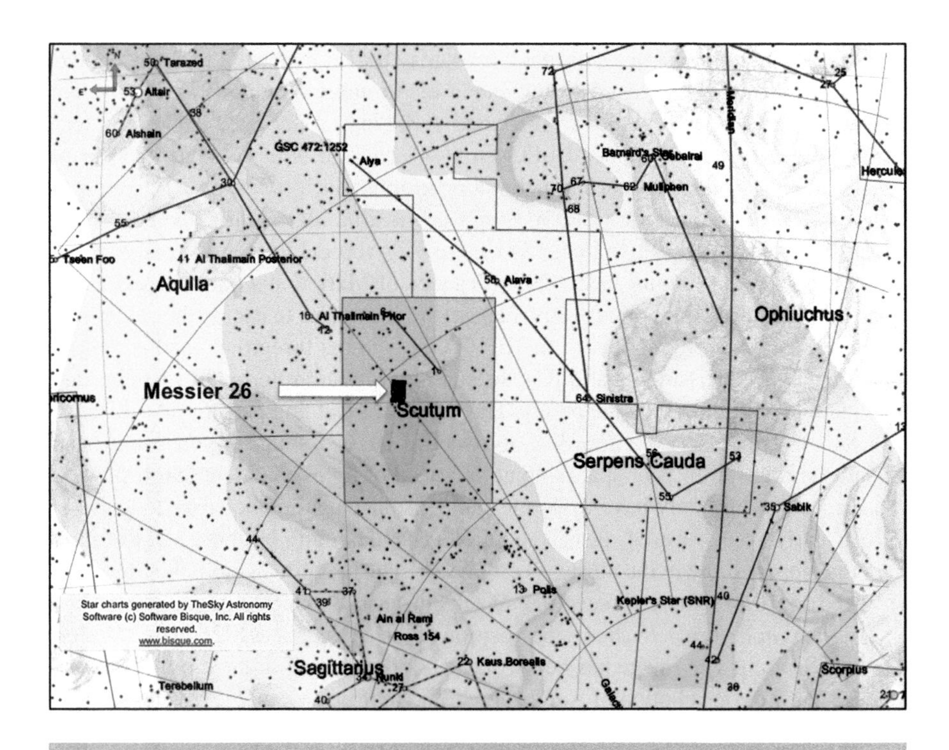

Fig. 3.78 Chart showing the location of M26

Serpens Cauda above and to the west. Scutum sits on top of the constellation of Sagittarius, so if you search for the well-known "teapot" of Sagittarius and look up – there is Scutum! If you are in the southern hemisphere, of course, you will need to look down.

The image was taken from Siding Spring at 04h 36m 13s local time on April 5. The Universal Time was 18h 36m 13s (April 4). The Right Ascension of M26 is 18h 46m 14s, and the Local Sidereal Time was 17h 25m 17s. The difference between the LST and the RA is 1h 20m 57s, so the cluster was east of the meridian and would cross it at 05h 57m 10s. The telescope used was a Fast Newtonian Astrograph with an aperture of 406 mm and a focal length of 1425 mm. The camera is an Apogee Aspen CG16070 with a Class 1 CCD.

Messier 26: Data relating to the image in Figure 3.76

REMOTELY IMAGED FROM NEW SOUTH WALES, AUSTRALIA
NGC 6694
OBJECT TYPE: Lenticular Galaxy

RA: 18h 46m 14s
DEC: -09° 21' 39"
ALTITUDE: +61° 08' 59"
AZIMUTH: 45° 01' 25"
FIELD OF VIEW: 1° 26' 35" x 0° 57' 32"
OBJECT SIZE: 15'
POSITION ANGLE: 86° 39' from North
EXPOSURE TIME: 300 s
DATE: 5th April
LOCAL TIME: 04h 36m 13s
UNIVERSAL TIME: 18h 36m 13s (4th April)
SCALE: 1.07 arcsec/pixel
MOON PHASE: 61.14% (waxing)

Telescope Optics
OTA: 16" 0.4m Astro Systeme Austria
Optical Design: Fast Newtonian Astrograph
Aperture: 400 mm
Focal Length: 1425 mm
F/Ratio: f/3.5
Guiding: Off-Axis Guiding
Mount: Paramount PME

Instrument Package
CCD: Apogee Aspen CG16070 Class 1 CCD
Pixel Size: 7.4μm square
Sensor: KAI – 16070 ABG
Cooling: -30°C default
Array: 4864 x 3232 pixels
FOV: 88.3 x 58.7 arcmin

Location
Observatory: Siding Spring
UTC +10.00 (Australia Daylight savings time is observed)
31° 16' 24" South
149 03' 52" East
Elevation: 1122 meters (3681 ft)

Messier 27

Messier 27, shown in Figure 3.79 is a planetary nebula, born from the death of a star at its center some 48,000 years ago (SEDS). Gas was blown outwards to form the expanding nebula. Its dimensions are approximately 8.0' x 5.7' of arc. It is located in the constellation of Vulpecula and was imaged at an altitude of just over 28°. These are Messier's notes on M27: "On July 12, 1764, I have worked on the research of the nebulae, and I have discovered one in the constellation Vulpecula, between the two forepaws, & very near the star of fifth magnitude, the fourteenth of that constellation, according to the catalog of Flamsteed: One sees it well in an

Fig. 3.79 Image of Messier 27, a planetary nebula in Vulpecula

ordinary refractor of three feet & a half [FL]. I have examined it with a Gregorian telescope, which magnified 104 times: it appears in an oval shape; it doesn't contain any star; its diameter is about 4 minutes of arc. I have compared that nebula with the neighboring star which I have mentioned above [14 Vul]; its right ascension has been concluded at 297d 21' 41", & its declination 22d 4' 0" north." The word "Vulpecula" translates to "Fox," hence Messier's reference to the "two forepaws."

M27 is shown on the negative image in Figure 3.80. The Field of View of the image is 0° 47' 06" x 0° 47' 06" as measured on the plate solution. North is at the top with east to the left, with a position angle of 358° 17'. Three stars are labeled on the image to give an idea of relative magnitudes. The brightest of these – HIP 98375 – has a magnitude of 5.68. It lies at a distance of 158 ly in comparison with M27, which is at a distance of 1,250 ly. The second brightest, HIP 98505, is magnitude 7.67 and is at 62.8 ly from us. The third star, HIP 98523, is magnitude 8.09 and lies at a distance of 784 ly.

M27 lies in the constellation of Vulpecula, as can be seen in Figure 3.81. Once more, the black rectangle is the actual image to scale, superimposed on the chart as a result of star matching. M27 is the only Messier Object in Vulpecula. It is located about 2° from the magnitude 4.57 star 13 Vulpeculae. There is a single Caldwell Object in Vulpecula – C37, which is an open cluster and lies about 5° away from

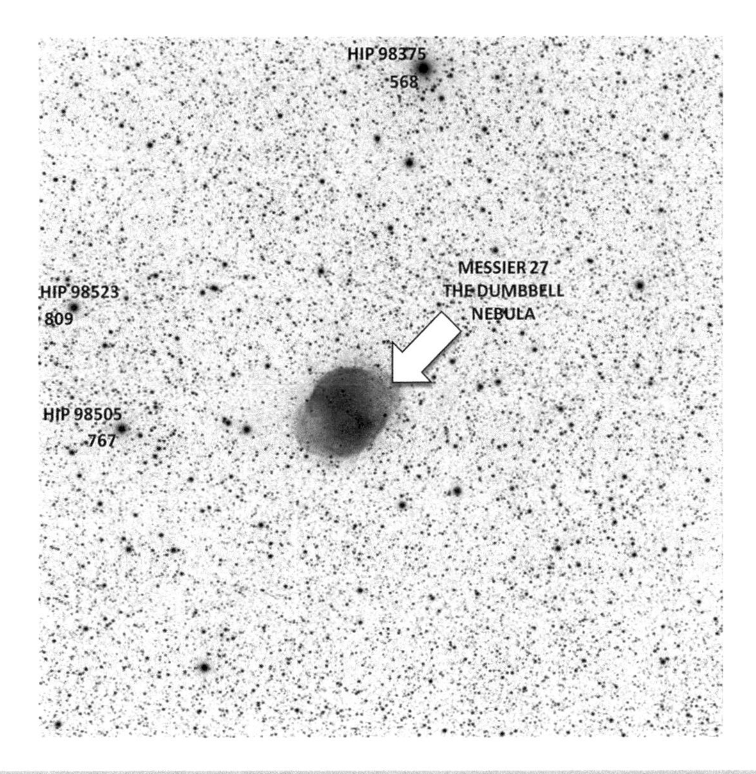

Fig. 3.80 Negative image of Messier 27

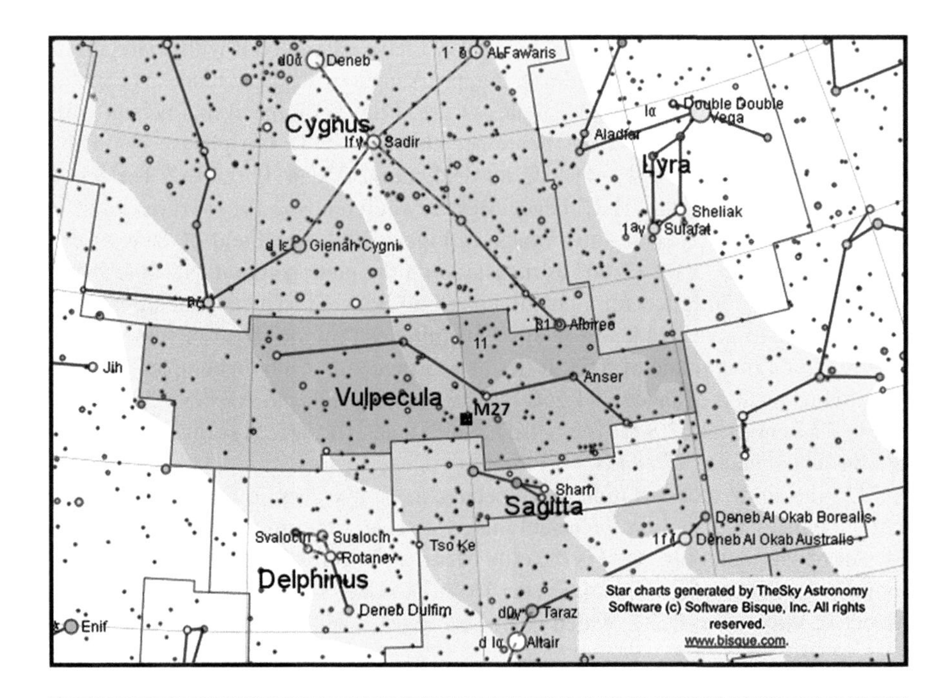

Fig. 3.81 Chart showing the location of M27

M27. Vulpecula's partial claim to fame comes from that fact that it is in this constellation that the first Pulsar was found by Irish astrophysicist Susan Jocelyn Bell. The constellation borders Cygnus, Lyra, Hercules, Sagitta, Delphinus and Pegasus, so there is plenty to look at in this part of the sky!

M27 was imaged from New Mexico on October 4, 50 minutes after midnight local time. The Universal Time was 06h 50m. The Right Ascension of M27 is 20h 01m 40.50s, and the Local Sidereal Time was 00h 0m 19s. The difference of about 4h indicates the length of time since M27 crossed the meridian. The altitude of M27 was low at about 28° when the image was taken. This is only just above the minimum altitude for this telescope of 25°, below which, the telescope would be pointing at the inside of the observatory wall! The telescope used to image M27 is a 150-mm apochromatic refractor with a focal length of 1,095 mm.

Messier 27: Data relating to the image in Figure 3.79

REMOTELY IMAGED FROM MAYHILL, NEW MEXICO
NGC6853 (DUMBBELL NEBULA)
OBJECT TYPE: Planetary Nebula

RA: 20h 00m 20s
DEC: +22° 46' 26"
ALTITUDE: +28° 20' 45"

AZIMUTH: 280° 07' 13"
FIELD OF VIEW: 0° 47' 04" x 0° 47' 04"
OBJECT SIZE: 8' x 5.7'
POSITION ANGLE: 358° 16' from North
EXPOSURE TIME: 600 s
DATE: 4th October
LOCAL TIME: 00h 50m 00s
UNIVERSAL TIME: 06h 50m 00s
SCALE: 1.38 arcsec/pixel
MOON PHASE: 9.67% (waxing)

TELESCOPE OPTICS
OTA: Takahashi TOA-150
Optical Design: Apochromatic Refractor
Aperture: 150 mm
Focal Length: 1095 mm
F/Ratio: f/7.3
Guiding: Internal
Mount: Paramount GTS

Instrument Package
SBIG ST-4000M One-Shot Color CCD
A/D Gain: 0.6e-/ADU
Pixel Size: 7.4um square
Resolution: 1.45 arcsec/pixel
Sensor: Frontlit
Cooling: -20°C Winter (-10°C Summer)
Array: 2048 x 2048 (8.3 Megapixels)
FOV: 49.6 x49.6 arcmin

Location
Observatory: New Mexico Skies
UTC Minus 7.00 (Daylight savings time is observed)
Minimum Target Elevation: Approx 25 – 45 Degrees
(N or S) 32° 54' Decimal: 32.9 North
(W or E) 105° 31' Decimal: 105.5 West
Elevation: 2225 meters (7298 ft)

Messier 28

Messier 28 (NGC 6626), shown in Figure 3.82, is a globular cluster in the constellation of Sagittarius. Although an interesting object, it cannot compare to the spectacular appearance of M13, or even M22, which is a very near neighbor. It has a diameter of 11.7' of arc, which is only 1/3 the size of Messier 22. The globular is

Fig. 3.82 Image of Messier 28, a globular cluster in Sagittarus

very concentrated towards the center, so it does not give that glittering effect that a globular with more evenly distributed stars displays. Messier noted on the night of July 26 and 27, 1764, "I have discovered a nebula in the upper part of the bow of Sagittarius, at about 1 degree from the star Lambda of that constellation, & little distant from the beautiful nebula which is between the head & the bow [M22]: that new one may be the third of the older one, & doesn't contain any star, as far as I have been able to judge when examining it with a good Gregorian telescope which magnifies 104 times: it is round, its diameter is about 2 minutes of arc; one sees it with difficulty with an ordinary refractor of 3 feet & a half of length." English astronomer W. H. Smyth commented, "It is not very bright; and is preceded by two telescopic stars in a vertical line. Messier, who enrolled it in 1764, describes it as a nebula without a star, and seen with difficulty in his 3 1/2 foot telescope. But Sir William Herschel resolved it, and placed it among the stellar clusters; his son [John] recommended it as a testing-object for trying the space-penetrating powers of telescopes."

The negative image in Figure 3.83 identifies two stars in the field. The first, HIP 89980, is magnitude 6.19 and is a cool M3 star, which lies at a distance of 1,164.85 ly in the Hipparcos Catalog data. The second star, HIP 90266, is magnitude 8.48 and spectral type B9, so it is a hot star at a distance of 2,038.48 ly. Messier 28 itself lies at a distance of 18,000 ly, so it is considerably further away than either of the

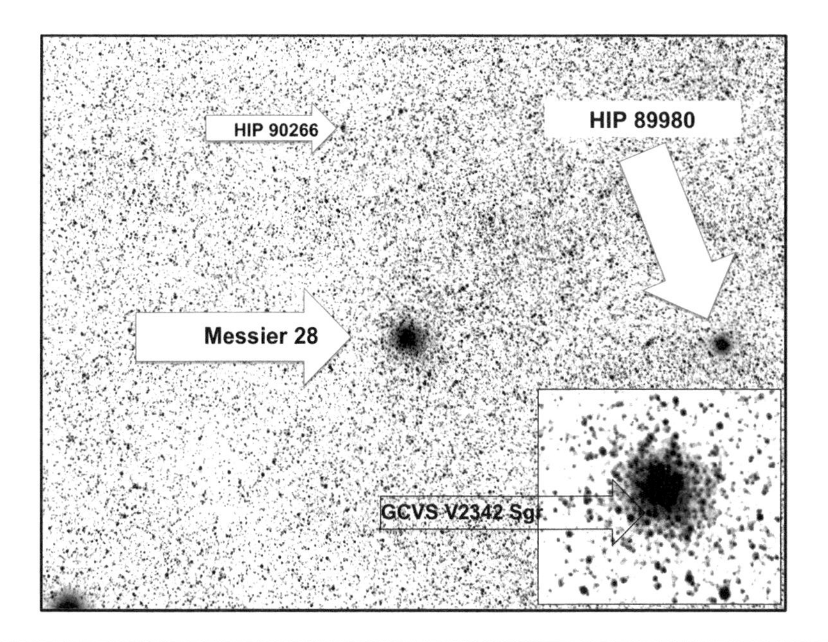

Fig. 3.83 Negative image of Messier 28

identified stars. In the enlarged view of M28 itself, a variable star is identified. The star is GCVS V2342 Sgr, where GCVS stands for the General Catalog of Variable Stars. This star has a variation of 0.7 magnitudes, 13.1 at its brightest and 13.8 at its faintest. It has a period of 92 days. In 1987, the Jodrell Bank radio telescope in England discovered a millisecond pulsar in M28. This is a neutron star that is rotating in the region of 1000 times a second. It has the name PSR 1937+21A, the A signifying that it was the first to be discovered in M28. This pulsar was the first to be discovered in any globular cluster. There are now a dozen known millisecond pulsars in Messier 28. On to the bottom left of the image is in fact Kaus Borealis, or 22-Lambda Sagittarii, the knob on the top of the Sagittarius teapot lid!

The chart in Figure 3.84 shows north at the top and east to the left. M28 lies just above the well-known teapot asterism. The rectangular image reflects the ratio of the pixels on the chip in the camera. In this case, the ratio is 1,600 X 1,200 pixels. The plate solution gave a figure for the Position Angle of 357° 19' from north. This means that the camera is aligned to within 2° 41' of north which is acceptable. Also on the chart is the arrowed meridian line west of the M28 globular cluster.

The image was taken at 03h 04m 22s on April 2. The Universal Time was 17h 04m 22s. Messier 28 has a Right Ascension of 18h 25m 36s, and the Local Sidereal Time was 15h 41m 21s, so the difference was 2h 44m 15s. Adding this to the local time gives a meridian transit time of 05h 48m 37s. The telescope, shown in the photograph, used to take the image was the 90-mm refractor at Siding Spring with a single-shot color camera.

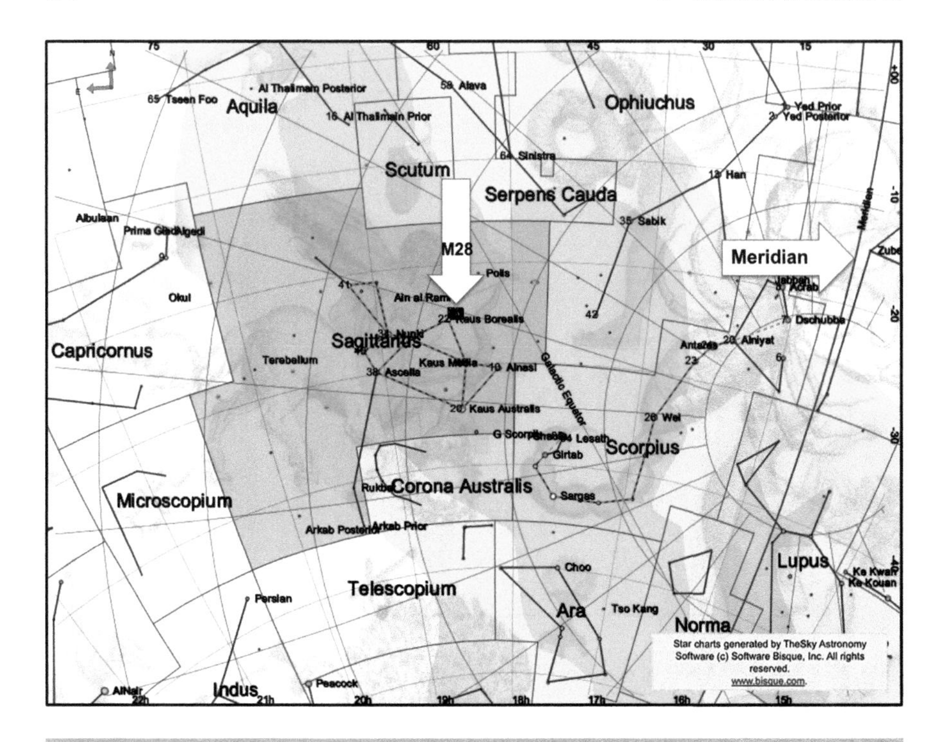

Fig. 3.84 Chart showing the location of M28

Messier 28: Data relating to the image in Figure 3.82

REMOTELY IMAGED FROM NEW SOUTH WALES, AUSTRALIA
NGC 6626
OBJECT TYPE: Globular Cluster

RA: 18h 25m 36s
DEC: -24° 51' 24"
ALTITUDE: +53° 24' 47"
AZIMUTH: 90° 26' 27"
FIELD OF VIEW: 1° 37' 44" x 1° 13' 18"
OBJECT SIZE: 11.2’
POSITION ANGLE: 357° 19' from North
EXPOSURE TIME: 300 s
DATE: 2nd April
LOCAL TIME: 03h 04m 22s
UNIVERSAL TIME: 17h 04m 22s
SCALE: 3.66 arcsec/pixel
MOON PHASE: 27.21% (waxing)

Telescope Optics
OTA: Takahashi Sky 90
Optical Design: Apochromatic Refractor
Aperture: 90 mm
Focal Length: 417 mm
F/Ratio: f/5.6
Guiding: None
Mount: Paramount PME

Instrument Package
CCD: SBIG ST2000 XMC
Colour CMOS
Pixel Size: 7.4μm square
Sensor: Frontlit
Cooling: CNA
Array: 1600 x 1200 pixels
FOV: 60.5 x 80.7 arcmin

Location
Observatory: Siding Spring
UTC +10.00 (Australia Daylight savings time is observed)
31° 16' 24" South
149 03' 52" East
Elevation: 1122 meters (3681 ft)

Messier 29

Messier 29, shown in Figure 3.85, is an open cluster located in the Constellation of Cygnus. It is about 10' of arc in size. Messier reported on July 29, 1764, "A cluster of 7 or 8 very small stars, which are below Gamma Cygni, which one sees with an ordinary telescope of 3.5-foot in the form of a nebula." The cluster was at an altitude of just over 46°, peeping over the wall of the observatory with its declared minimum altitude of 35° to 45°. American astronomer Robert Burnham Jr. described M29 as "a small and visually rather indistinguished star cluster located in a rich and crowded area of the Cygnus Milky Way about 1.7° SSE from Gamma Cygni." It is conveniently located near the central star of the cross of Cygnus, which makes it relatively easy to find. It is also known as NGC 6913.

A circle of roughly 10' diameter has been drawn on the negative image in Figure 3.86 to give an idea of scale. Four stars with distances taken from the Hipparcos and Tycho star catalogs are identified. The cluster itself is listed by various sources as being at a distance of somewhere between 4,000 ly to 7,200 ly. A 2014 paper by Lithuanian astronomer V. Straizys et al entitled "The Enigma Of The Open Cluster M29 (NGC 6913) Solved" puts the distance at 1.54 ± 0.15

Fig. 3.85 Image of Messier 29, an open cluster in Cygnus

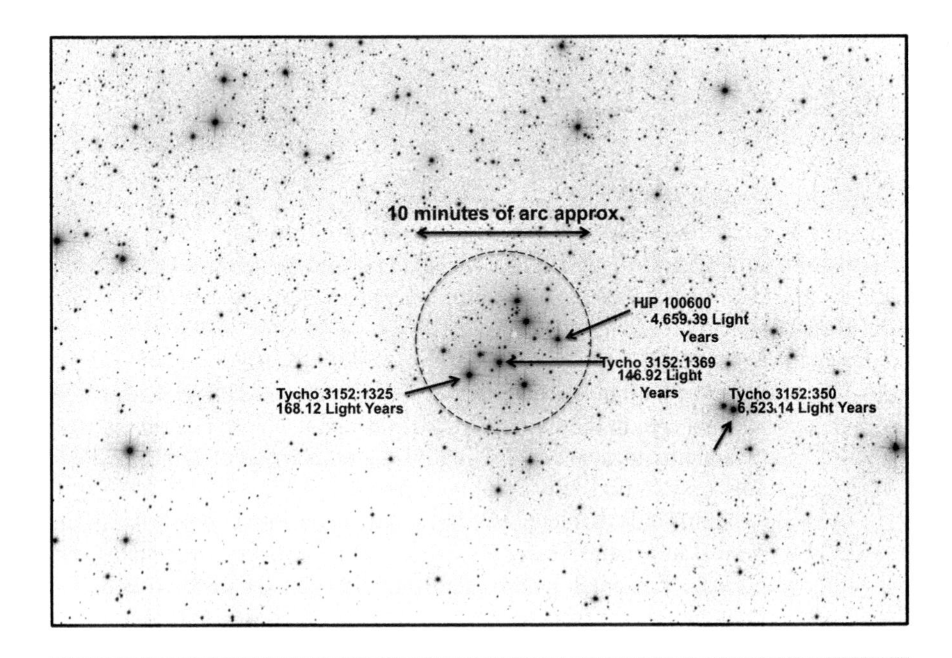

Fig. 3.86 Negative image of Messier 29

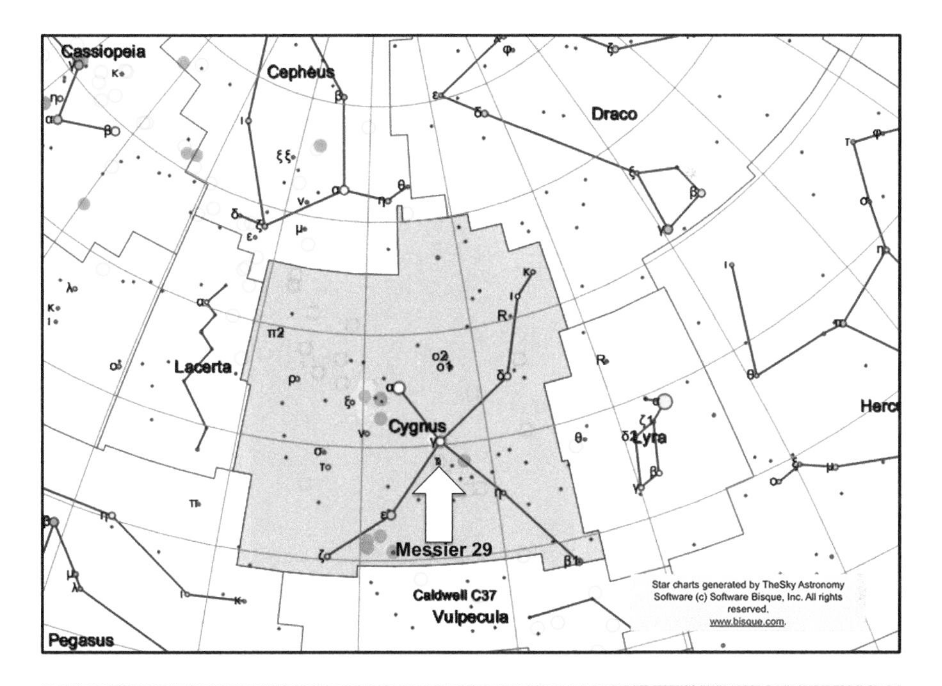

Fig. 3.87 Chart showing the location of M29

kiloparsecs – equivalent to 4,928 ly. Tycho 3152 :1325 is magnitude 8.92 and is only 168 ly distant from the Tycho data. HIP 100600 is magnitude 9.36 and is placed at a distance of 4,659 ly from the Hipparcos Catalog, which is closer to the Straizys value. Burnham noted that M29 is greatly affected by interstellar absorption – in other words, there is a large amount of obscuring dust between us and the cluster. He points out that perhaps this would be a very striking cluster if it were not for this obscuring dust.

The chart in Figure 3.87 shows the exact location of M29 in Cygnus, marked by the tiny superimposed image. There is in fact another Messier object in Cygnus – M39 – that is also an open cluster. There are seven Caldwell objects: are C12 (Spiral Galaxy), C15 (Planetary Nebula), C19 (Open Cluster and Nebula), C20 (Nebula) , C27 (Nebula), C33 (Nebula) and C34 (Nebula). Cygnus is surrounded by the constellations Lacerta, Cepheus, Draco, Lyra, Vulpecula and Pegasus. The open cluster in C19 is very different to M29 in that it is embedded in a nebula known as the Cocoon Nebula.

The image was taken on October 6 in Spain at 01h 17m 54s local time. The Universal Time was 23h 17m 54s (October 5). The Local Sidereal Time was 00h 10m 30s, so the First Point of Aries had crossed the meridian some 10 minutes earlier. The Right Ascension of M29 is 20h 24m 36s, so it had crossed the meridian just less than 4 hours earlier. The altitude of roughly 46° gave an Air Mass of 1.39. The 10-minute image was taken using a telescope with 431 mm of aperture.

Messier 29: Data relating to the image in Figure 3.85

REMOTELY IMAGED FROM NERPIO, SPAIN
NGC6913
OBJECT TYPE: Open Cluster

RA: 20h 24m 36s
DEC: +38° 33' 34"
ALTITUDE: +46° 05' 48"
AZIMUTH: 289° 56' 12"
FIELD OF VIEW: 0° 42' 13" x 0° 28' 09"
OBJECT SIZE: 10'
POSITION ANGLE: 272° 11' from North
EXPOSURE TIME: 600 s
DATE: 6th October
LOCAL TIME: 01h 17m 54s
UNIVERSAL TIME: 23h 17m 54s
SCALE: 0.63 arcsec/pixel
MOON PHASE: 20.96% (waxing)

Telescope Optics
OTA: Planewave 17" CDK
Optical Design: Corrected Dall – Kirkham Astrograph
Aperture: 431 mm
Focal Length: 2929 mm
F/Ratio: f/6.8
Guiding: External
Mount: Paramount ME

Instrument Package
SBIG STL - 1100M
A/D Gain: 2.2 e-/ADU
Pixel Size: 9μm square
Resolution: 0.63 arcsec/pixel
Sensor: Frontlit
Cooling: -20°C default
Array: 4008 x 2672 (10.7 Megapixels)
FOV: 28.2 x 42.3 arcmin

Location
Observatory: AstroCamp – MPC Code – I89
UTC +1.00 (Daylight savings time is observed)
Minimum Target Elevation: Approx 35 – 45 Degrees
North: 38° 09'
West 0002° 19'
Elevation: 1650 meters (5413 ft)

Messier 30

Messier 30 (NGC 7099), shown in Figure 3.88, is a globular cluster in the constellation of Capricornus. Messier noted, "Nebula discovered below the tail of Capricorn, very near to the star 41 of that constellation, of 6th magnitude, according to Flamsteed. One sees it with difficulty with an ordinary telescope of 3.5-foot [FL]. It is round & contains no star; its position determined from Zeta Capricorni." As usual, Messier was unable to observe the true nature of a globular cluster, as he could not resolve any stars with the equipment available to him on the date of his observation in 1764. He used a range of instruments, but his refractors were only equivalent to a small aperture instrument, and his reflectors were poor, as they used speculum metal for mirrors. He cataloged the object on August 3, 1764. M30 has a Declination of -23° 06' 03", low in the sky for northern observers.

The negative image in Figure 3.89 shows M30 and two nearby stars for comparison. The brighter star is HIP 107128 (41 Capricorni), which has a magnitude of 5.2, and the fainter is HIP 107037, which has a magnitude of 8.36. The brighter star lies at a distance of 75.8725 Parsecs, which is equivalent to 247.46 ly. M30 is at a distance of 26,100 ly – 1/4 of the size our Galaxy (SEDS). Messier 30 subtends about 12 arcmin – equivalent to a diameter of about 90 ly. This globular does seem very

Fig. 3.88 Image of Messier 30, a globular cluster in Capricornus

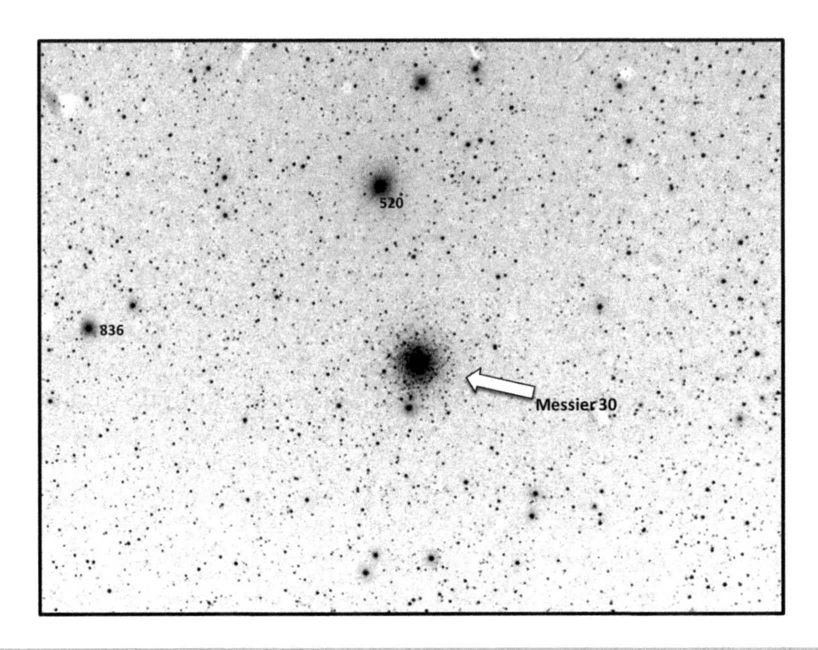

Fig. 3.89 Negative image of Messier 30

compressed in comparison with others such as M13 or M15. Burnham considers that M30 was first resolved into stars by Sir William Herschel in 1783. He quoted both William's son, John Herschel, who regarded M30 as having an elliptical shape, and Admiral Smyth, who called it a "fine, pale white cluster. ...bright and from the struggling streams of stars on its north edge has an elliptical aspect with a central blaze; few other stars in the field."

The 10-minute image of M30 can be seen superimposed on the chart (Figure 3.90) in the constellation of Capricornus. It is the only Messier Object in this constellation. In fact, Patrick Moore in his *Data Book of Astronomy* made the comment that "All in all, Capricornus is a rather barren constellation redeemed by the presence of the bright globular cluster M30." This is in stark contrast to its neighbor Sagittarius, which contains 15 Messier Objects! The difference is that Capricornus skirts the edge of the Milky Way, whereas Sagittarius runs directly through it. Like all globular clusters, M30 is very old, having been around for some 12.93 billion years (Forbes)!

The image was taken at 22h 26m 39s on October 6. The Universal Time was 11h 26m 39s. By taking the image from Siding Spring, New South Wales, it was possible to do so with M30 at an altitude of just over 77°. The lunar phase was about 25%, but the Moon was over 68°away from M30. M30 has a Right Ascension of 21h 41m 19s, and the Local Sidereal Time was 22h 24m 52s. Thus, the globular had crossed the meridian 1h 43m 33s earlier. If you refer back to the chart of M30, you can see that the vertical meridian line in red is about an hour and a quarter to the east (left) of the black rectangle image of M30. The telescope used to take the image had a single-shot color camera mounted on a 3-inch refractor, with an exposure time of 10 minutes.

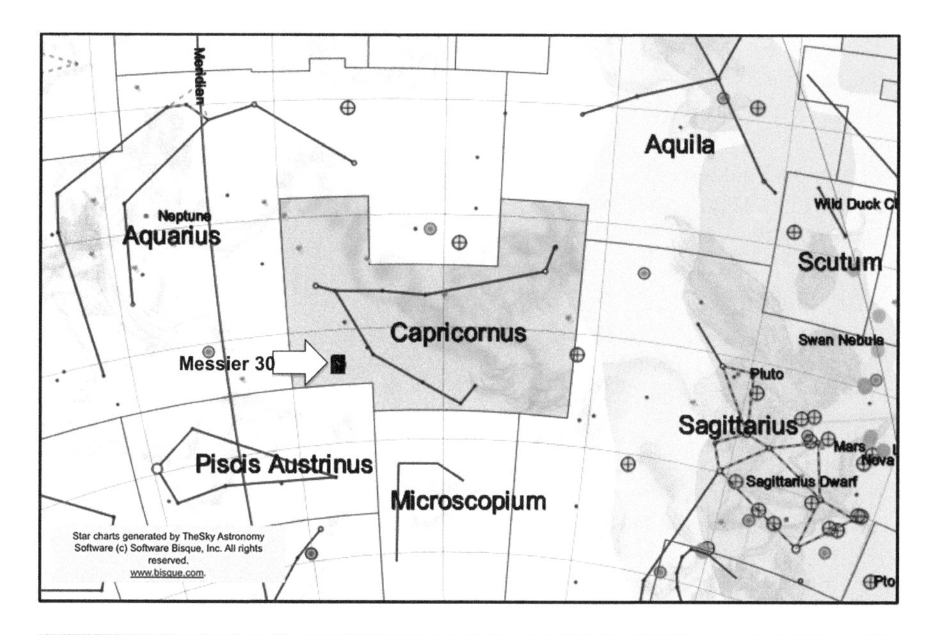

Fig. 3.90 Chart showing the location of M30

Messier 30: Data relating to the image in Figure 3.88

REMOTELY IMAGED FROM NEW SOUTH WALES, AUSTRALIA
NGC 7099
OBJECT TYPE: Globular Cluster

RA: 21h 41m 19s
DEC: -23° 06' 03"
ALTITUDE: +77° 21' 40"
AZIMUTH: 307° 25' 50"
FIELD OF VIEW: 1° 37' 42" x 1° 13' 17"
OBJECT SIZE: 12'
POSITION ANGLE: 90° 08' from North
EXPOSURE TIME: 600 s
DATE: 6th October
LOCAL TIME: 22h 26m 39s
UNIVERSAL TIME: 11h 26m 39s
SCALE: 3.66 arcsec/pixel
MOON PHASE: 24.99% (waxing)

Telescope Optics
OTA: Takahashi Sky 90
Optical Design: Apochromatic Refractor
Aperture: 90 mm
Focal Length: 417 mm

F/Ratio: f/5.6
Guiding: None
Mount: Paramount PME

Instrument Package
CCD: SBIG ST2000 XMC
Colour CMOS
Pixel Size: 7.4μm square
Sensor: Frontlit
Cooling: CNA
Array: 1600 x 1200 pixels
FOV: 60.5 x 80.7 arcmin

Location
Observatory: Siding Spring
UTC +10.00 (Australia Daylight savings time is observed)
31° 16' 24" South
149 03' 52" East
Elevation: 1122 meters (3681 ft)

Messier 31

The spectacular nearby galaxy Messier 31 (NGC 224), shown in Figure 3.91, is a must for all amateur astronomers, as it is easily visible with the naked eye on a clear dark night. It is disappointing through smaller telescopes but comes alive in short exposure images. The original image, which had a field of view of 3° 53' 16" x 2° 35' 31", has been cropped to just fit in the Andromeda galaxy. As it is a naked-eye object, it is difficult to say who first discovered it, but it was recorded by the Persian astronomer Al-Sufi in 964 and was apparently known to Persian astronomers before then. The Bavarian astronomer, Simon Marius, noted of M31: "Of these observations the first is, that since December 15, 1612, I have discovered & observed a star or a fixed star, as I could not find elsewhere in all the sky. It is situated near the third & northern star in the girdle of Andromeda. Without instrument there can be seen something like a nebula; but with the telescope no stars can be seen, as in the nebula in Cancer [M44] & other nebulous stars, but only glimmering rays, which are the brighter the closer they are to the center. In the center is a faint & pale glow, which occupies a diameter of about a quarter of a degree." Messier referred to M31 as "The beautiful nebula of the belt of Andromeda, shaped like a spindle."

The negative image in Figure 3.92 shows the Great Nebula in Andromeda and identifies its two companion galaxies M32 and M110. SEDS lists the distance to Messier 31 as 2.9 million ly, with a size of 178' x 63'. This would give M31 an actual size of 150,000 ly x 53,000 ly. (A distance of 2.3 million ly would give an actual size of roughly 120,000 ly x 42,000 ly.) The companion galaxies to M31, M110 and M32, are identified but are dealt with elsewhere in the book. You can get a lot of interest out of a simple 3-minute image of M31. I have identified the bright

Fig. 3.91 Image of Messier 31, a galaxy in Andromeda

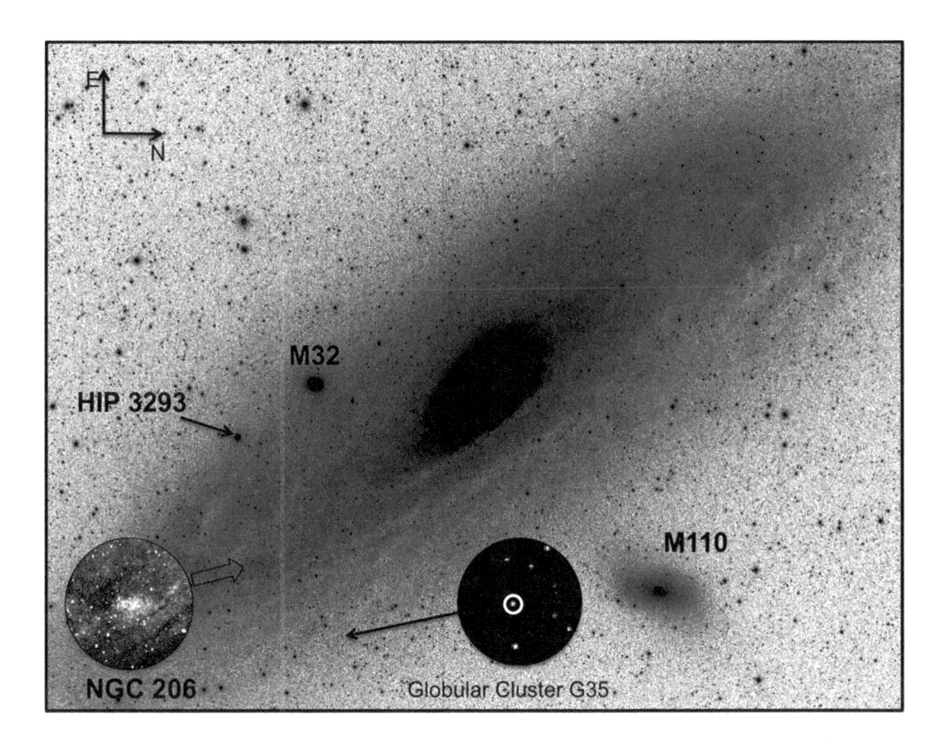

Fig. 3.92 Negative image of M31

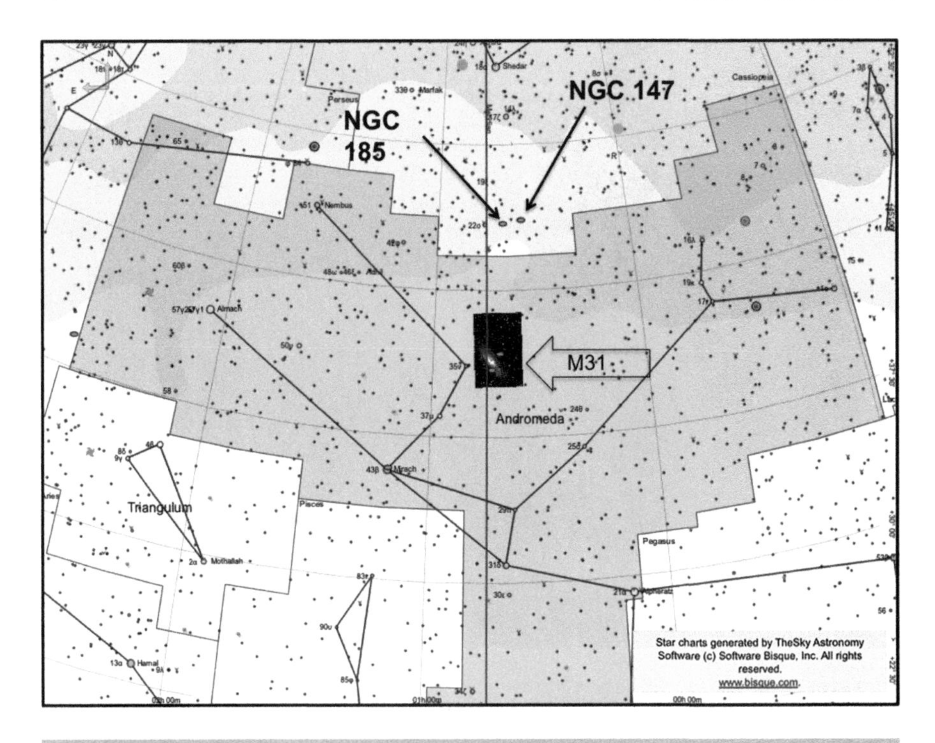

Fig. 3.93 Chart showing the location of M31

star cloud NGC 206 and enlarged it from the original image. Another thing that you can do is to identify globular clusters in the Andromeda galaxy itself. I tracked down one of these known as G35 and pointed it out with an enlargement. If you take your own image you will be able to enlarge it and search for other globulars. You will need a chart of these – over 500 have been catalogued. You will soon find a chart by googling for it.

The chart in Figure 3.93 shows the location of M31 in the constellation of Andromeda. Notice the vertical line that is just to the left of the M31 core in the uncropped embedded image. This is the meridian, so M31 has just crossed it very recently (see below). Two galaxies, NGC 147 and NGC 185, can also be seen. Although they are in a different constellation (Cassiopeia), they are in fact companion galaxies in the M31 group. So, together with M32 and M110, there are four dwarf elliptical galaxies that are gravitationally connected to M31.

The image was taken on September 24th at 01h 34m 01s local time in New Mexico. This corresponded to a Universal Time of 07h 34m 01s. The Right Ascension of M31 is 00h 43m 41s, and the Local Sidereal Time was 00h 45m 01s. The LST is higher than the RA, so M31 must have crossed the meridian from east to west already, in fact, just over a minute prior. Subtracting one from the other gives a difference of 0h 01m 20s. This means that the meridian transit time of M31 was 01h 32m 41s. The telescope used to take the image of M31 was a Takahashi FSQ apochromatic astrograph with an aperture of 106 mm and an exposure time of 180 seconds.

Messier 31: Data relating to the image in Figure 3.91

REMOTELY IMAGED FROM MAYHILL, NEW MEXICO
NGC 224
OBJECT TYPE: Spiral Galaxy

RA: 00h 43m 41s
DEC: +41° 21' 38"
ALTITUDE: +81° 25' 07"
AZIMUTH: 358° 19' 10"
FIELD OF VIEW: 3° 53' 16" x 2° 35' 31"
OBJECT SIZE: 178' X 63'
POSITION ANGLE: 92° 16' from North
EXPOSURE TIME: 180s
DATE: 24th September
LOCAL TIME: 01h 34m 01s
UNIVERSAL TIME: 07h 34m 01s
SCALE: 3.49 arcsec/pixel
MOON PHASE: 40.22% (waning)

Telescope Optics
OTA: Takahashi FSQ Fluorite
Optical Design: Petzval Apochromat Astrograph
Aperture: 106 mm
Focal Length: 530 mm
F/Ratio: f/5.0
Guiding: External
Mount: Paramount GTS-1100S

Instrument Package
SBIG STL-11000M
A/D Gain: 2.2e-/ADU
Pixel Size: 9um square
Resolution: 3.5 arcsec/pixel
Sensor: Frontlit
Cooling: -15°C default
Array: 4008 x 2672 (10.7 Megapixels)
FOV: 155.8 x 233.7 arcmin

Location
Observatory: New Mexico Skies
UTC Minus 7.00 (Daylight savings time is observed)
Minimum Target Elevation: Approx. 25 – 45 Degrees
(N or S) 32° 54' Decimal: 32.9 North
(W or E) 105° 31' Decimal: 105.5 West
Elevation: 2225 meters (7298 ft)

Messier 32

Messier 32 (NGC 221) shown in Figure 3.94, is an elliptical galaxy and a close companion of M31. The fact that it is overshadowed by its massive neighbor makes it a difficult object to image. The size of M32 is only 8' x 6' of arc, in comparison with M31, which is 178' x 63'. It has a rough diameter of 8,000 ly – small, compared with our own Galaxy of around 100,000 ly. The image taken was with a 17" telescope based at AstroCamp in Spain. M32 lies at a distance of 2.9 million ly. Messier reports: "Small nebula without stars, below & at some minutes from that of the belt of Andromeda. This small nebula is round, its light fainter than that of the belt. M. le Gentil has discovered it on October 29, 1749. M. Messier saw it, for the first time, in 1757, & he has not found any change (Diam. 2')." Messier catalogued the object on August 3, 1764.

The negative image in Figure 3.95 shows the proximity of M32 to M31 – all in all not a very exciting target. A magnitude 7 star is pointed out for comparison purposes. This star (HIP 3293) is only 160 ly from us, in comparison to the 2.9 Million ly of M32. It is difficult to visualize the enormous range difference on the image. The field of view of this image is roughly 42' x 28', so the full Moon would almost match the shorter side. The position angle is 272°, so to get north at the top, the image must be rotated just less than 90° clockwise. The circled area on the negative image shows the image in relation to the nearby M31 galaxy, which is not apparent from the actual image.

The chart in Figure 3.96 shows the position of M32 in the constellation of Andromeda. M32 lies fairly centrally in the constellation. Andromeda lies between

Fig. 3.94 Image of Messier 32, an elliptical galaxy in Andromeda

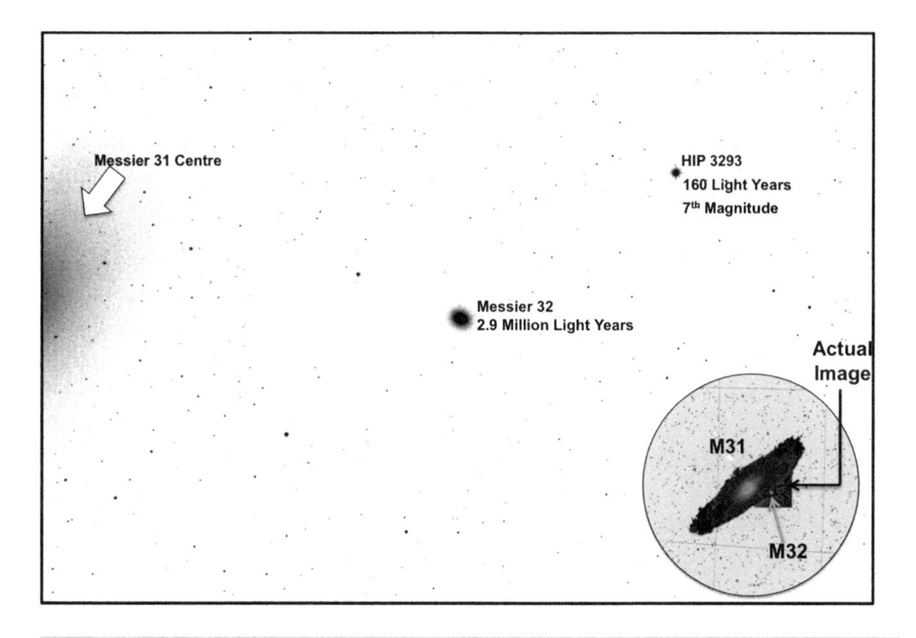

Fig. 3.95 Negative Image of Messier 32

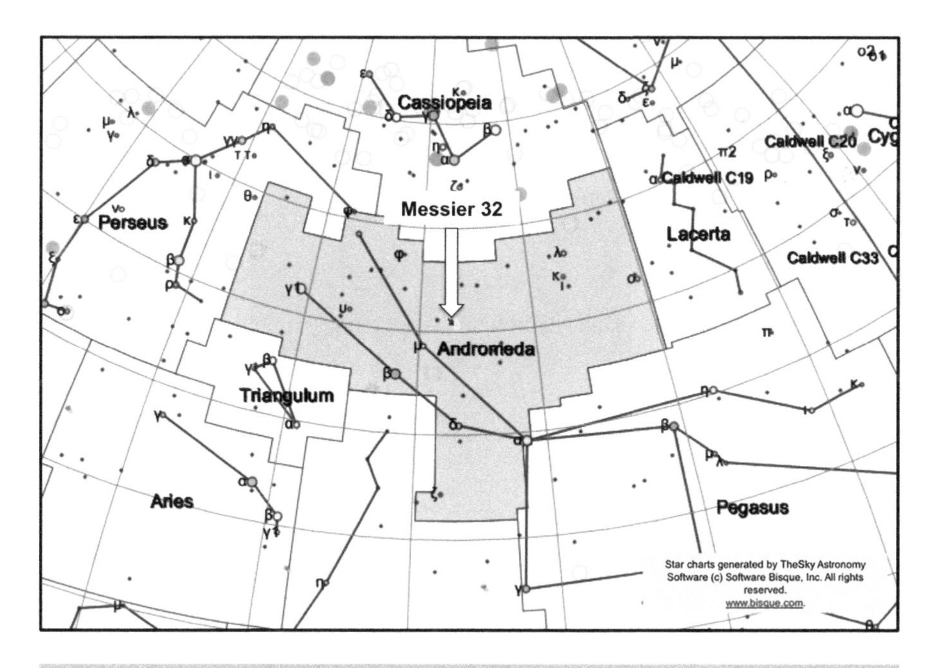

Fig. 3.96 Chart showing the location of M32

the constellations of Perseus and Lacerta, above the constellation of Pegasus to the right and Triangulum to the left. The constellation Cassiopeia lies directly above Andromeda in the chart.

The image shown was taken on September 25 at 23h 00m 12s local time. The Moon was no problem, as it was over 97° away with a phase of 24.60% (waning). Universal Time was 21h 00m 12s. The Right Ascension of M32 is 00h 43m 39s, and the Local Sidereal Time was 21h 03m 00s. Thus, the galaxy had a few hours to go before reaching the meridian. This corresponded to an azimuth of 67° 41' 44". The telescope used was a corrected Dall-Kirkham astrograph with an aperture of 431 mm.

Messier 32: Data relating to the image in Figure 3.94

REMOTELY IMAGED FROM SPAIN
NGC 221
OBJECT TYPE: Elliptical Galaxy

RA: 00h 43m 39s
DEC: +40° 57' 28"
ALTITUDE: +49° 32' 05"
AZIMUTH: 67° 41' 44"
FIELD OF VIEW: 0° 42' 13" x 0° 28' 09"
OBJECT SIZE: 8' X 6'
POSITION ANGLE: 272° 10' from North
EXPOSURE TIME: 300 s
DATE: 25th September
LOCAL TIME: 23h 00m 12s
UNIVERSAL TIME: 21h 00m 12s
SCALE: 0.63 arcsec/pixel
MOON PHASE: 24.60% (waning)

Telescope Optics
OTA: Planewave 17" CDK
Optical Design: Corrected Dall – Kirkham Astrograph.
Aperture: 431 mm
Focal Length: 2929 mm
F/Ratio: f/6.8
Guiding: External
Mount: Paramount ME

Instrument Package
SBIG STL - 1100M
A/D Gain: 2.2 e-/ADU
Pixel Size: 9μm Square
Resolution: 0.63 arcsec/pixel
Sensor: Frontlit

Cooling: -20°C default
Array: 4008 x 2672 (10.7 Megapixels)
FOV: 28.2 x 42.3 arcmin

Location
Observatory: AstroCamp – MPC Code – I89
UTC +1.00 (Daylight savings time is observed)
Minimum Target Elevation: Approx 35 – 45 Degrees
North: 38° 09'
West 0002° 19'
Elevation: 1650 meters (5413 ft)

Messier 33

Messier 33, shown in Figure 3.97, is a spectacular galaxy in the constellation of Triangulum. Anglo-Irish astronomer William Parsons identified this galaxy as a spiral. The Galaxy is about 3,000,000 ly distant, which is further away than M31.

Fig. 3.97 image of Messier 33, a galaxy in Triangulum

The major axis of the galaxy is about 73' of arc. This corresponds to roughly 60,000 ly – smaller than our own Galaxy. The galaxy is sometimes known as the Pinwheel Galaxy, or NGC 598. It is a member of our local group of galaxies. Messier discovered the galaxy on August 25 1764: "Nebula discovered between the head of the Northern Fish [Pisces] & the great Triangle, a bit distant from a star of 6th magnitude: The nebula is of a whitish light of almost even density, however a little brighter along two-third of its diameter, & contains no star. One sees it with difficulty with an ordinary telescope of 1-foot. Its position was determined from Alpha Trianguli. Seen again September 27, 1780 (Diam. 15')" (SEDS).

The negative image of M33 in Figure 3.98 shows a number of objects in the field. Firstly, there is magnitude 8.11 star HIP 7403 of type K5, lying 1,269 ly away from Earth. NGC 604 is a giant diffuse nebula and region of star formation that was discovered by William Herschel on September 11, 1784. NGC 595 is another star-forming nebula (HII region) that was discovered on October 1, 1864 by Heinrich Ludwig d'Arrest. Another star-forming nebula is NGC 588, which is shown in the

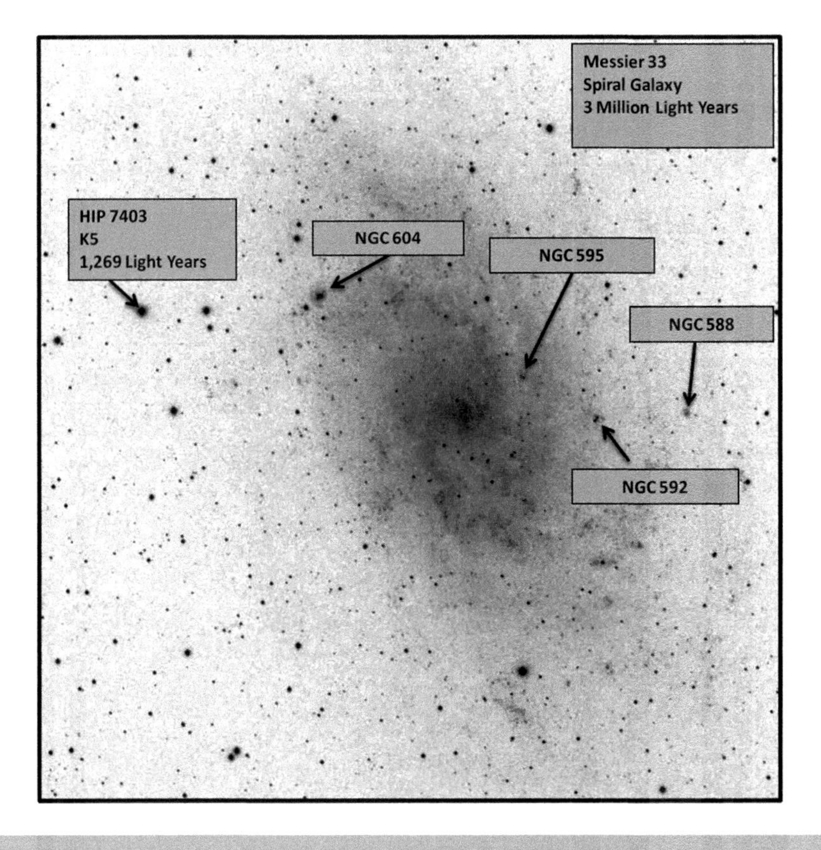

Fig. 3.98 Negative image of Messier 33

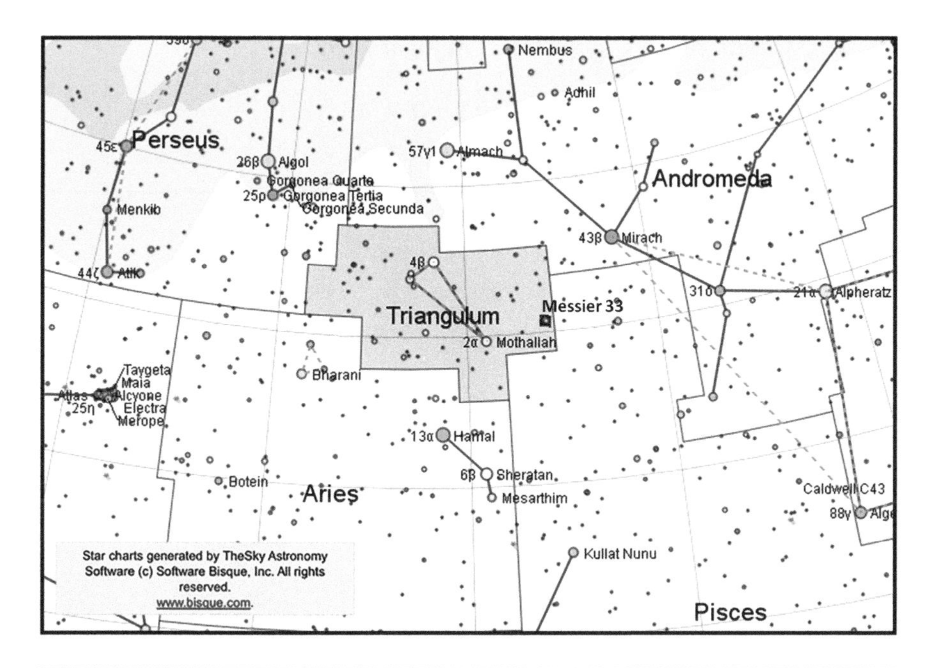

Fig. 3.99 Chart showing the location of M33

negative image to the right-hand side. The SIMBAD database describes NGC 588 as a giant HII region, which contains an ionizing embedded cluster of stars. The final object identified in the negative image is NGC 592 – another HII region and star cluster.

The chart in Figure 3.99 shows the location of M33 in Triangulum. Again, the black square is the actual image taken. M33 is the only Messier object in Triangulum. There are no Caldwell objects in this constellation. Triangulum shares a border with Andromeda, so the great galaxy M31 is not too far away in the sky. M33 is part of the local group of galaxies, together with M31 and its companions M32 and M110. There are over total 50 galaxies regarded as being in the local group. The small constellation of Triangulum also borders Aries, Perseus and Pisces. Aries does not contain any Messier objects, but Perseus has the open cluster M34 and planetary nebula M76 and Pisces has the spiral galaxy M74. Perseus also includes the Caldwell Objects C14 and C34.

The image was taken in New Mexico on September 24 at 12h 11m 27s. The Universal Time was 06h 11m 27s. M33 has a Right Ascension of 01h 36m 01s, and the Local Sidereal Time was 23h 22m 14s. Thus, M33 was east of the meridian, corresponding to an azimuth of just less than 86°. It was at an altitude of just less than 62°. The moon was below the horizon. M33 was imaged from New Mexico using a 150-mm telescope.

Messier 33: Data relating to the image in Figure 3.97

REMOTELY IMAGED FROM MAYHILL, NEW MEXICO
NGC 598
OBJECT TYPE: Galaxy

RA: 01h 34m 50s
DEC: +30° 44' 35"
ALTITUDE: +61° 51' 18"
AZIMUTH: 85° 07' 06"
FIELD OF VIEW: 0° 47' 04" x 0° 47' 04"
OBJECT SIZE: 73' X 45'
POSITION ANGLE: 177° 58' from North
EXPOSURE TIME: 300 s
DATE: 24th September
LOCAL TIME: 12h 11m 27s
UNIVERSAL TIME: 06h 11m 27s
Scale: 1.38 arcsec/pixel
MOON PHASE: 100%

Telescope
OTA: Takahashi TOA-150
Optical Design: Apochromatic Refractor
Aperture: 150 mm
Focal Length: 1095 mm
F/Ratio: f/7.3
Guiding: Internal
Mount: Paramount GTS

Instrument Package
SBIG ST-4000M One-Shot Color CCD
A/D Gain: 0.6e-/ADU
Pixel Size: 7.4um square
Resolution: 1.45 arcsec/pixel
Sensor: Frontlit
Cooling: -20°C Winter (-10°C Summer)
Array: 2048 x 2048 (8.3 Megapixels)
FOV: 49.6 x49.6 arcmin

Location
Observatory: New Mexico Skies
UTC Minus 7.00 (Daylight savings time is observed)
Minimum Target Elevation: Approx 25 – 45 Degrees (N or S)
32° 54' Decimal: 32.9 North
(W or E) 105° 31' Decimal: 105.5 West
Elevation: 2225 meters (7298 ft)

Messier 34

Messier 34 (NGC 1039) is an open cluster located in the constellation of Perseus. The 10-minute color image in Figure 3.100 really brings the cluster stars alive. It has a diameter of about 35' of arc -- larger than the full Moon at about 30'! It is believed to have been discovered by Giovanni Batista Hodierna before 1654 and rediscovered by Charles Messier on August 25, 1764: "Cluster of small stars, between the head of Medusa [Algol] & the left foot of Andromeda, a little below the parallel of Gamma [Andromedae]: with an ordinary telescope of 3 foot one can distinguish the stars. Its position has been determined from Beta [Persei], the head of Medusa (Diam. 15')" (SEDS). The cluster lies about 1,450 ly away from us so with an angular size of 35'. The diameter of M34 is about 15 ly (if you convert the

Fig. 3.100 Image of Messier 34, an open cluster in Perseus

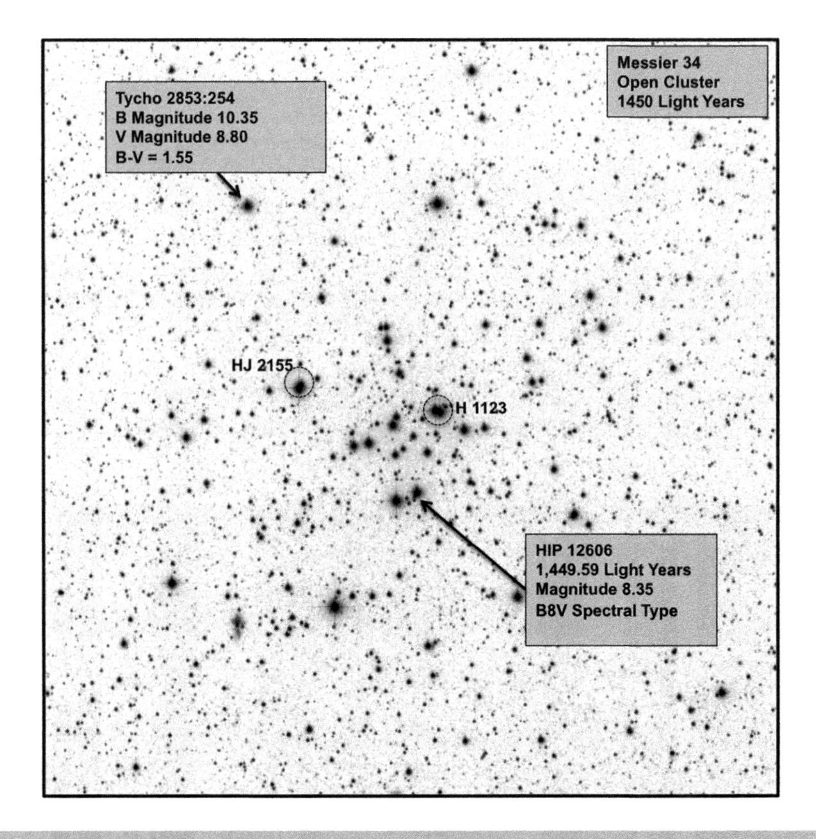

Fig. 3.101 Negative image of Messier 34

35' to radians and multiply by the distance in ly (small angle formula), it provides you with the size in ly). It is believed to have an age of around 180 to 225 million years old, making it older than the Pleiades and younger than the Hyades.

The negative image in Figure 3.101 highlights the cluster star HIP 12606 of magnitude 8.35 and spectral type B8V. The B-type star is hot and blue in color, as can be seen from the color image. The other star highlighted in the negative image is Tycho 2853:254. The Tycho catalog provides the B (Blue) and V (Visual or Green) magnitudes. Calculating B-V gives 1.55. A star with this value corresponds to an orange or red cool star of spectral type M (ignoring any reddening due to intervening dust). Looking back to the color image, the orange color is clearly shown. Two double stars are identified in the cluster, shown circled on the negative image. The first is H 1123. The components are magnitude 8.39 and 8.46 and are separated by 20.1". The second is HJ 2155 with components of mage 8.26 and 10.27 (Washington Double Star Catalog).

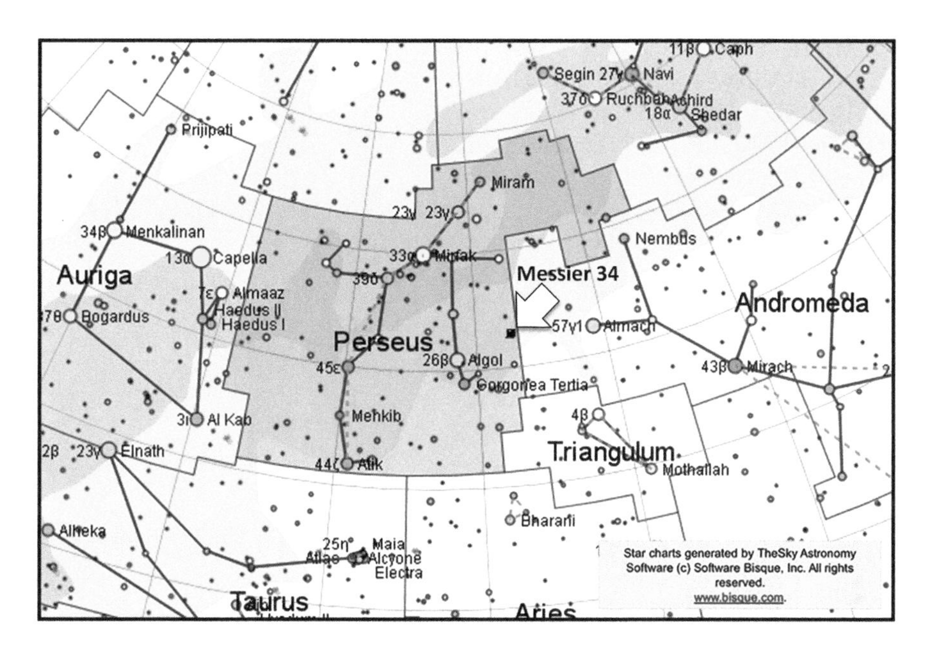

Fig. 3.102 Chart showing the location of M34

The image of M34 can be seen superimposed on the chart in Figure 3.102 showing its location to the west side of the constellation of Perseus. There is one other Messier object in Perseus – M76 -- known as the Little Dumbbell Nebula, or the Cork Nebula. There is an interesting Caldwell object, C23, very close to M34, but in the adjacent constellation of Andromeda. Perseus itself contains the Caldwell object C14, better known as the Perseus Double Cluster, and C24, an active galaxy. Perseus is also adjacent to the constellation of Triangulum and so not far from the spectacular M33 galaxy. It shares borders with Cassiopeia, Camelopardalis, Auriga, Taurus, and Aries.

Messier 34 was imaged from New Mexico at local time 23h 41m 11s on October 1. The Universal Time was 05h 41m 11s. M34 has a Right Ascension of 02h 43m 11s, and the Local Sidereal Time 23h 23m 25s. It was east of the meridian with an azimuth value of 60° 37' 21". The altitude of the object was +49° 53' 52", just above my recommended minimum limit of 45°. The equipment used to take the image was a 6-inch telescope with a single-shot color camera.

Messier 34: Data relating to the image in Figure 3.100

REMOTELY IMAGED FROM MAYHILL, NEW MEXICO
OBJECT TYPE: Open Cluster

RA: 02h 43m 11s
DEC: +42° 49' 49"
ALTITUDE: +49° 53' 52"

AZIMUTH: 60° 37' 21"
FIELD OF VIEW: 0° 47' 04" x 0° 47' 04"
OBJECT SIZE: 35' X 35'
POSITION ANGLE: 177° 56' from North
EXPOSURE TIME: 600 s
DATE: 1st October
LOCAL TIME: 23h 41m 11s
UNIVERSAL TIME: 05h 41m 11s (2nd October)
SCALE: 1.38 arcsec/pixel
MOON PHASE: 1.48% (waxing)

Telescope Optics
OTA: Takahashi TOA-150
Optical Design: Apochromatic Refractor
Aperture: 150 mm
Focal Length: 1095 mm
F/Ratio: f/7.3
Guiding: Internal
Mount: Paramount GTS

Instrument Package
SBIG ST-4000M One-Shot Color CCD
A/D Gain: 0.6e-/ADU
Pixel Size: 7.4um square
Resolution: 1.45 arcsec/pixel
Sensor: Frontlit
Cooling: -20°C Winter (-10°C Summer)
Array: 2048 x 2048 (8.3 Megapixels)
FOV: 49.6 x49.6 arcmin

Location
Observatory: New Mexico Skies
UTC Minus 7.00 (Daylight savings time is observed)
Minimum Target Elevation: Approx. 25 – 45 Degrees
(N or S) 32° 54' Decimal: 32.9 North
(W or E) 105° 31' Decimal: 105.5 West
Elevation: 2225 meters (7298 ft)

Messier 35

Messier 35 (NGC 2168), shown in Figure 3.103, is an open cluster in the constellation of Gemini. The cluster is easy to observe with binoculars and small telescopes. It has a fainter companion cluster NGC 2158. Credit for discovery is given

Fig. 3.103 Image of Messier 35, an open cluster in Gemini

to Philippe Loys de Chéseaux in 1745/6. It appears in John Bevis's *Uranographia Britannica* printed in 1750. Messier observed in 1764, "Cluster of very small stars, near the left foot of Castor, at a little distance from the stars Mu & Eta of that constellation [Gemini]." He later reported, "When examining this star cluster with an ordinary refractor of 3 feet, it seemed to contain nebulosity; but having examined it with a good Gregorian telescope which magnified 104 times, I have noticed that it is nothing but a cluster of small stars, among which there are some which are of more light; its extension may be 20 minutes of arc."

The negative image in Figure 3.104 shows Messier 35 and points out NGC 2158. The distance to M35 is 2,800 ly, and it has an apparent diameter of 28' of arc. These give M35 an actual diameter of approximately 23 ly. NGC 2158 is at a distance of about 16,000 ly, much further away than M35. It has an apparent diameter of about 5 arcmin, which corresponds to a diameter of 23 ly. It would seem that both clusters are the same actual size. The WEBDA cluster website gives a log age for M35 of 7.979, meaning that it has an age of 95.3 million years. This is actually young for a cluster, and stars are still hot, blue and burning hydrogen. The log age for NGC 2158 is given by WEBDA as 9.023, meaning that it is over 1 billion years old. Thus, NGC 2158 contains orange and red stars.

The location of Messier 35 in the constellation of Gemini is shown in the chart in Figure 3.105. M35 is not far from the star 7-Eta Geminorum (Propus or HIP

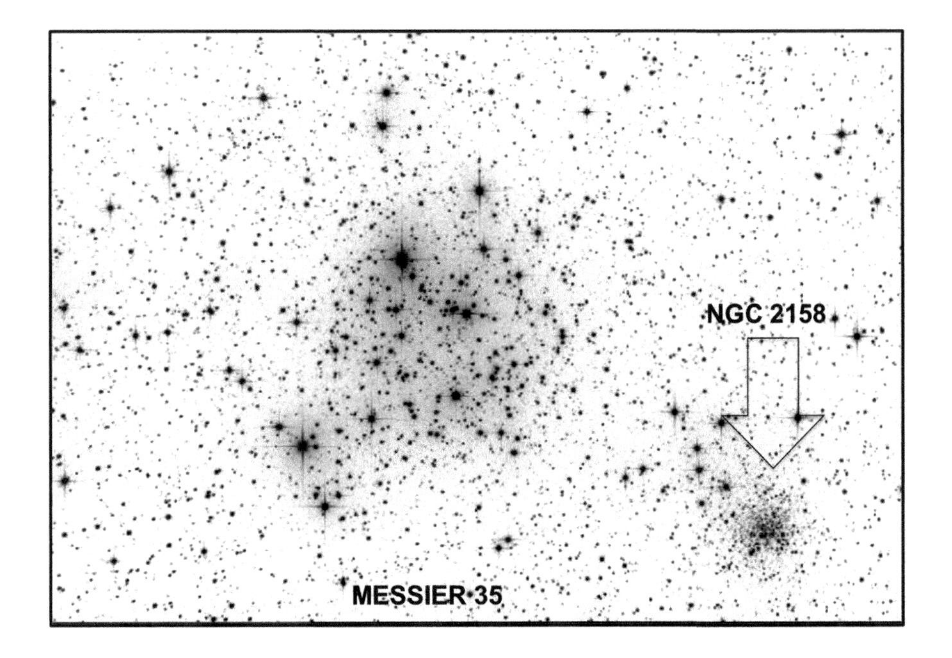

Fig. 3.104 Negative image of Messier 35

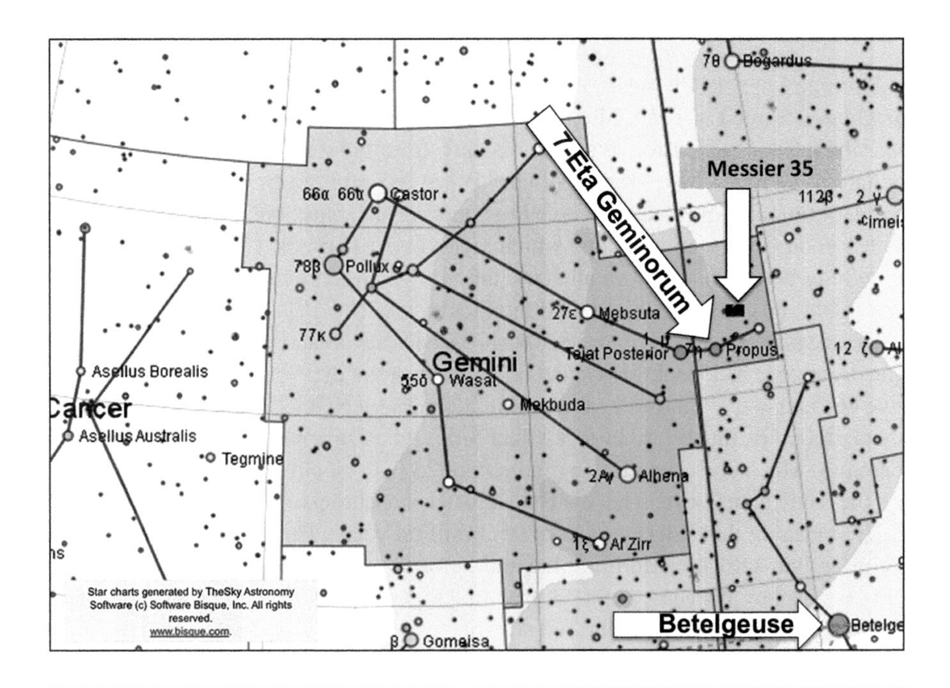

Fig. 3.105 Chart showing the location of M35

29655). This is magnitude 3.31 and is a cool M3 star. It lies at a distance of 349.20 ly (Hipparcos Data). The angular separation between M35 and Propus is approximately 2° 17'. Gemini is adjacent to the constellation of Orion. The angular separation between M35 and the star Betelgeuse is 17° 15'. Betelgeuse is also a cool M star but has a magnitude of 0.45. It is at a distance of 427.47 ly. Betelgeuse makes a good starting point if you are trying to locate M35.

The image was taken on November 14 at 02h 48m 16s local time in New Mexico. This corresponds to a Universal Time of 09h 48m 16s. The Right Ascension of M35 is 06h 09m 49s, and the Local Sidereal Time was 06h 20m 36s. The LST is higher than the RA, so M31 must have crossed the meridian from east to west already. Subtracting one from the other gives a difference of 0h 10m 47s. Subtracting this from the local time when the image was taken gives a meridian transit time of 02h 37m 29s. The telescope used to take the image of M35 was a Takahashi Epsilon Astrograph with an aperture of 250 mm.

Messier 35: Data relating to the image in Figure 3.103

REMOTELY IMAGED FROM MAYHILL, NEW MEXICO
NGC 2168
OBJECT TYPE: Open Cluster

RA: 06h 09m 49s
DEC: +24° 20' 43"
ALTITUDE: +81° 14' 14"
AZIMUTH: 196° 20' 14"
FIELD OF VIEW: 1° 00' 28" x 0° 40' 45"
OBJECT SIZE: 28'
POSITION ANGLE: 357° 30' from North
EXPOSURE TIME: 60s
DATE: 14th November
LOCAL TIME: 02h 48m 16s
UNIVERSAL TIME: 09h 48m 16s
RESOLUTION: 1.66 arcsec/pixel
MOON PHASE: 0.36% (waxing)

Telescope Optics
OTA: Takahashi Epsilon 250
Optical Design: Hyperbolic Flat-Field Astrograph
Aperture: 250 mm
Focal Length: 850 mm
F/Ratio: f/3.4
Guiding: External
Mount: Paramount ME

Instrument Package
CCD: SBIG ST-10XME
Pixel Size: 6.8μm square
Resolution: 1.65 arcsec/pixel
Sensor: Frontlit
Cooling: -15°C default
Array: 2184 x 1472 (3.2 Megapixels)
FOV: 40.4 x 60 arcmin

Location
Observatory: New Mexico Skies
UTC Minus 7.00 (Daylight savings time is observed)
Minimum Target Elevation: Approx. 25 – 45 Degrees
(N or S) 32° 54' Decimal: 32.9 North
(W or E) 105° 31' Decimal: 105.5 West
Elevation: 2225 meters (7298 ft)

Messier 36

Messier 36 (NGC 1960), shown in Figure 3.106, is one of three distinctive open clusters in the constellation of Auriga, (M36, M37, M38), which are favorite targets for amateur astronomers. It has a diameter of 12' of arc – less than half the diameter of the full Moon. The cluster was discovered by Sicilian Giovanni Batista Hodierna in the 17th Century. Messier made the following comments after his observation of M36 in 1764: "In the night of September 2 to 3, 1764, I have determined the position of a star cluster in Auriga, near the star Phi of that constellation. With an ordinary refractor of 3 feet & a half, one has difficulty to distinguish these small stars; but when employing a stronger instrument, one sees them very well; they don't contain between them any nebulosity: their extension is about 9 minutes of arc. I have compared the middle of this cluster with the star Phi Aurigae, & I have determined its position; its right ascension was 80d 11' 42", & its declination 34d 8' 6" north."

Messier 36, shown in Figure 3.107, lies at a distance of 4,100 ly. Assuming a diameter of 12', it appears to have an actual diameter of about 14 ly, using the small angle formula. It is described as a young open cluster with an age of about 16 million years. One study (Sanner et al 2000) investigated 404 stars in the region down to magnitude 14, and by studying the proper motion of these stars, determined that only 178 of these were actual cluster members. One star that is not a cluster member is highlighted in the negative image. HIP 26354, a magnitude 9 star, which lies at a distance of 911 ly and is a foreground star that is less than a quarter of the distance of the M36 cluster itself. The central circle (arrowed) on the negative image shows the double star WDS BKO 269, which has components of magnitude 9.38 and 13.00.

Fig. 3.106 Image of Messier 36, an open cluster in Auriga

The chart in Figure 3.108 shows the square image of M36 in its correct location. The other Messier objects, M37 and M38 have already been mentioned. There is only one Caldwell object in Auriga, C31, otherwise known as the "flaming star" nebula (IC 405). M36 lies between the stars Castor and Pollux in Gemini to the east and the star Atik in Perseus to the west. Within Auriga itself, M36 lies below a line joining the stars Bogardus (Theta Aurigae) and Al Kab (Iota Aurigae), and almost centrally to a line joining M37 and M38.

The image of Messier 36 was taken at local time 12h 39m 48s, just after midnight on November 25 from New Mexico. The Universal Time was 07h 39m 48s. The cluster has a Right Ascension of 05h 37m 26s, and the Local Sidereal Time was 04h 55m 16s. M36 was therefore east of the meridian with an azimuth of 78° 17' 40". The altitude of M36 was +81° 06' 14", so it was high in the sky. The 5-minute color image was taken with a 6-inch telescope based in New Mexico.

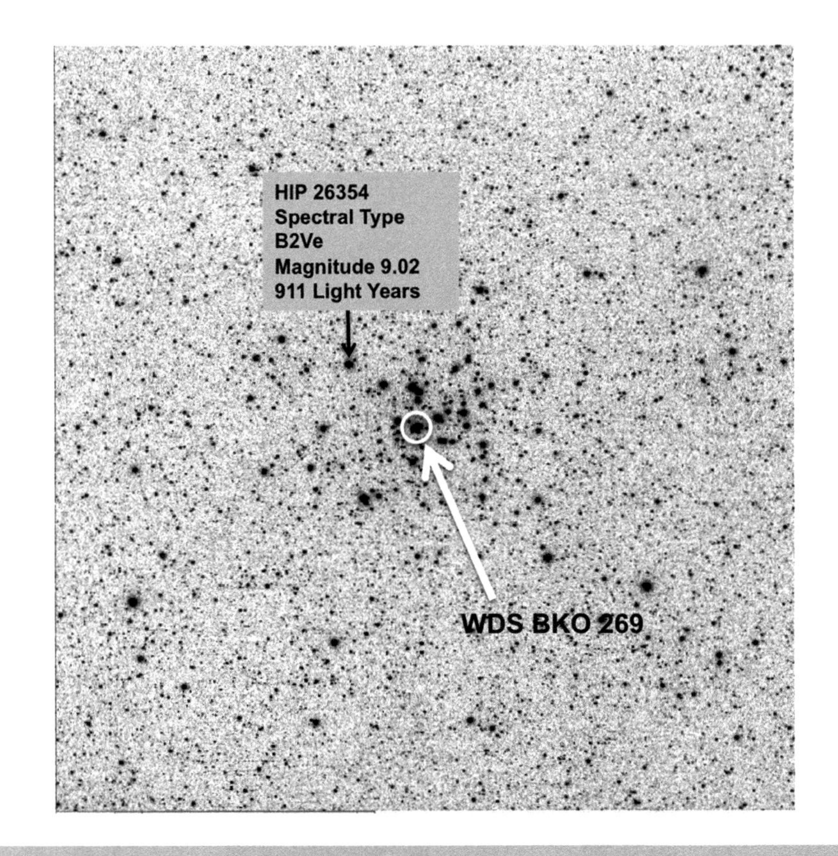

Fig. 3.107 Negative image of Messier 36

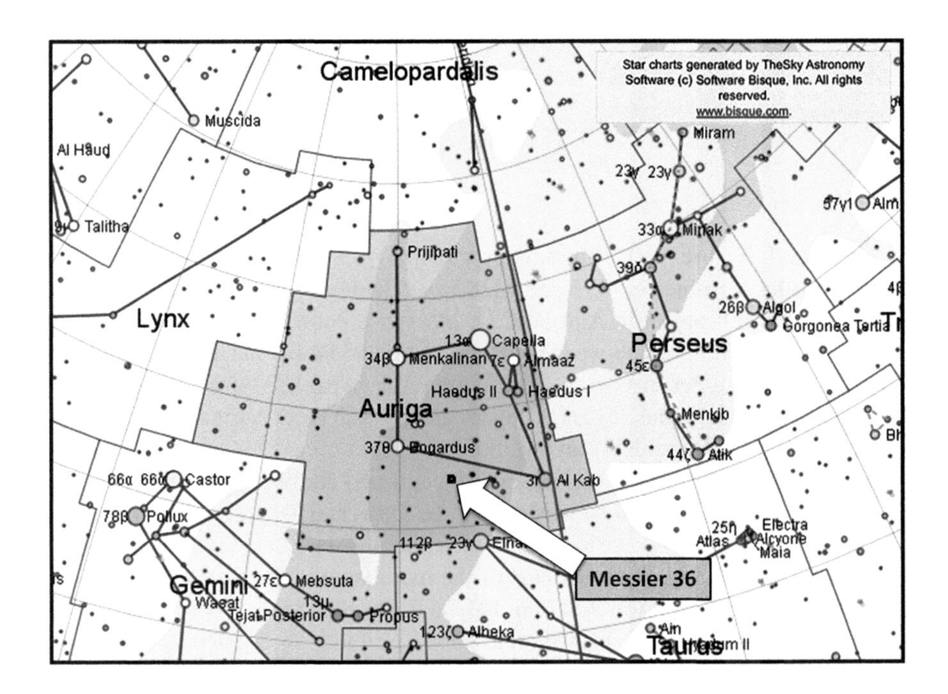

Fig. 3.108 Chart showing the location of M36

Messier 36: Data relating to the image in Figure 3.106

REMOTELY IMAGED FROM MAYHILL, NEW MEXICO
NGC 1960
OBJECT TYPE: Open Cluster

RA: 05h 37m 26s
DEC: +34° 08' 51"
ALTITUDE: +81° 06' 14"
AZIMUTH: 78° 17' 40"
FIELD OF VIEW: 0° 47' 08" x 0° 47' 08"
OBJECT SIZE: 12' X 12'
POSITION ANGLE: 177° 50' from North
EXPOSURE TIME: 300 s
DATE: 25th November
LOCAL TIME: 12h 39m 48s
UNIVERSAL TIME: 07h 38m 49s
SCALE: 1.38 arcsec/pixel
MOON PHASE: 15.10% (waning)

Telescope Optics
OTA: Takahashi TOA-150
Optical Design: Apochromatic Refractor
Aperture: 150 mm
Focal Length: 1095 mm
F/Ratio: f/7.3
Guiding: Internal
Mount: Paramount GTS

Instrument Package
SBIG ST-4000M One-Shot Color CCD
A/D Gain: 0.6e-/ADU
Pixel Size: 7.4um Square
Resolution: 1.45 arcsec/pixel
Sensor: Frontlit
Cooling: -20°C Winter (-10°C Summer)
Array: 2048 x 2048 (8.3 Megapixels)
FOV: 49.6 x49.6 arcmin

Location
Observatory: New Mexico Skies
UTC Minus 7.00 (Daylight savings time is observed)
Minimum Target Elevation: Approx. 25 – 45 Degrees
(N or S) 32° 54' Decimal: 32.9 North
(W or E) 105° 31' Decimal: 105.5 West
Elevation: 2225 meters (7298 ft)

Chapter 4

Messier 37 to Messier 74

Messier 37

Messier 37 (NGC 2099), shown in Figure 4.1, is another open cluster in the constellation of Auriga, as mentioned in the section on M36. The size of the cluster is approximately 24' of arc across, which is double the diameter of Messier 36. As in the case of M36, it was originally discovered by Sicilian astronomer Giovanni Batista Hodierna. M37 is the brightest of the three clusters in Auriga – M36, M37 and M38. American astronomer Robert Burnham Jr. described M37 as "one of the finest galactic star clusters" and went on to say "A superb galactic star cluster for telescopes of all sizes, usually considered the finest of the three Messier open clusters in Auriga, and apparently first observed by Messier himself in 1764." Messier reported, "In the same night [as observing M36], I have observed a second cluster of small stars which were not very distant from the preceding, near the right leg of Auriga & on the parallel of the star Chi of that constellation: the stars there are smaller than that of the preceding cluster: they are also closer to each other, & contain a nebulosity. With an ordinary refractor of 3 feet & a half, one has difficulty to see these stars; but one distinguishes them with an instrument of greater effectivity. I have determined the position for this cluster, which may have an extension of 8 to 9 minutes of arc: its right ascension was 84d 15' 12", & its declination 32d 11' 51" north."

The distance to M37, shown in Figure 4.2, is 4,400 ly. Based on this information, my simple calculation gives the cluster a diameter of about 31 ly. Burnham stated that the cluster contains about 150 stars down to magnitude 12.5, and that the total population may be over 500 stars. M37 is regarded as being much older than M36.

L. Adam, *Imaging the Messier Objects Remotely from Your Laptop*, The Patrick Moore Practical Astronomy Series, https://doi.org/10.1007/978-3-319-65385-3_4

Fig. 4.1 Image of Messier 37, an open cluster in Auriga

It contains a significant number of Red Giant stars signifying age. The age of M37 is estimated to be about 300 million years. On the negative image in Figure 4.2, the star that appears bright orange in the color image is pointed out. This star is HIP 27661 a cooler M-type star – however this is not part of the cluster, as it lies at a distance of only 580 ly (Hipparcos Catalog).

The chart in Figure 4.3 shows the position of M37 within the constellation of Auriga, adjacent to M36 and M38. The camera used to take the image contains 1,600 X 1,200 pixels so the chip is not "square." You can see that in the shape of the dark rectangle shaped image superimposed on Figure 4.3. The camera used to image M36 had 2,400 X 2,400 pixels, so the image is exactly square. The red line to the right of M37 is the meridian.

M37 was imaged at the local time of 22h 19m 28s in New South Wales on January 17. Universal Time was 11h 19m 28s. The Right Ascension of M37 is 05h 53m 26s, and the Local Sidereal Time was 05h 03m 46s. Thus, the cluster was roughly 50 minutes away from the meridian. The transit time of M37 was 23h 13m. The telescope gave a wide field of view of 1° 37' 38" x 1° 13' 13", as you can see roughly three times the size of the cluster on the long side. The color image shown was taken with a 90-mm telescope and an exposure of 300 seconds.

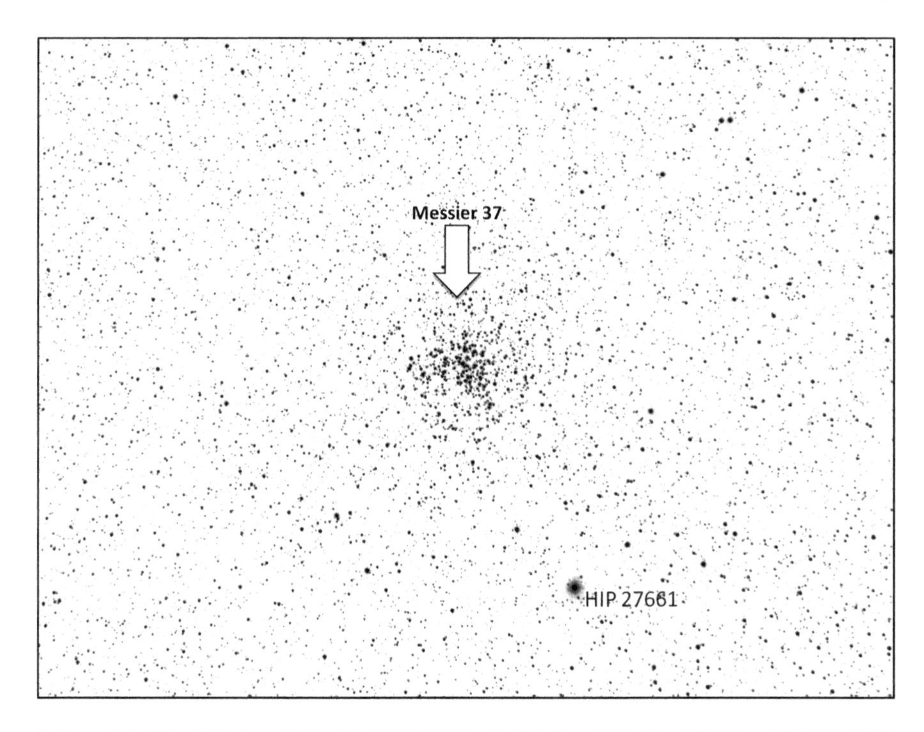

Fig. 4.2 Negative image of Messier 37

Messier 37: Data relating to the image in Figure 4.1

REMOTELY IMAGED FROM NEW SOUTH WALES, AUSTRALIA
NGC 2099
OBJECT TYPE: Open Cluster

RA: 05h 53m 26s
DEC: +32° 33' 15"
ALTITUDE: +25° 08' 02"
AZIMUTH: 11° 33' 03"
FIELD OF VIEW: 1° 37' 38" x 1° 13' 13"
OBJECT SIZE: 24' X 24'
POSITION ANGLE: 356° 39' from North
EXPOSURE TIME: 300 s
DATE: 17th January
LOCAL TIME: 22h 19m 38s
UNIVERSAL TIME: 11h 19m 38s
RESOLUTION: 3.66 arcsec/pixel
MOON PHASE: 73.19% (waning)

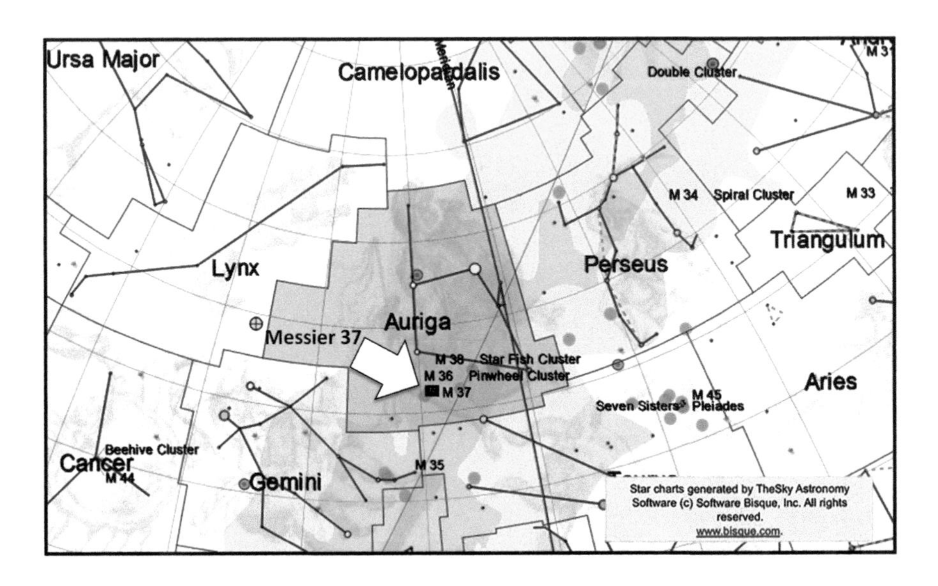

Fig. 4.3 Chart showing the location of M37

Telescope Optics
OTA: Takahashi Sky 90

Optical Design: Apochromatic
Refractor Aperture: 90 mm
Focal Length: 417 mm
F/Ratio: f/5.6
Guiding: None
Mount: Paramount P ME

Instrument Package
CCD: SBIG ST2000 XMC
Colour CMOS
Pixel Size: 7.4μm square
Sensor: Frontlit
Cooling: CNA
Array: 1600 x 1200 pixels
FOV: 60.5 x 80.7 arcmin

Location
Observatory: Siding Spring
UTC +10.00 (Australia Daylight savings time is observed)
31° 16’ 24” South
149 03’ 52” East
Elevation: 1122 meters (3681 ft)

Messier 38

Messier 38 (NGC 1912) is one of three similar open clusters in the constellation of Auriga, which is always a popular choice for visual observing. The cluster is about 15' of arc across -- about 1/2 the diameter of the full moon. There were two factors that could have significantly affected the final image, shown in Figure 4.4. Firstly, the altitude of M38 was very low at 21°, so I was lucky to be able to point the telescope over the edge of the runoff roof shed that houses the telescope. Secondly, the moon had a phase of 60.68% (waxing) just over 46° away from M38. The image was saved and opened in IRIS, then saved as a TIFF file. I opened that file in Photoshop PS2 and increased the saturation level to bring out the colors. I then added some spikes using the ProDigital Software option on the filter menu. Lastly, I used PS2 to make a copy, converted the copy to monochrome and inverted the colors to make the reference image.

The monochrome reference image of M8 in Figure 4.5 shows the magnitudes of some of the stars, ranging from magnitude 8.4 down to magnitude 14.13. The SEDS database gives the distance of the cluster as 4,200 ly. Of course, there are many foreground stars in this image. For example, the star labeled magnitude 8.84 is

Fig. 4.4 Image of Messier 38, an open cluster in Auriga

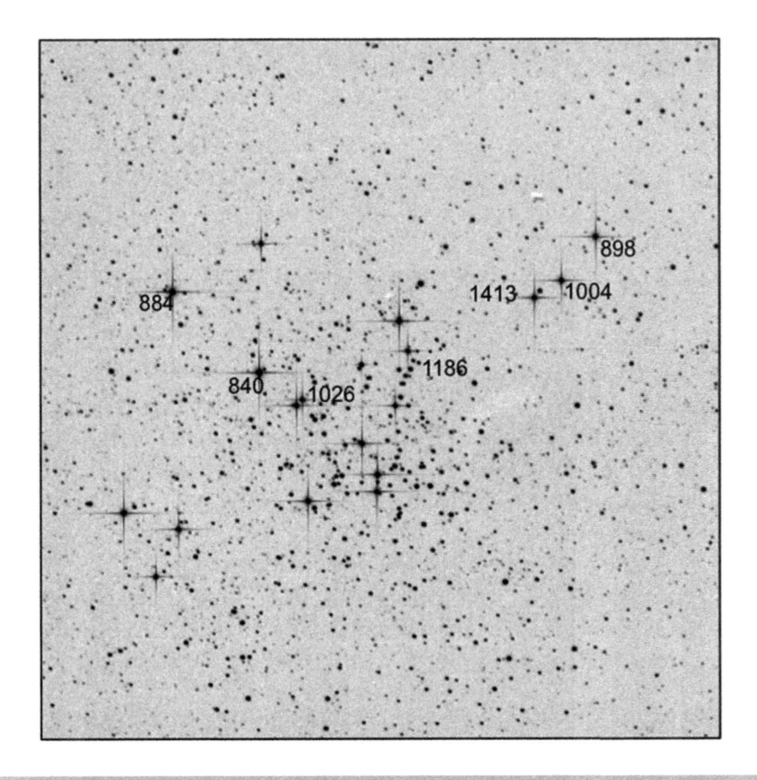

Fig. 4.5 Negative image of Messier 38

given a distance of 1,630.78 ly in the Tycho catalog. The star with magnitude 10.26, with the label just to the right of the star, is given a distance of only 472.69 ly in the Tycho catalog. My star chart shows a faint planetary nebula above the center of M38, but at magnitude 18.9, it is much fainter than the limiting magnitude of my image, which is about magnitude 15.

M38, shown in Figure 4.6, is located in the constellation of Auriga – fairly close to the similar cluster M36. With a diameter of about 15', you need to choose a camera with an appropriate field of view. The telescope/camera combination used is listed as having a field of view of 49.6' square. The square shape is simply a result of the chip in the camera being square. You will notice that the field of view (FOV) actually achieved is just over 47'. The original FITS image was plate-solved to provide the actual FOV. Of course, the best place to image any object is when it is at its highest.

The image of M38 was taken at the local time in New Mexico of 22h 38m 54s. The local time at my location was 05h 38m 54s, so it was an early start for me. The RA of M38 is roughly 5h 30m, and the Local Sidereal Time was about 11h 10m. Thus, M38 was about 6 hours away from the meridian. Of course, it also needs to be dark! Ideally, you should choose the time of year that the object you are targeting is near the meridian, at a time also convenient to yourself. Of course, with a shared

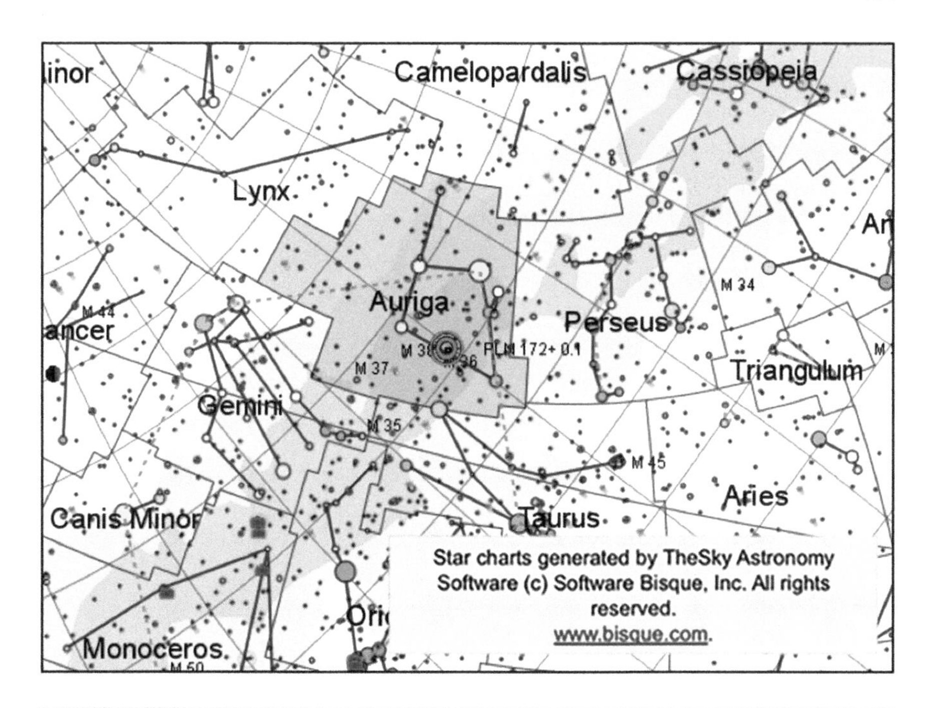

Fig. 4.6 Chart showing the location of M38

remote telescope, unless you have booked in advance, you will find that when you need to image the object the telescope is fully booked! Clearly, my timing was not good on this one, and the image will have suffered by having to go through quite a thickness of atmosphere.

Messier 38: Data relating to the image in Figure 4.4

REMOTELY IMAGED FROM MAYHILL, NEW MEXICO
NGC 1912
OBJECT TYPE: Open Cluster

RA: 05h 29m 47s
DEC: +35° 51’ 56”
ALTITUDE: +21° 57’ 03”
AZIMUTH: 299° 27’ 28”
FIELD OF VIEW: 47’ 06”
OBJECT SIZE: 15’ X 15’
POSITION ANGLE: 11° 38’
EXPOSURE TIME: 600 s
DATE: 14th April
LOCAL TIME: 22h 38m 54s
UNIVERSAL TIME: 04h 38h 54s

RESOLUTION: 2048 X 2048
MOON PHASE: 60.68%

Telescope Optics
OTA: Takahashi TOA-150
Optical Design: Apochromatic Refractor
Aperture: 150 mm
Focal Length: 1095 mm
F/Ratio: f/7.3
Guiding: Internal
Mount: Paramount GTS

Instrument Package
SBIG ST-4000M One-Shot Color CCD
A/D Gain: 0.6e-/ADU
Pixel Size: 7.4um Square
Resolution: 1.45 arcsec/pixel
Sensor: Frontlit
Cooling: -20°C Winter (-10°C Summer)
Array: 2048 x 2048 (8.3 Megapixels)
FOV: 49.6 x49.6 arcmin

Location
Observatory: New Mexico Skies
UTC Minus 7.00 (Daylight savings time is observed)
Minimum Target Elevation: Approx. 25 – 45 Degrees
(N or S) 32° 54’ Decimal: 32.9 North
(W or E) 105° 31’ Decimal: 105.5 West
Elevation: 2225 meters (7298 ft)

Messier 39

Messier 39 is an open cluster in the constellation of Cygnus. M39 is about 32’ of arc in size, and the image in Figure 4.7 captures most of the cluster. If you compare this with the M37 cluster of 24’, it seems quite large – even more so when you think of M39 being slightly larger than the full moon. M39 also has the New General Catalog designation NGC 7092. As you can see, it is not a particularly dense cluster, with American astronomer Robert Burnham Jr. noting that it has 30 actual cluster members and describing it as “a large but very loose structured galactic star cluster.” Messier commented, “In the night of October 24 to 25, 1764, I observed a cluster of stars near the tail of Cygnus: One distinguishes them with an ordinary [nonachromatic] refractor of 3 and a half feet; they don't contain any nebulosity; its extension can occupy a degree of arc. I have compared it with the star Alpha Cygni,

Fig. 4.7 Image of Messier 39, an open cluster in Cygnus

& I have found its position in right ascension of 320d 57' 10", & its declination of 47d 25' 0" north."

M39 is about 800 ly distant (SEDS). The WEBDA page for M39 gives the distance as 326 parsecs. As one parsec is about 3.26 ly, this is equivalent to 1063 ly. As always, consulting different catalogs will give different values. With an assumed diameter of 32' and a distance of 800 ly, the cluster has a diameter of just less than 8 ly. That makes M39 only about 1/4 of the actual size of M37, about 2/3 the actual size of M36 and about 1/3 the actual diameter of M38. In the negative image in Figure 4.8, you will notice that the stars are either spectral type A or B – indicating hot stars. The distances to the selected stars are shown, drawn from Hipparcos data. Note that the image has north to the right and east at the top.

The chart in Figure 4.9 shows the location of M39 in the constellation of Cygnus, represented by the linked image of the object. Note that the orientation of the chart matches that of the image, so east is up and north is to the right. In the description of the other Messier object in Cygnus (M29), I referred to the seven Caldwell objects in Cygnus. M29 is only 10' across but is almost double the actual size of M39 (15 ly). Cygnus is adjacent to the constellation of Vulpecula, which means that the Dumbbell Nebula M27 is close by, the constellation of Lyra, which

Fig. 4.8 Negative image of Messier 39

contains M57, the Ring Nebula and globular cluster M56, the constellation of Draco, containing the Spindle Galaxy M102 and Cepheus, the constellation Lacerta, which is completely bereft of Messier objects and finally the constellation of Pegasus, which contains the globular cluster of M15.

The image was taken from Auberry, California at local time 18h 13m 17s on December 22. Universal Time was 02h 13m 17s on December 23. M39 has a Right Ascension of 21h 31m 58.903s, and the Local Sidereal Time was 00h 23m 55s. The azimuth angle is 303° 58' 40". M39 had transited the meridian some hours earlier at 15h 22m. The image was taken with a 24-inch telescope and an exposure time was 300 seconds.

Messier 39: Data relating to the image in Figure 4.7

REMOTELY IMAGED FROM AUBERRY, CALIFORNIA
NGC 7092
OBJECT TYPE: Open Cluster

RA: 21h 32m 33s
DEC: +48° 30' 02"

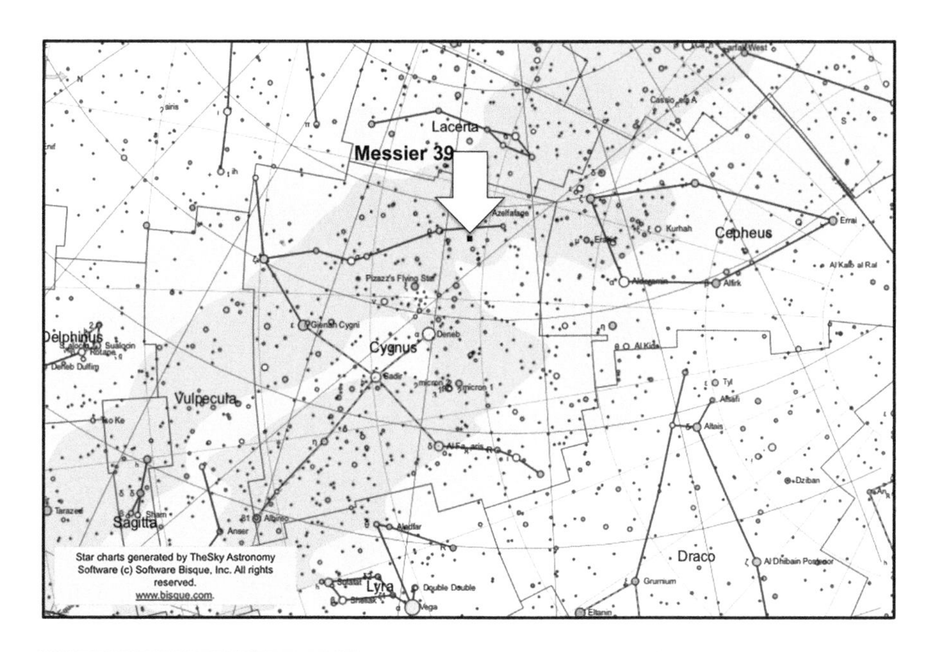

Fig. 4.9 Chart showing the location of M39

ALTITUDE: +57° 02' 55"
AZIMUTH: 304° 04' 19"
FIELD OF VIEW: 0° 32' 08" x 0° 32' 08"
OBJECT SIZE: 32' X 32'
POSITION ANGLE: 88° 29' from North
EXPOSURE TIME: 300 s
DATE: 22nd December
LOCAL TIME: 18h 13m 17s
UNIVERSAL TIME: 02h 13m 17s
SCALE: 0.63 arcsec/pixel
MOON PHASE: 31.26% (waning)

Telescope Optics
OTA: Planewave 24" (0.61m)
Optical Design: Corrected Dall-Kirkham Astrograph
Aperture: 610 mm
Focal Length: 3962 mm
F/Ratio: f/6.5
Guiding: Active Guiding Disabled
Mount: Planewave Ascension 200HR

Instrument Package
FLI-PL09000 CCD
A/D Gain: e-/ADU

Pixel Size: 12 µm square
Array: 3056 x 3956
Sensor: Frontlit
Cooling: -35°C default
FOV: 31.8 x 31.8 arcmin

Location
Observatory: Sierra Remote Observatory – MPC U69 UTC
Minus -8.00 (Daylight savings time is observed)
Minimum Target Elevation: Approx 25 Degrees
37.07°N, 119.4W
Elevation: 1405 meters (4610 ft)

Messier 40

Messier 40, shown in Figure 4.10, should not have been included in Messier's list, as it is simply a double star in the constellation of Ursa Major. It is known as Winnecke 4. On October 24, 1764, Messier noted, "Two stars very close together & very small, placed at the root of the tail of the Great Bear." It is believed that under poor observing conditions, the double star may have appeared as being a little fuzzy or comet-like, which led to him including them as potential objects to be mistaken for a comet. The two stars are similar in brightness. It was the German Astronomer Friedrich August Theodor Winnecke who first listed the double star in his 1869 catalog. The abbreviated number for the double star is WNC 4. It was the later the American amateur astronomer John Mallas who identified WNC 4 as M40 in a letter to *Sky and Telescope* magazine in August 1966. Mallas had used coordinates from Messier's observation and determined the precession to 1950 coordinates, which confirmed the position as that of Winnecke 4.

The Tycho catalog lists the brightest of the pair (Tycho 3840:1031) as having a magnitude of 9.64. It lists the fainter component (Tycho 3840:564) at magnitude 10.09. Distance measurements have confirmed that the two stars are not physically associated and are just a line of sight effect. From the solved image in Figure 4.11, a separation of about 52" of arc can be measured. The negative image identifies the bright star HIP 60212, which has a magnitude of 5.54, and two nearby galaxies in the same field. NGC 4290 is a spiral galaxy with a major axis of 2.2' and a minor axis of 1.5'. The other galaxy is NGC 4284, with a major axis of 1.9' and a minor axis of 0.9'.

The location of M40 is shown on the chart in Figure 4.12. It lies between Megrez and Alioth – two bright stars in the Big Dipper asterism in Ursa Major – but is a lot closer to Megrez. There are six other Messier objects in the constellation of Ursa

Fig. 4.10 Image of Messier 40, a double star in Ursa Major

Major. These are M81 and M82, M97, M101, M108 and M109. Surprisingly, there are no Caldwell objects in Ursa Major, but I suppose the objects of most interest had already been listed by Messier when Patrick Moore was searching for candidates for the 109 Caldwell object list. However, there are almost 400 cataloged NGC objects, so there is a lot more to be found.

The image was taken at the local time of 05h 07m 38s on December 27 from New Mexico. Universal Time was 12h 07m 38s. M40 has a Right Ascension of 12h 22m 43s, and the Local Sidereal Time was 11h 29m 59s. M40 would transit the meridian in about 54 minutes at 06h 01m. M40 had an altitude of +63° 14' 14" at the time the image was taken. The azimuth of the object was 15° 41' 16". At the transit time, M40 would be at an altitude of almost 65°. The image was taken with a 6-inch telescope.

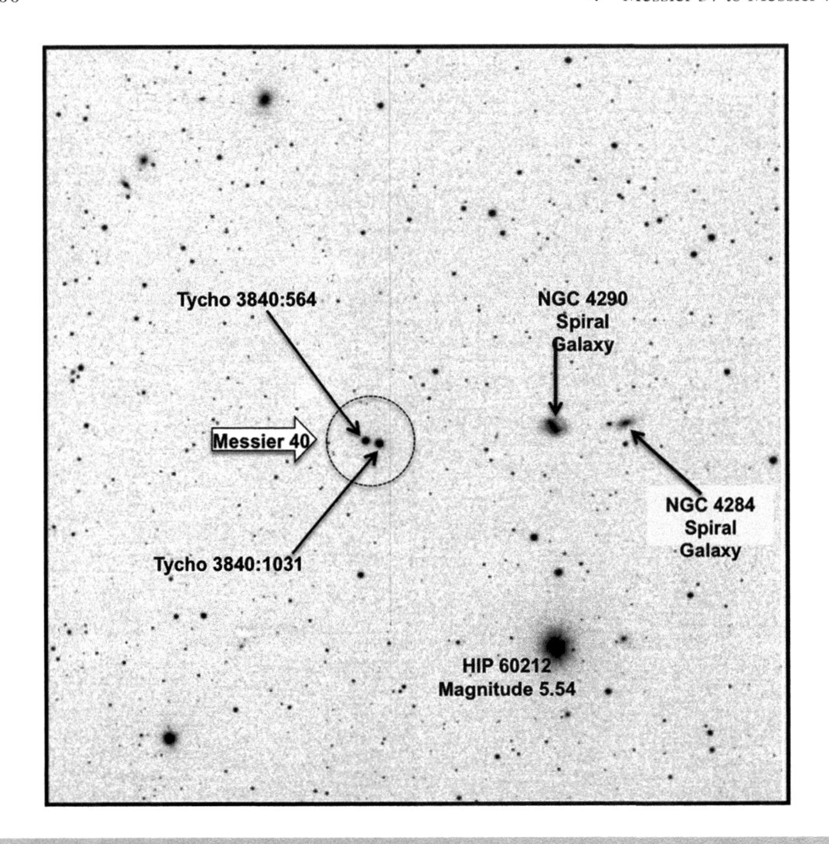

Fig. 4.11 Negative image of Messier 40

Messier 40: Data relating to the image in Figure 4.10

REMOTELY IMAGED FROM MAYHILL, NEW MEXICO
Winnecke 4
OBJECT TYPE: Double Star

RA: 12h 23m 06s
DEC: +57° 59' 16"
ALTITUDE: +63° 14' 14"
AZIMUTH: 15° 41' 16"
FIELD OF VIEW: 0° 47' 08" x 0° 47' 08"
OBJECT SIZE: 0
POSITION ANGLE: 177° 54' from North
EXPOSURE TIME: 300 s
DATE: 27th December
LOCAL TIME: 05h 07m 38s
UNIVERSAL TIME: 12h 07m 39s
SCALE: 1.38 arcsec/pixel
MOON PHASE: 3.16% (waning)

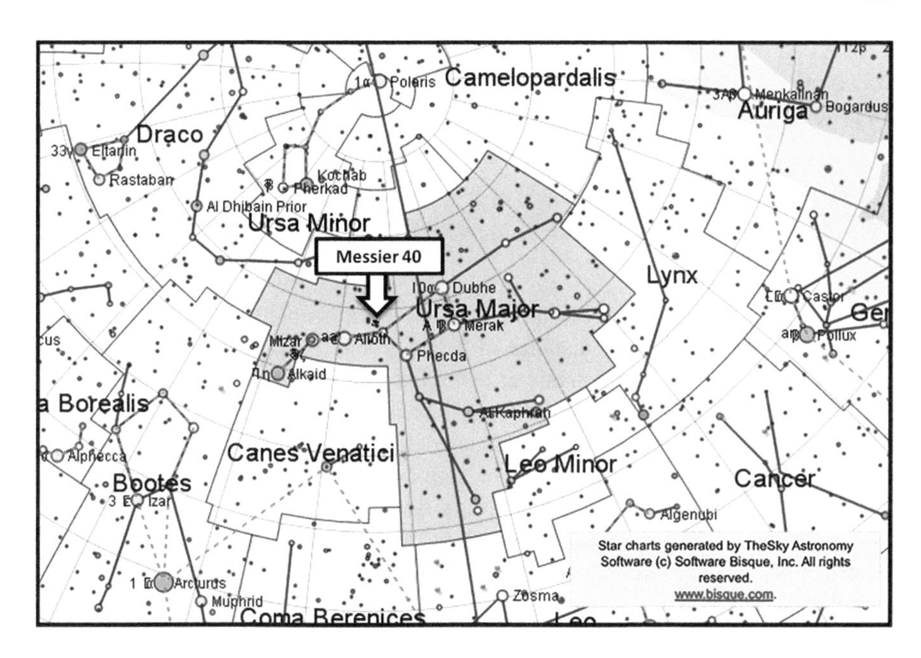

Fig. 4.12 Chart showing the location of M40

Telescope Optics
OTA: Takahashi TOA-150
Optical Design: Apochromatic Refractor
Aperture: 150 mm
Focal Length: 1095 mm
F/Ratio: f/7.3
Guiding: Internal
Mount: Paramount GTS

Instrument Package
SBIG ST-4000M One-Shot Color CCD
A/D Gain: 0.6e-/ADU
Pixel Size: 7.4um Square
Resolution: 1.45 arcsec/pixel
Sensor: Frontlit
Cooling: -20°C Winter (-10°C Summer)
Array: 2048 x 2048 (8.3 Megapixels)
FOV: 49.6 x49.6 arcmin

Location
Observatory: New Mexico Skies
UTC Minus 7.00 (Daylight savings time is observed)
Minimum Target Elevation: Approx. 25 – 45 Degrees

(N or S) 32° 54’ Decimal: 32.9 North
(W or E) 105° 31’ Decimal: 105.5 West
Elevation: 2225 meters (7298 ft)

Messier 41

Messier 41 (NGC 2287), shown in Figure 4.13, is an open cluster in the constellation of Canis Major. It has a distance of 2,300 ly and an apparent diameter of 38’ of arc. Calculating the actual diameter using the small angle formula would give it a value of about 25 ly. It is believed to have been originally discovered by Sicilian Giovanni Battista Hodierna and is included in his catalog of nebulous objects. Charles Messier observed the cluster on January 16, 1765, commenting, “Cluster of stars below *Sirius*, near Rho Canis Majoris; this cluster appears nebulous in an ordinary telescope of one foot; it is nothing more than a cluster of small stars.” American astronomer Robert Burnham Jr. regarded M41 as a “fine galactic star cluster.”

Fig. 4.13 Image of Messier 41, an open cluster in Canis Major

The negative image in Figure 4.14 identifies three stars in the field. The first one is HIP 32406, which is a magnitude 6.91 K3 star, usually referred to as the main star of the cluster. This giant cool star is distinctly orange, however, John Herschel commented: "The chief star, of 8th magnitude, is red." This star is placed at 1,331.25 ly in the Hipparcos Catalog. Another orange star nearby is HIP 32393, a K0 star. This one is fainter at magnitude 7.44 with a Hipparcos distance of 1,496.13 ly. The third star is HIP 32504 with a magnitude of 6.07. As you can see in the color image, this star is blue and hot. It is type B7 and has a designation of 12 Canis Majoris. It has a Hipparcos distance of 668.35 ly. Based on the Hipparcos data, all three stars would be line of sight foreground stars if the cluster distance is accepted as 2,300 ly.

The chart in Figure 4.15 shows the location of M41 in the constellation of Canis Major. North is up and east is to the west. This constellation contains the magnitude -1.44 Sirius to the north of M41, and the angular separation between Sirius and the cluster can be measured at just over 4°. Roughly to the east of M41 is the star 15 Canis Majoris. This is also known as HIP 33092 and is a magnitude 4.82 Type B1 star. It is just less than 2° away from Messier 41 which is the only Messier object in Canis Major. There are two Caldwell objects. The first is C58 or NGC 2360, an

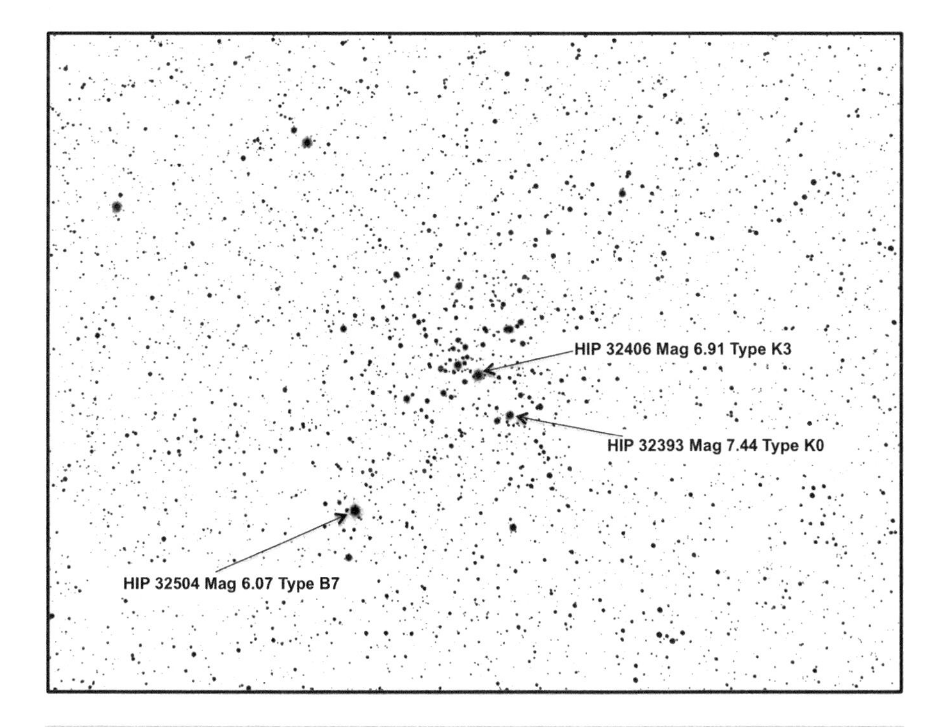

Fig. 4.14 Negative image of Messier 41

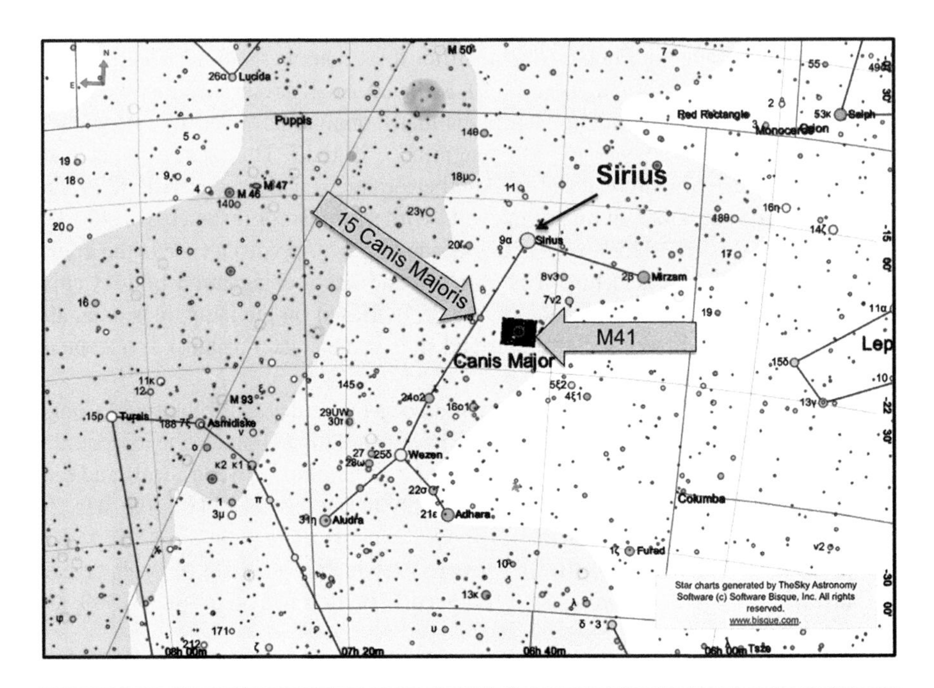

Fig. 4.15 Chart showing the location of M41

open cluster, which was discovered by Caroline Herschel (William's sister!), and the second is C64 or NGC 2362, an open cluster and nebula.

The image of Messier 41 was taken at 23h 39m 23s on December 19. The corresponding Universal Time was 11h 39m 23s. The Right Ascension of M41 is 06h 46m 45s, and the Local Sidereal Time when the image was taken was 03h 29m 43s. Thus, M46 had 3h 17m 2s to go before reaching the meridian, making the meridian transit time 02h 56m 25s. The single-color image camera was used with a 90-mm refractor to obtain the image.

Messier 41: Data relating to the image in Figure 4.13

REMOTELY IMAGED FROM NEW SOUTH WALES, AUSTRALIA
NGC 2287
OBJECT TYPE: Open Cluster

RA: 06h 46m 44s
DEC: -20° 46' 17"
ALTITUDE: +44° 49' 04"
AZIMUTH: 88° 58' 07"
FIELD OF VIEW: 1° 37' 42" x 1° 13' 17"
OBJECT SIZE: 38'
POSITION ANGLE: 357° 05' from North
EXPOSURE TIME: 300 s

DATE: 19th December
LOCAL TIME: 23h 39m 23s
UNIVERSAL TIME: 11h 39m 23s
SCALE: 3.66 arcsec/pixel
MOON PHASE: 65.89% (waning)

Telescope Optics
OTA: Takahashi Sky 90
Optical Design: Apochromatic Refractor
Aperture: 90 mm
Focal Length: 417 mm
F/Ratio: f/5.6
Guiding: None
Mount: Paramount PME

Instrument Package
CCD: SBIG ST2000 XMC
Colour CMOS
Pixel Size: 7.4μm square
Sensor: Frontlit
Cooling: CNA
Array: 1600 x 1200 pixels
FOV: 60.5 x 80.7 arcmin

Location
Observatory: Siding Spring
UTC +10.00 (Australia Daylight savings time is observed)
31° 16' 24" South
149 03' 52" East
Elevation: 1122 meters (3681 ft)

Messier 42

Messier 42 (NGC 1976), shown in Figure 4.16, must surely be one of the best-known objects in the night sky. It is known as the Great Nebula in Orion and forms part of the sword of Orion. Robert Burnham Jr. commented, "This is generally considered the finest example of a diffuse nebula in the sky, and one of the most wonderfully beautiful objects in the heavens." He also said, "In a moderately large telescope its appearance is impressive beyond words, and draws exclamations of delight and astonishment from all who view it. The great glowing irregular cloud, shining by the gleaming light of the diamond-like stars entangled in it, makes a marvelous spectacle which is unequalled anywhere in the sky." Messier noted, "Here is what I have reported about that nebula in the Journal of my Observations.

Fig. 4.16 Image of Messier 42, the "Great Nebula" in Orion

On March 4, 1769, the sky was perfectly serene, Orion was going to pass the meridian, I have directed to the nebula of this constellation a Gregorian telescope of 30 inches focal length, which magnified 104 times; one saw it perfectly well, & I drawed the extension of the nebula, which I compared consequently to the drawings which M. le Gentil has given of it, I found some differences. This nebula contains eleven stars; there are four near its middle, of different magnitudes & strongly compressed to each other; they are of an extraordinary brilliance."

The negative image in Figure 4.17 identifies objects in the field. A detached section of the nebula was listed by Messier as M43 and is just above M42 on the image which has north at the top. The bright star Na'ir al Saif is magnitude 2.75. This is 44-Iota Orionis, a hot O-type star otherwise known as HIP 26241. Star HIP 26314 is magnitude 6.56, Tycho 4778:1369 is magnitude 7.17, Tycho 4774:809 is magnitude 8.02 and Tycho 4774:849 is magnitude 8.56. At the bottom of the image is the open cluster NGC 1980 with 44 – Iota. There is nebulosity associated with this cluster, but is difficult to bring out on the same image without overexposing the central region of M42 even further. The diameter of M42 is 85' x 60' of arc, and it lies at a distance of 1,300 ly. That makes the real size of M42 roughly 32 ly X 23 ly.

The chart in Figure 4.18 shows the position of M42 in the constellation of Orion. The plate-solved image is superimposed on the chart providing the precise position. The sword hangs from Orion's Belt formed by the three stars Alnitak, Alnilam and

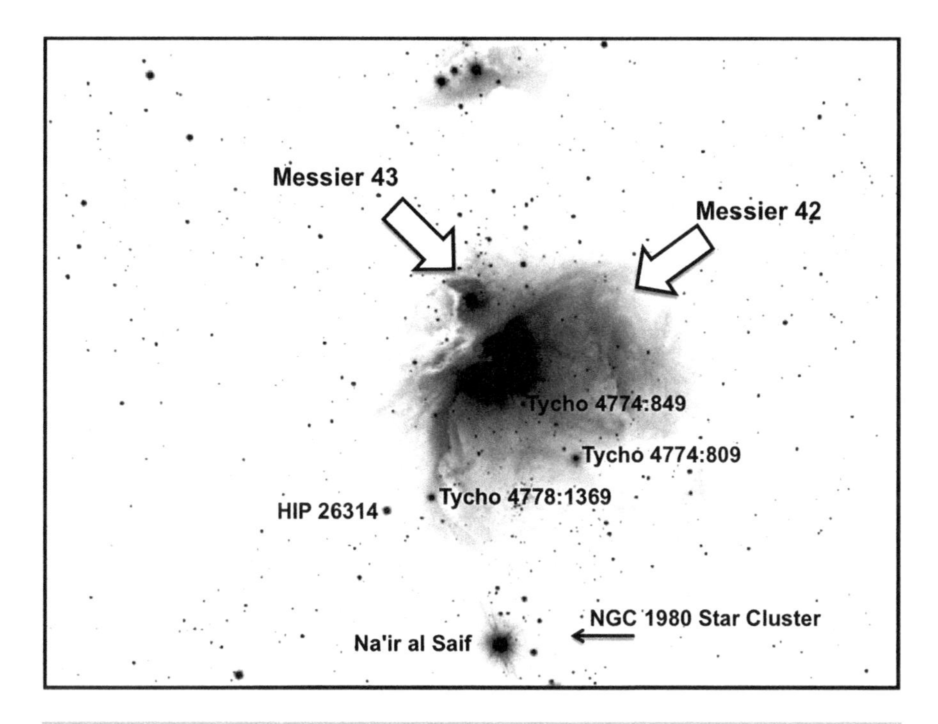

Fig. 4.17 Negative image of Messier 42

Mintaka. Below M42 are the two bright stars Saiph and Rigel. A well-known feature in Orion is the Horsehead Nebula, which is close to the star Alnitak, as is the Flame Nebula. These are two objects that you could easily capture using a remote telescope.

The image of M42 in Figure 4.16 was taken at local time 02h 05m 07s on December 29. The Universal Time was 15h 05m 07s (December 28). The Right Ascension of M42 is 05h 36m 08s, and the Local Sidereal Time was 07h 31m 10s. The RA is less than the LST, so M42 must have been west of the meridian. M42 was on the meridian at 0h 10m 5s. The image was taken with a 90-mm refractor from Siding Spring in Western Australia. The single-exposure color image shows M42 well but is overexposed in the central bright region.

Messier 42: Data relating to the image in Figure 4.16

REMOTELY IMAGED FROM NEW SOUTH WALES, AUSTRALIA
NGC 1976 (GREAT NEBULA)
OBJECT TYPE: Star-Forming Nebula

RA: 05h 36m 08s
DEC: -05° 23' 00"
ALTITUDE: +52° 38' 40"

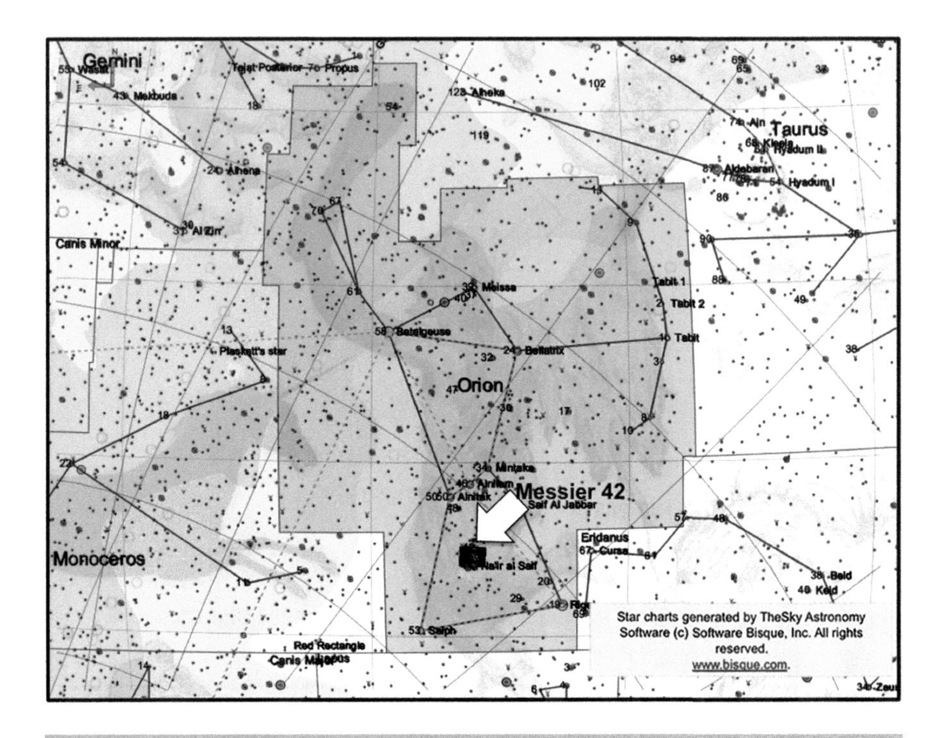

Fig. 4.18 Chart showing the location of M42

AZIMUTH: 307° 51' 54"
FIELD OF VIEW: 1° 37' 39" x 1° 13' 14"
OBJECT SIZE: 85' X 60'
POSITION ANGLE: 176° 51' from North
EXPOSURE TIME: 600 s
DATE: 29th December
LOCAL TIME: 02h 05m 07s
UNIVERSAL TIME: 15h 05m 07s
SCALE: 3.66 arcsec/pixel
MOON PHASE: 0.58% (waning)

Telescope Optics
OTA: Takahashi Sky 90
Optical Design: Apochromatic Refractor
Aperture: 90 mm
Focal Length: 417 mm
F/Ratio: f/5.6
Guiding: None
Mount: Paramount PME

Instrument Package
CCD: SBIG ST2000 XMC
Colour CMOS
Pixel Size: 7.4μm square
Sensor: Frontlit
Cooling: CNA
Array: 1600 x 1200 pixels
FOV: 60.5 x 80.7 arcmin

Location
Observatory: Siding Spring
UTC +10.00 (Australia Daylight savings time is observed)
31° 16' 24" South
149 03' 52" East
Elevation: 1122 meters (3681 ft)

Messier 43

Messier 43 (NGC 1982), shown in Figure 4.19, is really part of the Great Nebula in Orion but is separated from M42 by a dark nebula. M43 is called de Mairan's Nebula, first described by French astronomer Jean-Jacques d'Ortous de Mairan in 1731. Its dimensions are 20' X 15' of arc, and it lies at a distance of 1,300 ly. M43 is believed to get its energy from the central star, which is called Bond's Star. Following his observations on March 4, 1769, Messier noted, "The star which is above, & has little distance from that nebula, & of which is spoken in the Traité de l'Aurore boréale by M. de Mairan is surrounded, & equally by a very thin light; the star doesn't have the same brilliance as the four of the great nebula: its light is pale, & it appears covered by fog." The star is a variable, "GCVS NU Ori" with a range from magnitude 6.8 to 6.93. The monochrome image shown includes M42.

The negative image in Figure 4.20 pinpoints M43, and the enlargement at top left identifies four stars in the nebula. Star A is the previously mentioned central star, which is GCVS NU Ori. Star B is GCVS NQ Ori, another variable, which varies between magnitude 12.1 and 14.1. Stars C and D are also variables. C is GCVS MU Ori, varying between magnitude 13.8 and 15.8. Star D is GCVS MS Ori, which varies between magnitude 13.8 and 16.6. The star cluster (NGC 1977) has been pointed out at the top of the image. Only the lower part of the cluster has been captured on the image. NGC 1977 is also known as the running man nebula.

The chart in Figure 4.21 shows the position of M43 in the constellation of Orion. Orion sits with Lepus to the south, Eridanus to the west, with Taurus to the west and north, Gemini to the north and east and Monoceros to the east. Orion includes M42 as well as another Messier object, M78, which is a reflection nebula.

The image was taken at local time 21h 35m 53s on December 29 in New South Wales. The Universal Time was 10h 35m 53s. The Right Ascension of M43 is 05h

Fig. 4.19 Image of Messier 43, a nebula in Orion

36m 22s, and the Local Sidereal Time was 03h 05m 09s. As the RA is greater than the LST, this means that M43 was to the east of the meridian when the image was taken. This is confirmed by the Azimuth angle of 61° 37' 49". The difference between RA and LST is 2h 31m 13s. Adding this to the local time gives the transit time as 24h 07m 06s.

Messier 43: Data relating to the image in Figure 4.19

REMOTELY IMAGED FROM NEW SOUTH WALES, AUSTRALIA
NGC 1982 (MAIRAN'S NEBULA)
OBJECT TYPE: Bright Nebula

RA: 05h 36m 22s
DEC: -05° 15' 38"
ALTITUDE: +46° 04' 24"
AZIMUTH: 61° 37' 49"

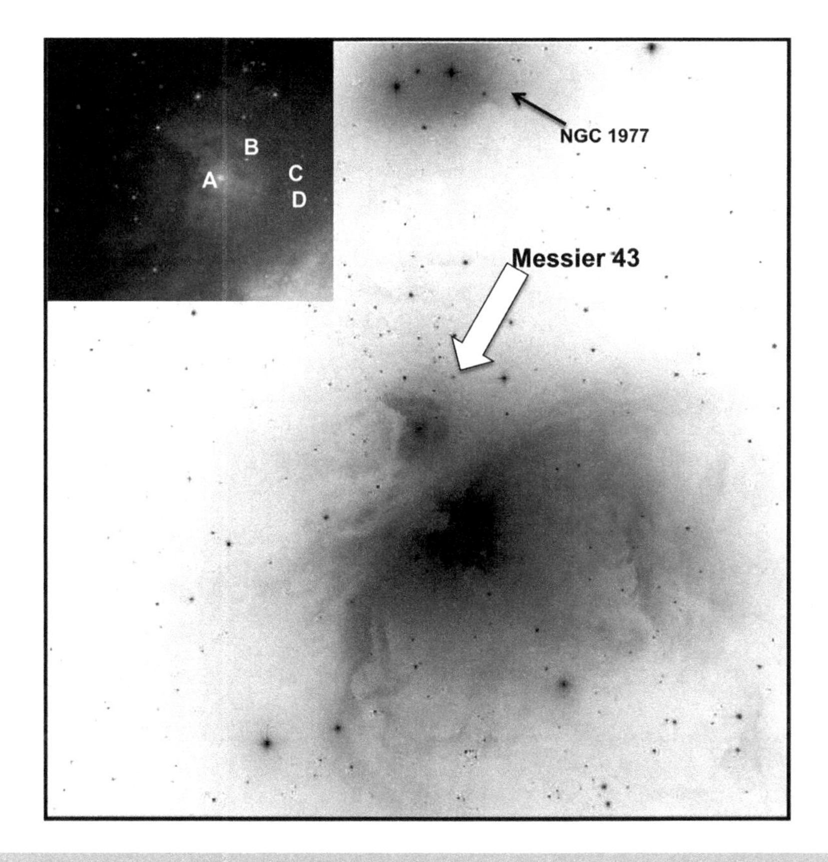

Fig. 4.20 Negative image of Messier 43

FIELD OF VIEW: 0° 55' 43" x 0° 55' 43"
OBJECT SIZE: 20' X 15'
POSITION ANGLE: 90° 51' from North
EXPOSURE TIME: 300 s
DATE: 29th December
LOCAL TIME: 21h 35m 53s
UNIVERSAL TIME: 10h 35m 53s
SCALE: 1.09 arcsec/pixel
MOON PHASE: 0.15% (waxing)

Telescope Optics
OTA: Planewave 20" (0.51m) CDK
Optical Design: Corrected Dall-Kirkham Astrograph
Aperture: 510 mm

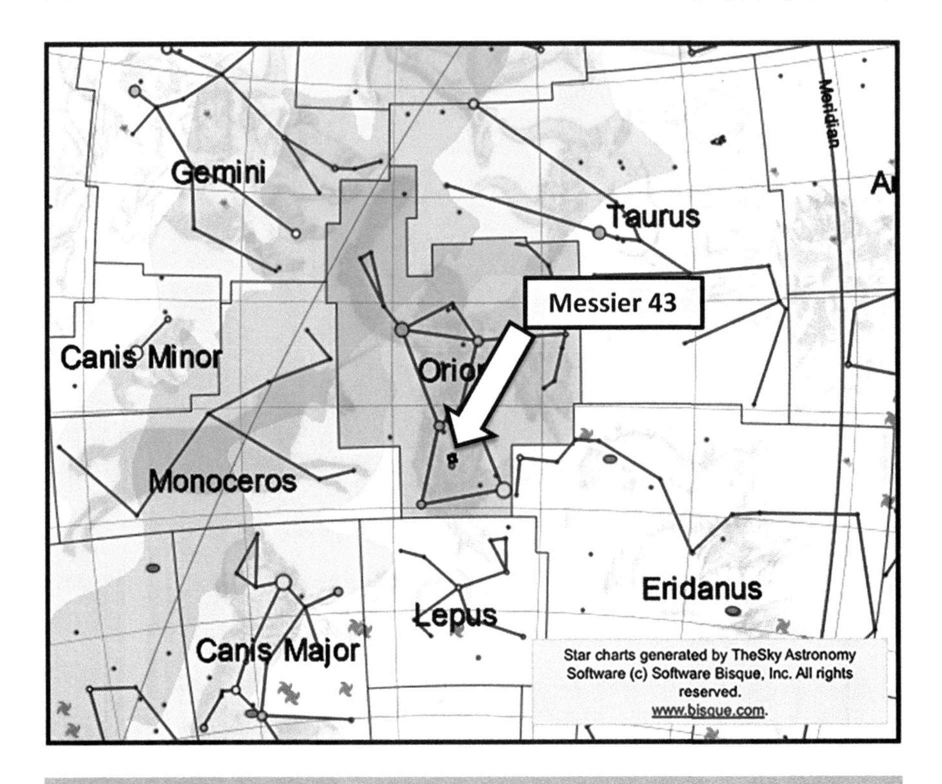

Fig. 4.21 Chart showing the location of M43

Focal Length: 2259 mm (focal reducer)
F/Ratio: f/4.4
Guiding: Active Guiding Disabled
Mount: Planewave Ascension 200HR

Instrument Package
CCD: PL09000 CCD Camera
Pixel Size: 12μm square
Sensor: Frontlit
Cooling: -30°C default
Array: 3056 x 3056 pixels
FOV: 55.9 x 55.9 arcmin

Location
Observatory: Siding Spring
UTC +10.00 (Australia Daylight savings time is observed)
31° 16' 24" South
149 03' 52" East
Elevation: 1122 meters (3681 ft)

Messier 44

Messier 44, shown in Figure 4.22, is in the constellation of Cancer and is popularly known as the Beehive or Praesepe Cluster. It is an easy naked-eye object, and it was Galileo who first turned his telescope to the "nebula" and resolved it into stars. English astronomer W.H. Smyth said, "the Praesepe, metamorphically rendered Bee-hive, is an aggregation of small stars which has long borne the name of a nebula, its components not being separately distinguishable by the naked eye; indeed, before the invention of the telescope, it was the only recognized one - the group is rather scanty in number, but splendid from the comparative magnitude of its constituents, which renders it a capital object for trying the light of a telescope. Yet Galileo discovered 36 small stars, when it was supposed that there were only three *nebulous* stars, which emitted the peculiar light."

The negative image in Figure 4.23 shows the cluster and identifies two bright stars in the frame. The first star is 47-Delta Cancri, also known as Asellus Australis or HIP 42911. This star is magnitude 3.94 with a spectral type K0. It is at a distance of 136.07 ly. The second star is 43-Gamma Cancri, also known as Asellus Borealis or HIP 42806. This star is magnitude 4.66 and is spectral type A1. It lies at a distance of 158.48 ly. The original image was enlarged to highlight the very faint spiral galaxy known as IC 2390, pointed out on the negative image. This galaxy is about 209 million ly away in comparison with the distance to M44 itself, which is a mere 577 ly distant. Thus, we are looking through the M44 cluster to an object that is

Fig. 4.22 Image of Messier 44, an open cluster in Cancer

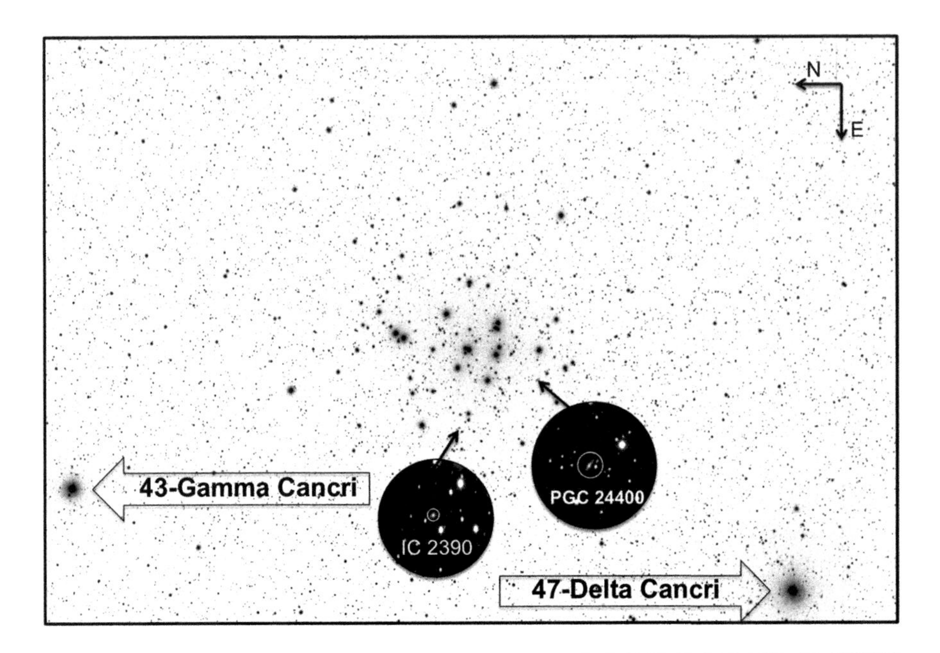

Fig. 4.23 Negative image of Messier 44

over 362,000 times further away than the cluster itself. The other highlighted spiral galaxy is PGC 24400. There are many more to find in the original image.

The chart in Figure 4.24 shows the location of Messier 44 towards the center of the constellation of Cancer. The star 65-Alpha Cancri is known as Acubens or HIP 44066. It is of magnitude 4.26 and is spectral type A5. It lies at a distance of 173.58 ly. There is an angular separation of almost 9° between Messier 44 and Acubens. The star 17-Beta Cancri is known as Tarf or HIP 40526. It is of magnitude 3.53 and spectral type K4. It lies at a distance of 290.43 ly. There is a distance of 11° 55' between Messier 44 and Tarf.

The image was taken on February 5 at 03h 11m 00s local time in New Mexico. This corresponded to a Universal Time of 10h 11m 00s. The Right Ascension of M44 is 08h 40m 45s, and the Local Sidereal Time was 12h 10m 38s. The LST is higher than the RA, so M44 must have crossed the meridian from east to west some time beforehand. Subtracting one from the other gives a difference of 3h 29m 53s. Subtracting this from the local time when the image was taken provides a meridian transit time of 23h 41m 07s on February 4. The telescope used to take the image of M44 was a Takahashi FSQ apochromatic astrograph with an aperture of 106 mm.

Messier 44: Data relating to the image in Figure 4.22

REMOTELY IMAGED FROM MAYHILL, NEW MEXICO
NGC 2632 (BEEHIVE OR PRAESEPE CLUSTER)

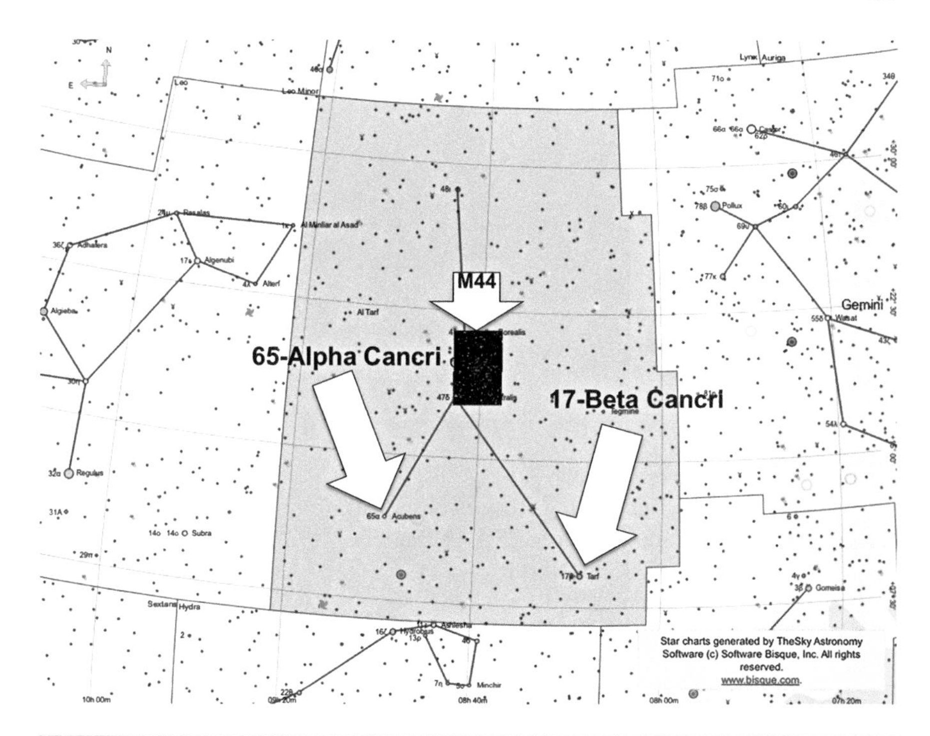

Fig. 4.24 Chart showing the location of M44

OBJECT TYPE: Open Cluster

RA: 08h 40m 45s
DEC: +19° 37' 19"
ALTITUDE: +41° 37' 13"
AZIMUTH: 267° 49' 39"
FIELD OF VIEW: 3° 53' 32" x 2° 35' 41"
OBJECT SIZE: 95'
POSITION ANGLE: 91° 58' from North
EXPOSURE TIME: 300 s
DATE: 5th February
LOCAL TIME: 03h 11m 00s
UNIVERSAL TIME: 10h 11m 00s
SCALE: 3.5 arcsec/pixel
MOON PHASE: 29.66% (waning)

Telescope Optics
OTA: Takahashi FSQ - Fluorite
Optical Design: Petzval Apochromatic Astrograph
Aperture: 106 mm
Focal Length: 530 mm

F/Ratio: f/5.0 Guiding: External
Mount: Paramount PME

Instrument Package
CCD: SBIG STL-11000M
Pixel Size: 9μm square
Sensor: Frontlit
Cooling: -15°C
Array: 4008 x 2672 pixels
FOV: 158.8 x 233.7 arcmin
Position Angle: 092.7°

Location
Observatory: New Mexico Skies
UTC Minus 7.00 (Daylight savings time is observed)
Minimum Target Elevation: Approx 25 – 45 Degrees
(N or S) 32° 54' Decimal: 32.9 North
(W or E) 105° 31' Decimal: 105.5
West Elevation: 2225 meters (7298 ft)

Messier 45

Messier 45 in the constellation of Taurus, shown in Figure 4.25, is better known as the Pleiades star cluster. It has been observed since ancient times, with the first recorded mention in 750 BCE by Homer and others. The 10-minute exposure taken from Spain shows the nebulosity involved with the cluster. M45 is also popularly known as the Seven Sisters. W.H. Smyth said, "The Pleiades constitute a celebrated group of stars, or miniature constellation, on the shoulders of Taurus; their popular influences have been said and sung for many years." Messier added the cluster to his list on March 4, 1769. 19th century Poet Laureate Alfred, Lord Tennyson, in his poem "Locksley Hall" referred to the Pleiades: "Many a night from yonder ivied casement, ere I went to rest, Did I look on great Orion sloping slowly to the West. Many a night I saw the Pleiads, rising thro' the mellow shade, Glitter like a swarm of fireflies tangled in a silver braid." It is difficult to understand why Messier would include the Pleiades if his list was solely of objects that could be mistaken for comets, as M45 is clearly a star cluster.

The negative image in Figure 4.26 identifies the major stars of the Pleiades. The cluster has a size of 110' of arc and lies at a distance of 440 ly. This indicates an actual size of about 14 ly. The stars are hot and blue, indicating a young cluster. The spectral type of the named stars is included on the image, which confirms this. WEBDA gives the log age of M45 as 8.13, which corresponds to roughly 135 million years. Most references, however, seem to give a lower age in the region of 100 million years. It may seem old initially, until you consider that the age of the

Fig. 4.25 Image of Messier 45, an open cluster in Taurus

Earth is about 5 billion years. The reflection nebula is caused by the young stars pumping energy into a surrounding dust cloud. This dust cloud has been shown to be independent of the cluster itself that just happens to be passing through the cloud. It is estimated that the stars in Messier 45 will have dispersed in about 250 million years.

The chart in Figure 4.27 shows the plate-solved image of the Pleiades superimposed exactly in the correct position. The Hyades is another well-known star cluster in Taurus and is located near the bright star Aldebaran. Aldebaran (87-Alpha Tauri or HIP 21421) is magnitude 0.87 of spectral type K7. It is not associated with the Hyades but is rather a foreground star in relation to that cluster. The Hyades are much closer to us than the Pleiades at a distance of 153 ly. The angular separation between Alcyone in the Pleiades and Aldebaran is roughly 13° 40'. The star Ain (74-Epsilon Tauri or HIP 20889) is a magnitude 3.53 star of spectral type K0, much fainter than Aldebaran and slightly nearer Messier 45. The angular separation between Alcyone and Ain is roughly 10° 45'.

The image was taken on September 30 from Nerpio, Spain at 06h 41m 48s. The corresponding Universal Time was 04h 41m 48s. The Right Ascension of M45 is 03h 47m 07s, and the Local Sidereal Time was 05h 09m 59s. So, M45 crossed the meridian 1h 22min 52s before, making its transit time 05h 18m 56s. A 6-inch Apochromatic Refractor with a focal length of 1,095 mm was used to take the image.

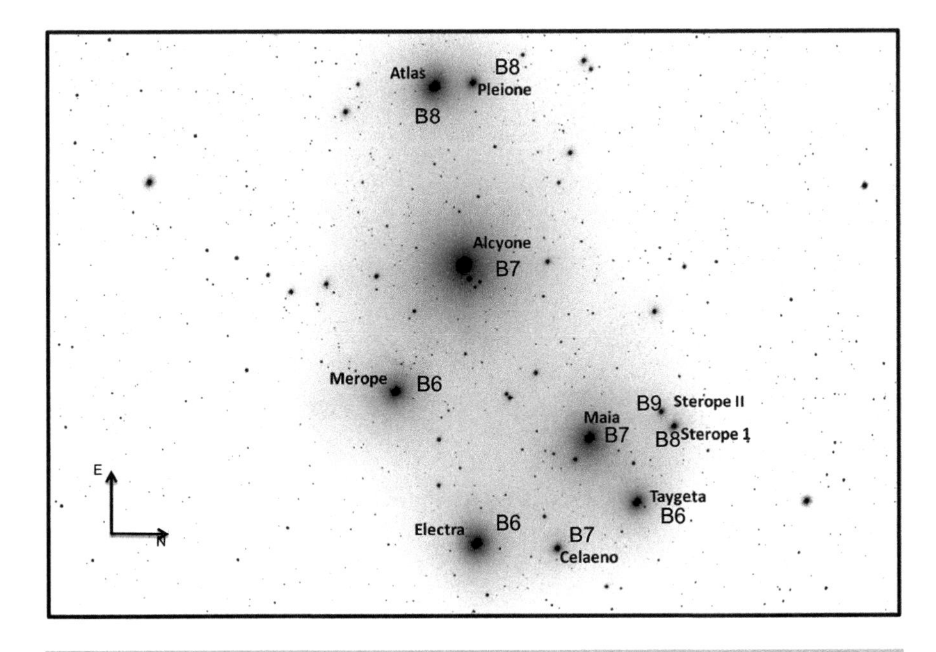

Fig. 4.26 Negative image of Messier 45

Messier 45: Data relating to the image in Figure 4.25

REMOTELY IMAGED FROM NERPIO, SPAIN
PLEIADES STAR CLUSTER
OBJECT TYPE: Open Cluster

RA: 03h 47m 07s
DEC: +24° 11' 47"
ALTITUDE: +67° 32' 00"
AZIMUTH: 237° 36' 41"
FIELD OF VIEW: 1° 51' 37" x 01° 14' 25"
OBJECT SIZE: 110'
POSITION ANGLE: 88° 15' from North
EXPOSURE TIME: 600 s
DATE: 30th September
LOCAL TIME: 06h 41m 48s
UNIVERSAL TIME: 04h 41m 48s
SCALE: 1.67 arcsec/pixel
MOON PHASE: 0.65% (waning)

Telescope Optics
OTA: Takahashi TOA-150
Optical Design: Apochromatic Refractor

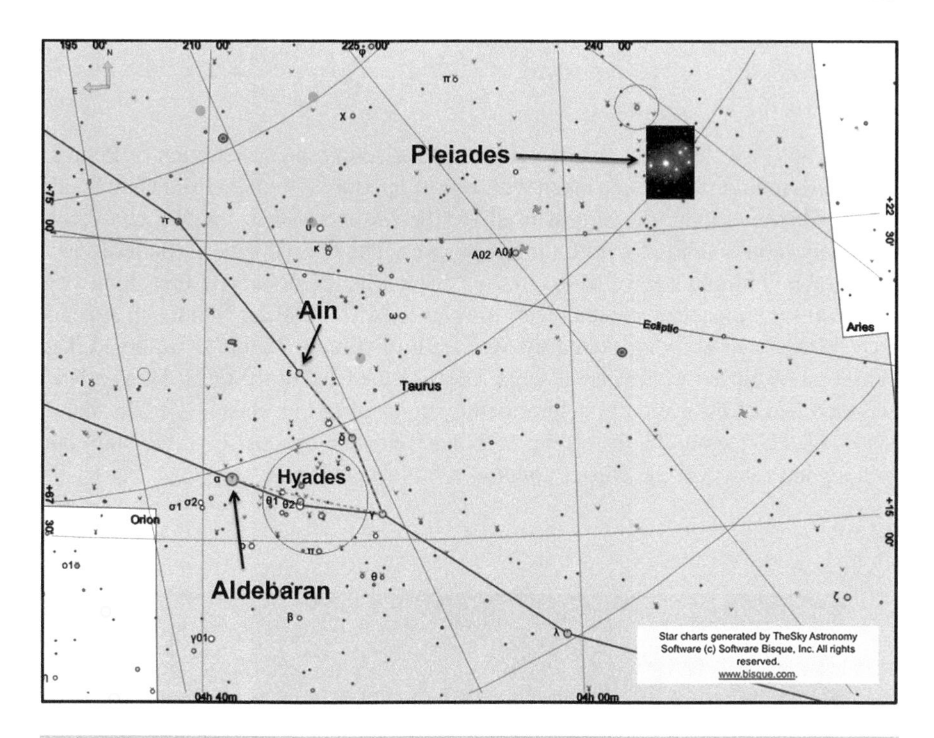

Fig. 4.27 Chart showing the location of M45

Aperture: 150 mm
Focal Length: 1095 mm
F/Ratio: f/7.3
Guiding: Internal
Mount: Paramount PME

Instrument Package
CCD: SBIG STL-11000M
Pixel Size: 9μm square
Sensor: Frontlit
Cooling: -20°C default
Array: 4008 x 2672 pixels
FOV: 75.4 x 113.1 arcmin

Location
Observatory: AstroCamp
UTC +1.00 (Madrid Daylight savings time is observed)
38° 09' North
02° 19' West
Elevation: 1650 meters (5413 ft)

Messier 46

Messier 46 (NGC 2437) is an open cluster in the southern constellation of Puppis. The image in Figure 4.28 was taken with a large aperture telescope from New South Wales. This telescope, the largest used for the Messier object images, can reach extremely faint magnitudes with long exposures. The telescope has a field of view comparable with the size of M46, so the cluster fills the field. The first thing you will spot is the planetary nebula – very similar to M57, the Ring Nebula, in appearance. Messier 46 was discovered by Messier on February 19 1771. He noted that M46 was, "A cluster of very small stars, between the head of the Great Dog and the two hind feet of the Unicorn, determined by comparing this cluster with the star 2 Navis, of 6th-magnitude, according to Flamsteed; one cannot see these stars but with a good refractor; the cluster contains a bit of nebulosity."

Fig. 4.28 Image of Messier 46, an open cluster in Puppis

The negative image in Figure 4.29 indicates the Planetary Nebula NGC 2438, which has a size of just over one arcmin at a distance of around 2,900 ly. Calculations show that the actual diameter of M46 will be about 0.85 of a light-year. The planetary nebula is the result of a star reaching the red giant phase when it has run out of fuel and collapsed, shedding the outer parts of the star until the high temperature core is revealed, which produces a solar wind, causes gaseous expansion, then provides energy to heat up the gas and make the nebula visible. M46 itself is at a distance of 5,400 ly, so with its apparent diameter of 27' of arc, it has an actual diameter of over 42 ly. WEBDA gives it a log age of 8.390, equivalent to over 245 million years. There are probably over 500 stars in M46. The brightest star in the cluster is HIP 37464, which has an apparent magnitude of 8.66.

The chart in Figure 4.30 shows the location of M46 in the constellation of Puppis. You will notice how tiny the 27-minute plate-solved square image is in relation to the size of the constellation. Images in this book range from almost 4° down to this image size of 27'. The magnitude -1.44 Sirius (9-Alpha Canis Majoris or HIP 32349) in Canis Major is not too far away towards the west. The angular separation between M46 and Sirius is 13° 51'. Sirius lies at a distance of a mere 8.6 ly.

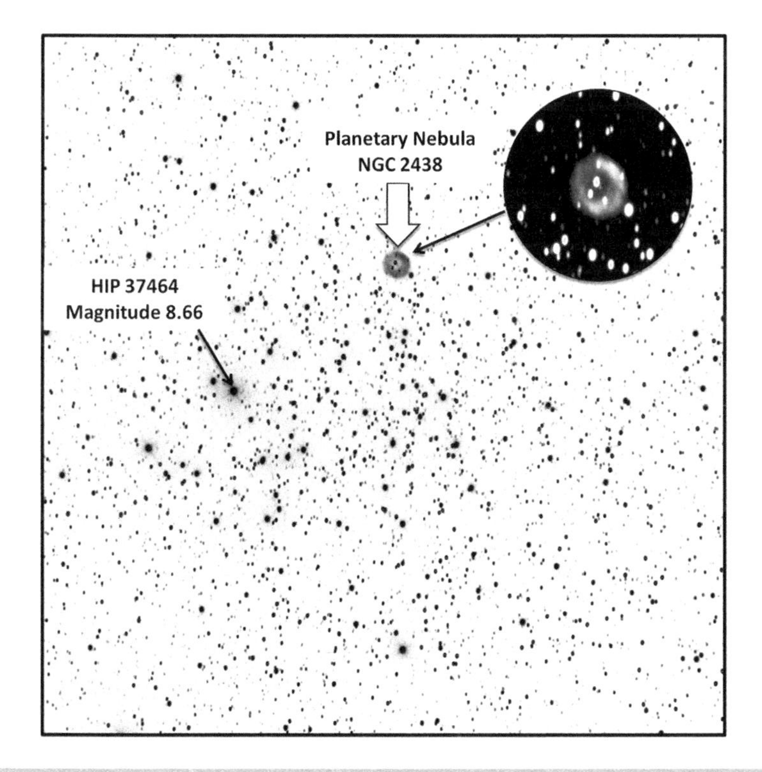

Fig. 4.29 Negative image of Messier 46

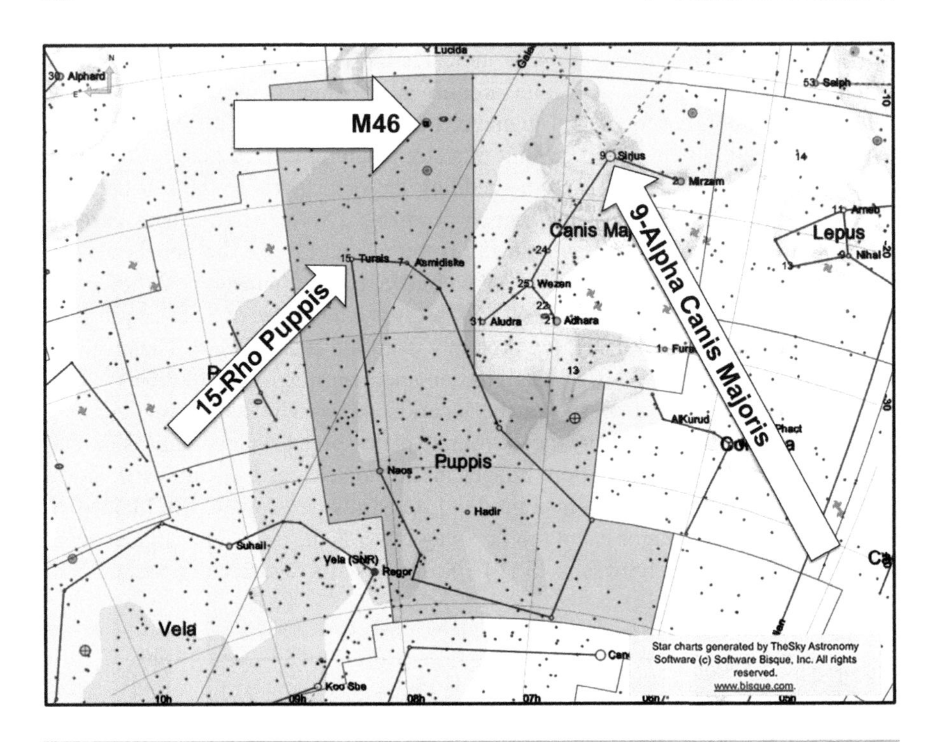

Fig. 4.30 Chart showing the location of M46

The star 15-Rho Puppis (Turais or HIP 39757) sits to the south west of Messier 46 at an angular distance of 11° 15'. This star has a magnitude of 2.83 and is at a distance of 62.73 ly.

The image was taken on November 9 from Siding Spring at 03h 02m 30s. The corresponding Universal Time was 16h 02m 30s (November 8). The Right Ascension of M46 is 07h 42m 33s, and the Local Sidereal Time was 05h 11m 35s. Thus, M46 would cross the meridian in 2h 30m 58s with a transit time of 05h 33m 28s. The telescope used to take the image of M46 was a Planewave 27" Corrected Dall-Kirkham Astrograph with an aperture of 700 mm and a focal length of 4,531 mm. The mount used was a Planewave Ascension Alt-Az.

Messier 46: Data relating to the image in Figure 4.28

REMOTELY IMAGED FROM NEW SOUTH WALES, AUSTRALIA
NGC 2437
OBJECT TYPE: Open Cluster

RA: 07h 42m 33s
DEC: -14° 50' 58"
ALTITUDE: +51° 51' 13"
AZIMUTH: 73° 19' 12"

FIELD OF VIEW: 0° 27' 49" x 0° 27' 49"
OBJECT SIZE: 27'
POSITION ANGLE: 179° 57' from North
EXPOSURE TIME: 300 s
DATE: 9th November
LOCAL TIME: 03h 02m 30s
UNIVERSAL TIME: 16h 02m 30s (8th November)
SCALE: 0.55 arcsec/pixel
MOON PHASE: 58.92% (waxing)

Telescope Optics
OTA: Planewave 27" (0.7m) CDK700WF
Optical Design: Corrected Dall-Kirkham Astrograph
Aperture: 700 mm
Focal Length: 4531 mm
F/Ratio: f/6.6
Guiding: Active Guiding Disabled
Mount: Planewave Alt-Az

Instrument Package
CCD: FLI PL09000
Pixel Size: 12μm square
Sensor: Frontlit
Cooling: -35°C default
Array: 3056 x 3056 pixels
FOV: 27.1 x 27.1 arcmin

Location
Observatory: Siding Spring
UTC +10.00 (Australia Daylight savings time is observed)
31° 16' 24" South
149 03' 52" East
Elevation: 1122 meters (3681 ft)

Messier 47

Messier 47 (NGC 2422), shown in Figure 4.31, is an open cluster in the constellation of Puppis. It lies close to M46 in the same constellation. There were a number of independent discoveries of this cluster, but the first is now believed to be Sicilian astronomer Giovanni Batista Hodierna in the 17th century. Messier independently discovered it on February 19, 1771. He commented that it was a cluster of stars, close to M46 but with brighter stars and that the cluster contained no nebulosity. W. H. Smyth made the comment that Messier 47 was "a very splendid field of large and small stars, disposed somewhat in lozenge shape." John Herschel described the

Fig. 4.31 Image of Messier 47, an open cluster in Puppis

cluster as "a very large, pretty rich, splendid cluster, which more than fills the field." He also referred to a fine double star at the center of the cluster.

The negative image in Figure 4.32 identifies some objects in the cluster. The bright star to the right (west) is HIP 36981 with a magnitude of 5.66. It is a spectral type B2 star lying at a distance of 1,655.62 ly (from Hipparcos data). The Messier 47 cluster itself is normally given a distance of about 1,600 ly, so it would appear that this star is part of the cluster. The magnitude 7.03 star HIP 37015 is of spectral type B3 and is at a distance of 1,689.93 ly, so would appear to be another confirmed cluster member. John Herschel referred to a "fine double star" at the center of the cluster. This star is Struve 1121.

The M47 image is shown embedded on the chart in Figure 4.33 in the constellation of Puppis. North is at the top and east is to the left. You can see just how close together the clusters of M47 and M46 are. M46 lies to the southeast of M47. There is an angular separation of just less than 78' of arc between them. There is a magnitude 4.83 star to the west of M47. This is HIP 36773, which is of spectral type A4 and lies at a distance of 3,362.44 ly. (Hipparcos). There is an angular separation of 41' between M47 and this star. To get a wider-angle perspective, look back at the chart for Messier 46.

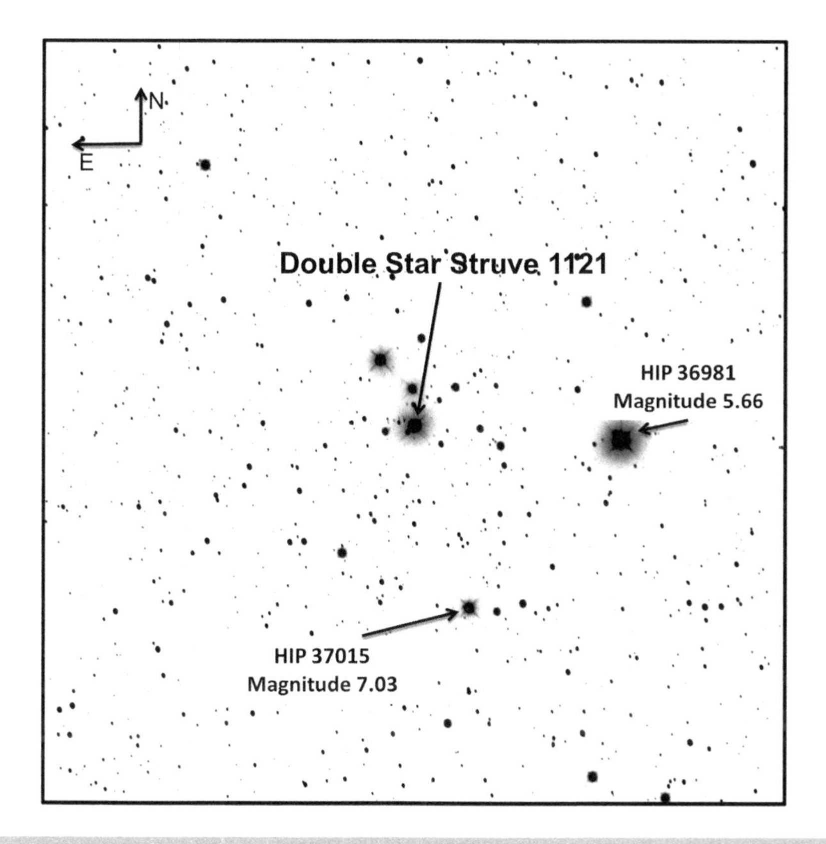

Fig. 4.32 Negative image of Messier 47

The image was taken on November 9 from Siding Spring at 04h 02m 45s. The corresponding Universal Time was 17h 02m 45s (November 8). The Right Ascension of M47 is 07h 37m 22s, and the Local Sidereal Time was 6h 12m 00s. M47 would cross the meridian in 1 h 25m 22s with a transit time of 05h 28m 07s. The telescope used to take the image of M47 was a Planewave Corrected Dall-Kirkham Astrograph with an aperture of 700 mm and a focal length of 4,531 mm. The mount used was a Planewave Ascension Alt-Az.

Messier 47: Data relating to the image in Figure 4.31

REMOTELY IMAGED FROM NEW SOUTH WALES, AUSTRALIA
NGC 2422
OBJECT TYPE: Open Cluster

RA: 07h 37m 22s
DEC: -14° 31' 03"
ALTITUDE: +64° 17' 02"
AZIMUTH: 54° 16' 57"

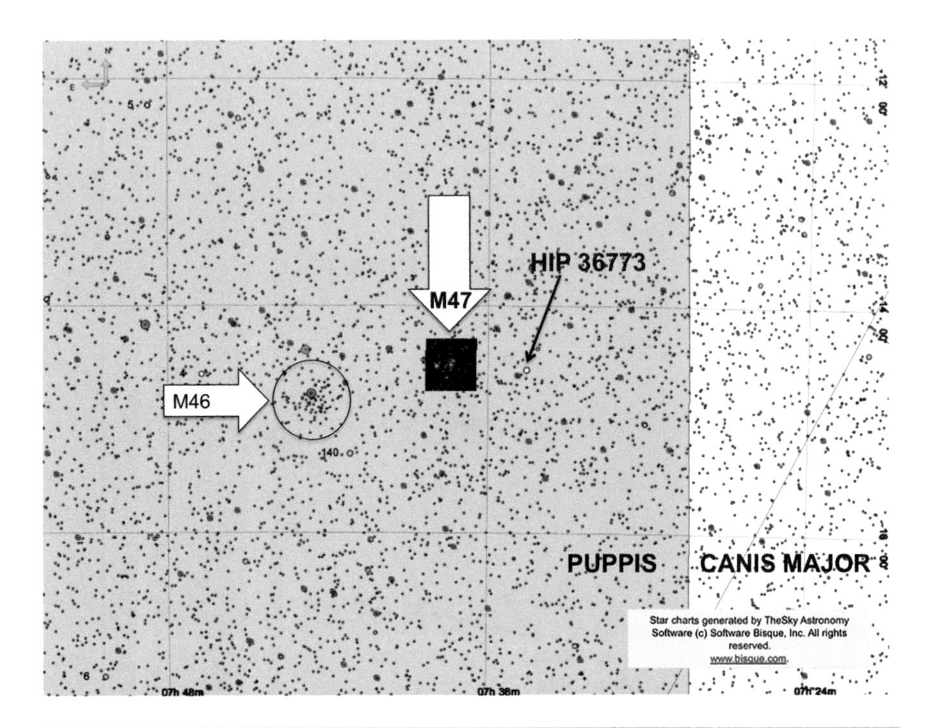

Fig. 4.33 Chart showing the location of M47

FIELD OF VIEW: 0° 27' 49" x 0° 27' 49"
OBJECT SIZE: 30'
POSITION ANGLE: 179° 57' from North
EXPOSURE TIME: 300 s
DATE: 9th November
LOCAL TIME: 04h 02m 45s
UNIVERSAL TIME: 17h 02m 45s (8th November)
SCALE: 0.55 arcsec/pixel
MOON PHASE: 59.36% (waxing)

Telescope Optics
OTA: Planewave 27" (0.7m) CDK700WF
Optical Design: Corrected Dall-Kirkham Astrograph
Aperture: 700 mm
Focal Length: 4531 mm
F/Ratio: f/6.6
Guiding: Active Guiding Disabled
Mount: Planewave Alt-Az

Instrument Package
CCD: FLI PL09000
Pixel Size: 12μm square
Sensor: Frontlit
Cooling: -35°C default
Array: 3056 x 3056 pixels
FOV: 27.1 x 27.1 arcmin

Location
Observatory: Siding Spring
UTC +10.00 (Australia Daylight savings time is observed)
31° 16' 24" South
149 03' 52" East
Elevation: 1122 meters (3681 ft)

Messier 48

Messier 48 (NGC 2548), shown in Figure 4.34, is an open cluster in the constellation of Hydra. It is quite large and is a little bigger than the field of view of the telescope used to take the image. Charles Messier discovered the cluster on February 19, 1771, but he miscalculated the coordinates, which took until the 20th Century to fully resolve. John Herschel observed the cluster in 1830 and commented that M48 was, "A superb cluster which fills the whole field; stars of 9th and 10th to the 13th magnitude -- and none below, but the whole ground of the sky on which it stands is singularly dotted over with infinitely minute points."

The image of Messier 48 has a field of view of 49' 06" x 32' 04" of arc, showing most of the cluster, which extends out as far as 54'. The distance to M48 is 1,500 ly, so the actual size of the full cluster will be about 24 ly. The image in Figure 4.35 shows the central concentration of stars with the frame covering roughly 21 ly x 14 ly. John Herschel was able to observe stars in the cluster with magnitude 9 and 10, down to magnitude 13. Four stars of decreasing brightness are identified on the negative image in Figure 4.35. The brightest star is HIP 40348 with a magnitude of 8.2. The next star towards the centre of the image is Tycho 4859:378 with a magnitude of 10.0. Going fainter still, I The star UCAC4 422:44127 at magnitude 12.48 is identified, along with the final GSC 4855:2551 with a magnitude of 15.27. Clearly, there are much fainter stars than this recorded on the image.

The chart in Figure 4.36 shows the position of M48 in the constellation of Hydra. North is at the top, and the solved plate is linked to the chart. M48 has an angular separation of roughly 18° 34' from the bright star 30-Alpha Hydrae, better known as Alphard or HIP 46390. This is located slightly to the south of east of M48. Alphard has a magnitude of 1.99 and is of spectral type K3, lying at a distance of 177.26 ly. The very bright star 10-Alpha Canis Minoris, better known as Procyon or HIP 37279, is roughly 13° 58' to the north east of M48. Procyon has a magnitude of 0.4 and is of spectral type F5, lying at a distance of a mere 11.41 ly.

Fig. 4.34 Image of Messier 48, an open cluster in Hydra

The image was taken on November 30 from New Mexico at 05h 09m 28s. The corresponding Universal Time was 12h 09m 28s. The Right Ascension of M48 is 08h 14m 34s, and the Local Sidereal Time was 09h 45m 23s. M48 crossed the meridian 1h 30m 49s ago, making the transit time of M48 3h 38m 39s. The telescope used to take the image of M48 was a Planewave Corrected Dall-Kirkham Astrograph with an aperture of 431 mm and a focal length of 1,940 mm. The mount used was a Planewave Ascension 200HR.

Messier 48: Data relating to the image in Figure 4.34

REMOTELY IMAGED FROM MAYHILL, NEW MEXICO
NGC 2548
OBJECT TYPE: Open Cluster

RA: 08h 14m 34s
DEC: -05° 48' 11"
ALTITUDE: +45° 47' 43"
AZIMUTH: 213° 25' 08"
FIELD OF VIEW: 0° 49' 06" x 0° 32' 44"
OBJECT SIZE: 54'
POSITION ANGLE: 273° 06' from North
EXPOSURE TIME: 300 s
DATE: 30th November
LOCAL TIME: 05h 09m 28s

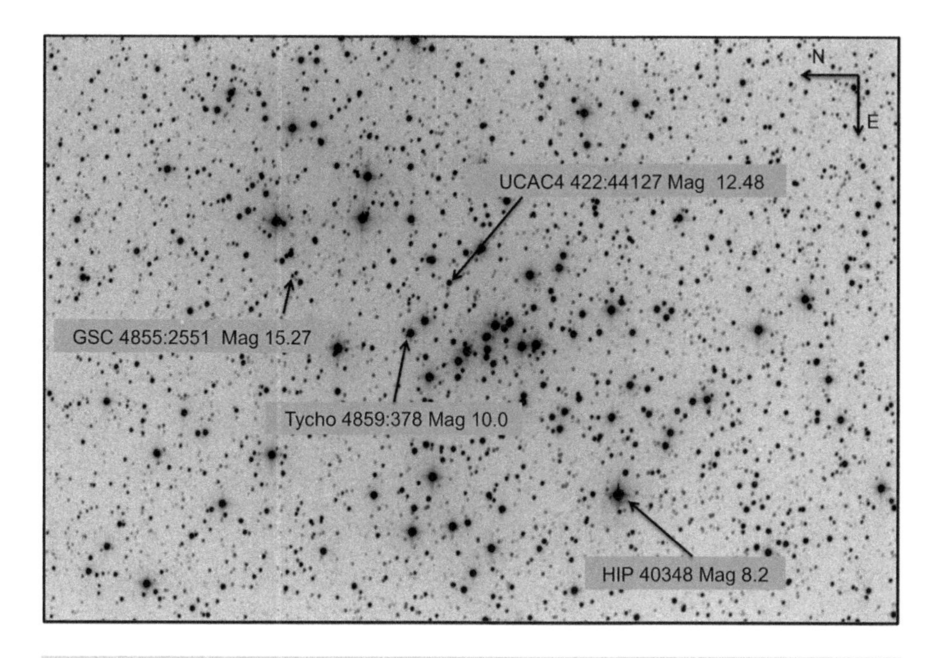

Fig. 4.35 Negative image of Messier 48

UNIVERSAL TIME: 12h 09m 28s
SCALE: 0.96 arcsec/pixel
MOON PHASE: 1.08% (waxing)

Telescope Optics
OTA: Planewave 17" CDK
Optical Design: Corrected Dall-Kirkham Astrograph
Aperture: 431 mm
Focal Length: 1940 mm
F/Ratio: f/4.5
Guiding: Active Guiding Disabled
Mount: Planewave Ascension 200HR

Instrument Package
CCD: FLI-PL6303E CCD Camera
Pixel Size: 9μm square
Resolution: 0.96 arcsec/pixel
Sensor: Frontlit
Cooling: -35°C default
Array: 3072 x 2048 (6.3 Megapixels)
FOV: 32.8 x49.2 arcmin

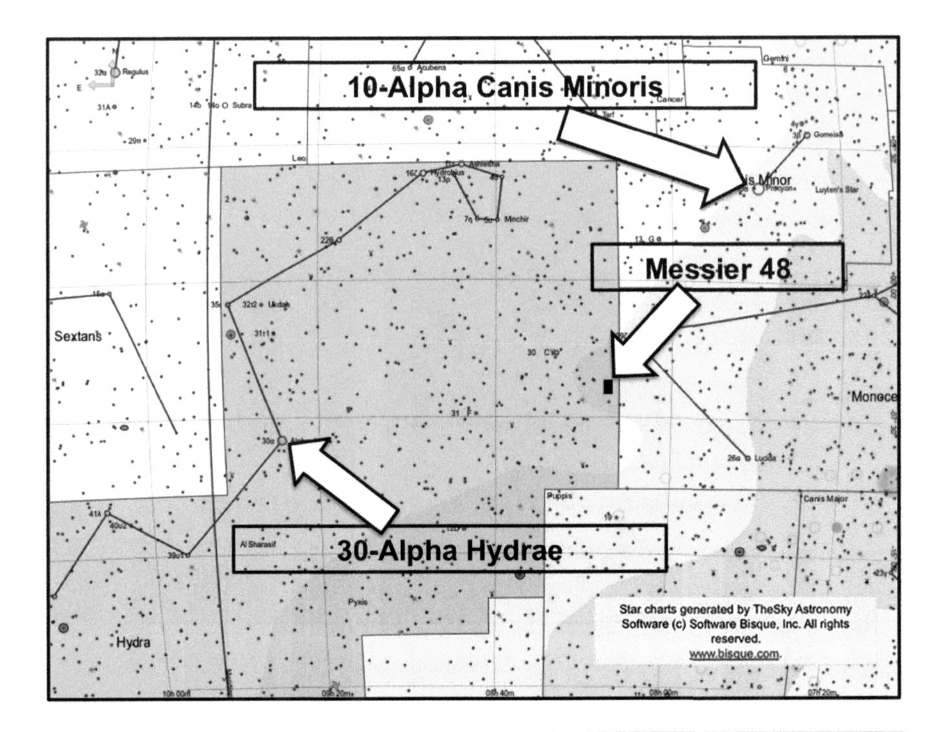

Fig. 4.36 Chart showing the location of M48

Location
Observatory: New Mexico Skies
UTC Minus 7.00 (Daylight savings time is observed)
Minimum Target Elevation: Approx. 25 – 45 Degrees
(N or S) 32° 54' Decimal: 32.9 North
(W or E) 105° 31' Decimal: 105.5 West
Elevation: 2225 meters (7298 ft)

Messier 49

Messier 49 (NGC 4472), shown in Figure 4.37, is an elliptical galaxy in the constellation of Virgo. Messier discovered this galaxy, which is part of the Virgo cluster. The image does not display any detail of the elliptical galaxy, but it does contain other galaxies of interest. M49 itself has a size of 9' x 7' of arc and lies at a distance of 60 million ly, which indicates an actual size of 157,000 x 122,000 ly, depending on its exact orientation, which is difficult to observe in an elliptical galaxy. The small star next to M49 does look as though it could be a supernova that has exploded in the galaxy, but in fact, it is just a star. Messier discovered M49 on

Fig. 4.37 Image of Messier 49, an elliptical galaxy in Virgo

February 19,1771, noting, "Nebula discovered near the star Rho Virginis. One cannot see it without difficulty with an ordinary telescope of 3.5-feet." W. H. Smyth commented that it is, "A bright, round, and well-defined nebula, on the Virgin's left shoulder; exactly on the line between Delta Virginis and Beta Leonis, 8 deg, or less than halfway, from the former star."

The negative image in Figure 4.38 shows Messier 49 at the center of the 97' x 73' frame. The locations of five other galaxies are indicated. NGC 4488 is a spiral galaxy 4.1' x 1.7' in apparent size. NGC 4492 is another spiral galaxy that is only 1.7' x 1.6'. NGC 4466 is also a spiral galaxy and measures 1.1' x 0.3'. NGC 4470, spiral galaxy number 4, is 1.3' x 0.9'. NGC 4416 is spiral number 5 and measures 1.7' x 1.5'. It is hard to comprehend the unthinkable numbers of stars and planets contained in these little smudges. The bright star HIP 61103 is magnitude 6.03 and spectral type K5. It lies at a distance of 795.50 ly – a long way for us, but nothing in comparison with M49's 60 million ly! I can understand why Messier would want to put M49 on his list of comet imposters -- it certainly could be mistaken for one.

The chart in Figure 4.39 shows the location of Messier 49 in the constellation of Virgo. You can see on the chart the many galaxies of the Virgo cluster. Smyth remarked that M49 was exactly on the line between Delta Virginis (Auva or HIP 63090) and Beta Leonis (Denebola or HIP 57632). You can confirm this on the

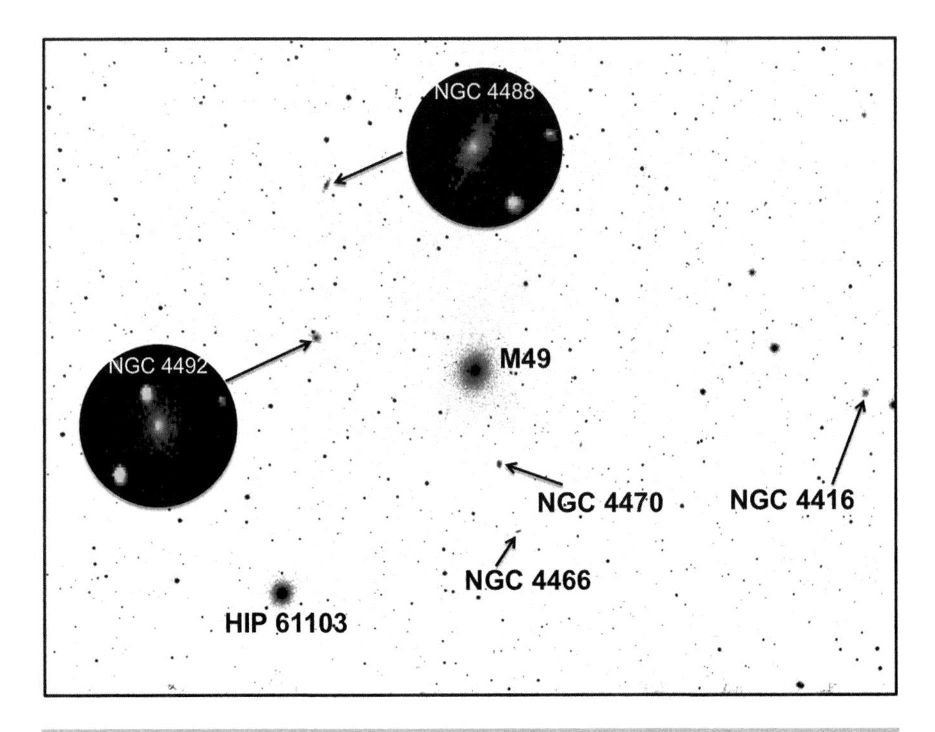

Fig. 4.38 Negative image of Messier 49

chart. He stated that the angular separation between M49 and Auva was 8°, which I measure at 7° 55' and that this was less than halfway to Denebola. I measure the distance between Auva and Denebola as 19° 49', so M49 is indeed less than halfway. Auva is an M3 star with a magnitude of 3.39 lying at a distance of 202.46 ly. Denebola is a mag 2.14 A4 star at a distance of 36.18 ly.

The image was taken on April 24 from Siding Spring at 00h 29m 04s. The corresponding Universal Time was 14h 29m 04s (April 24). The Right Ascension of M49 is 12h 30m 40s, and the Local Sidereal Time was 14h 32m 22s. So, M49 crossed the meridian 2h hours 1m 44s before with a transit time at 22h 27 20s. The single-shot color image camera was used with a 90-mm refractor to obtain the image.

Messier 49: Data relating to the image in Figure 4.37

REMOTELY IMAGED FROM NEW SOUTH WALES, AUSTRALIA
NGC 4472
OBJECT TYPE: Elliptical Galaxy

RA: 12h 30m 40s
DEC: +07° 54' 15"
ALTITUDE: +41° 13' 10"

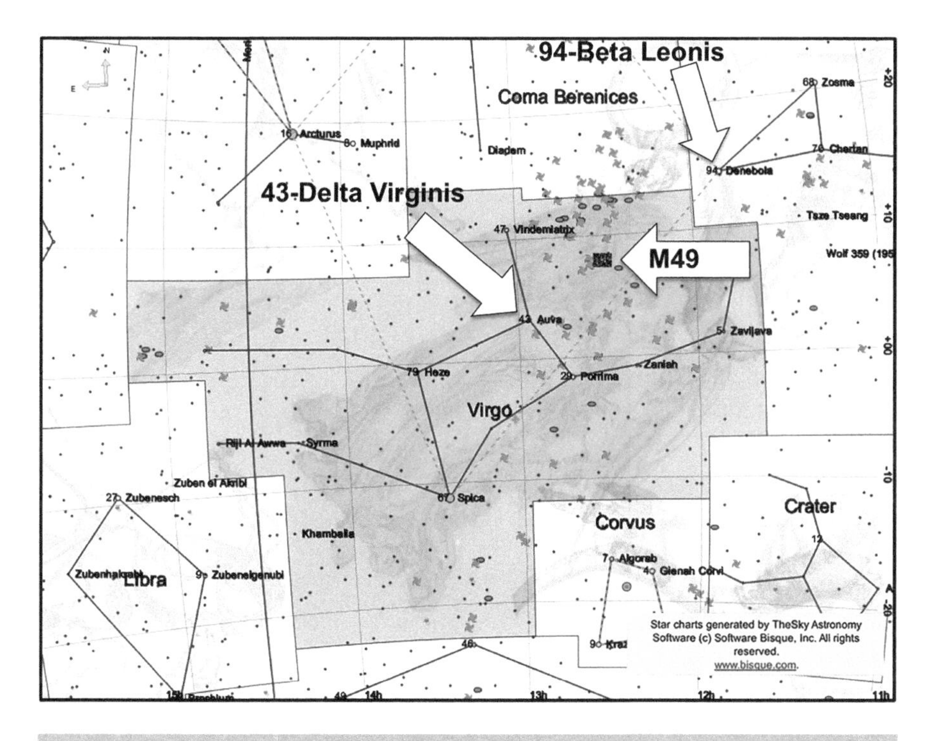

Fig. 4.39 Chart showing the location of M49

AZIMUTH: 318° 10' 37"
FIELD OF VIEW: 1° 37' 42" x 1° 13' 17"
OBJECT SIZE: 9' x 7.5'
POSITION ANGLE: 177° 02' from North
EXPOSURE TIME: 300 s
DATE: 24th April
LOCAL TIME: 00h 29m 04s
UNIVERSAL TIME: 14h 29m 04s
SCALE: 3.66 arcsec/pixel
MOON PHASE: 11.51% (waning)

Telescope Optics
OTA: Takahashi Sky 90
Optical Design: Apochromatic Refractor
Aperture: 90 mm
Focal Length: 417 mm
F/Ratio: f/5.6
Guiding: None
Mount: Paramount PME

Instrument Package
CCD: SBIG ST2000 XMC
Colour CMOS
Pixel Size: 7.4μm square
Sensor: Frontlit
Cooling: CNA
Array: 1600 x 1200 pixels
FOV: 60.5 x 80.7 arcmin

Location
Observatory: Siding Spring
UTC +10.00 (Australia Daylight savings time is observed)
31° 16' 24" South
149 03' 52" East
Elevation: 1122 meters (3681 ft)

Messier 50

Messier 50, shown in the monochrome image in Figure 4.40, is an open cluster in the constellation of Monoceros. It has a diameter of around 16' of arc and is believed to lie at a distance of 3,200 ly. At this distance, it would have an actual diameter of 15 ly. Messier commented on his observation of M50 on April 5, 1772, "I determined the position of a cluster of small stars, placed between the star Theta in the ear of Canis Major, & the right loins of Monoceros; I compared this cluster with the telescopic star, determined on April 3, & this one with a star of the seventh magnitude which was near the cluster." Messier was actually inspecting a comet on the night that he noticed the cluster, which had been found earlier by Jacques Montaigne of Limoges in France on March 8, 1772. In the cluster, there is a red star corresponding to the brightest star in the monochrome image, which contrasts starkly with the blue and white stars that you would see in a color image.

The negative image in Figure 4.41 gives an enlarged view of the cluster and identifies the red star mentioned previously. The star, HIP 33959, has a magnitude of 7.84 and is type K3. The Hipparcos Catalog places it at a distance of 3,977.52 ly. The star was referred to by Robert Burnham Jr. when he commented, "The cluster is a moderately compressed group about 10' in diameter with many outlying stragglers increasing the apparent size to about 20' X 15'. Curving arcs of stars give the perimeter a rather heart shaped outline, nicely highlighted by a reddish star some 7' south of the cluster centre." The reddish star is a K3 star with a surface temperature in the region of 4800 K. The star Tycho 5381:902 is also labeled. This star has a magnitude of 8.06. M50 is often referred to as the "Heart Cluster."

The chart in Figure 4.42 shows the location of Messier 50 in the constellation of Monoceros. Monoceros is to be found to the east of Orion, sandwiched between the

Fig. 4.40 Image of Messier 50, an open cluster in Monoceros

constellations of Canis Minor to the north and Canis Major to the south. The cluster is not too hard to find, as it lies just 10° away from the bright star Sirius. In his description of the location of M50, Messier referred to the star θ in Canis Major. This is marked on the chart just to the right of the arrow pointing to M50. Monoceros does not have any other Messier objects to view – only Messier 50. It does, however, contain four Caldwell objects. They are C46, otherwise known as NGC 2261 or Hubble's Variable Nebula, C49 (NGC 2237), which is the Rosette Nebula, C50 (NGC 2244), which is an open cluster embedded in the Rosette Nebula and C54 (NGC 2506), an open cluster originally discovered by William Herschel in 1791.

The image was taken at local time 22h 45m 00 on December 19 from Siding Spring. M50 has a Right Ascension of 07h 03m 38s. At the time the image was taken, the Local Sidereal Time was 03h 35m 02s. The difference between these is 3h 28m 36s. This puts the meridian crossing time at 02h 13m 36s the next morning. The altitude of M50 was +36° 30' 01" and the Azimuth angle 76° 20' 53". The telescope used to take the image was a large 510-mm Planewave Corrected Dall-Kirkham with a focal length of 2,259 mm.

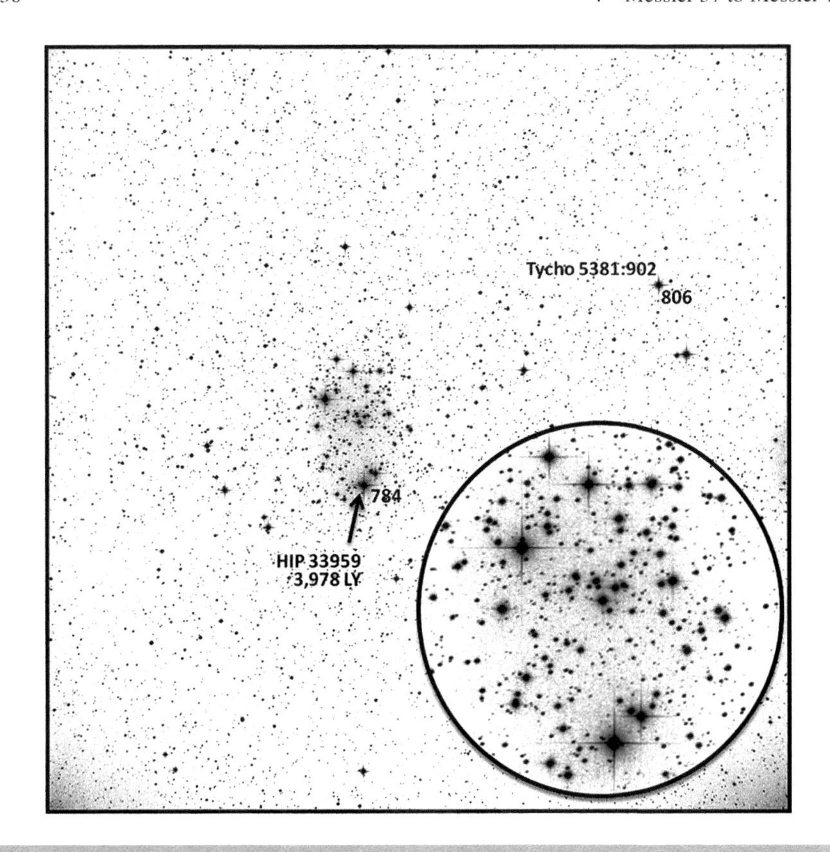

Fig. 4.41 Negative image of Messier 50

Messier 50: Data relating to the image in Figure 4.40

REMOTELY IMAGED FROM NEW SOUTH WALES, AUSTRALIA
NGC 2323
OBJECT TYPE: Open Cluster

RA: 07h 03m 38s
DEC: -08° 24' 39"
ALTITUDE: +36° 30' 01"
AZIMUTH: 76° 20' 53"
FIELD OF VIEW: 0° 55' 43" x 0° 55' 43"
OBJECT SIZE: 16'
POSITION ANGLE: 90° 47' from North
EXPOSURE TIME: 300 s
DATE: 19th December
LOCAL TIME: 22h 45m 00s
UNIVERSAL TIME: 11h 45m 00s
SCALE: 1.09 arcsec/pixel

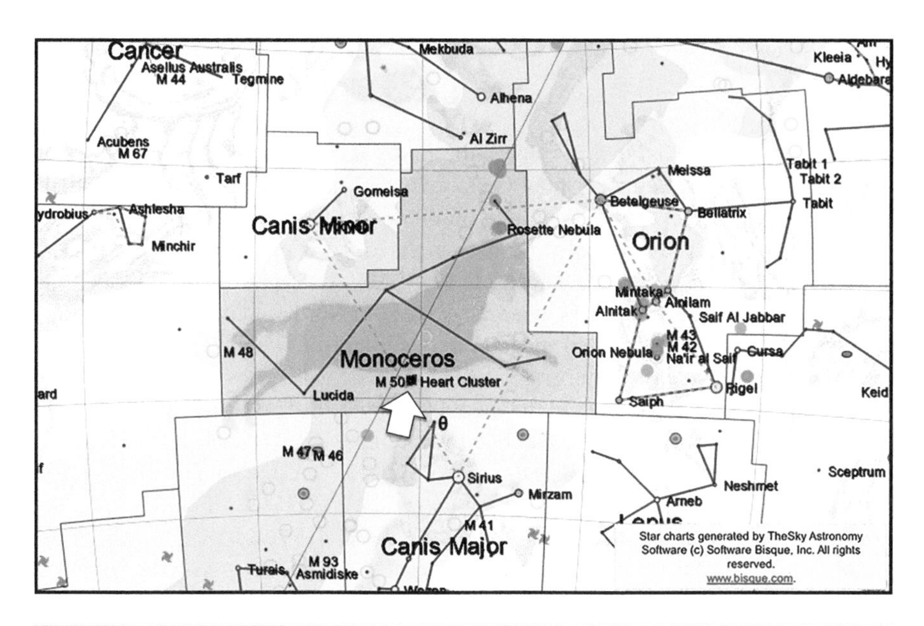

Fig. 4.42 Chart showing the location of M50

MOON PHASE: 65.85% (waning)

Telescope Optics
OTA: Planewave 20" (0.51m) CDK
Optical Design: Corrected Dall-Kirkham Astrograph
Aperture: 510 mm
Focal Length: 2259 mm (focal reducer)
F/Ratio: f/4.4
Guiding: Active Guiding Disabled
Mount: Planewave Ascension 200HR

Instrument Package
CCD: PL09000 CCD Camera
Pixel Size: 12μm square
Sensor: Frontlit
Cooling: -30°C default
Array: 3056 x 3056 pixels
FOV: 55.9 x 55.9 arcmin

Location
Observatory: Siding Spring
UTC +10.00 (Australia Daylight savings time is observed)
31° 16' 24" South
149 03' 52" East
Elevation: 1122 meters (3681 ft)

Messier 51

Messier 51, shown in Figure 4.43, was the first galaxy recognized to be spiral in form and is known as the Whirlpool Galaxy. It is 37 million ly distant. It was Irish astronomer Lord Rosse who made this observation with his large reflecting telescope in 1845. It had originally been discovered by Messier in October 1773. Regarding his observation of January 11, 1774, Messier reported that M51 was a: "Very faint nebula, without stars, near the eye of the Northern Greyhound [hunting dog], below the star Eta of 2nd magnitude of the tail of Ursa Major: M. Messier discovered this nebula on October 13, 1773, while he was watching the comet visible at that time. One cannot see this nebula without difficulties with an ordinary telescope of 3.5 foot." M51 is in fact two interacting galaxies. Each galaxy is given its own NGC number. The larger galaxy is NGC 5194 and the smaller NGC 5195. They are occasionally called M51A (NGC 5194) and M51B (NGC 5195).

The negative image in Figure 4.44 identifies NGC 5194 and NGC 5195. I used my software to measure the angular distance between the two centers. This was approximately 4' 30" of arc. Measuring the full length of M51 in the long axis gave about 11' to 15' -- it is very difficult to precisely define the start and end points. If I take the distance as 37 million ly and assume 11', I calculate an actual size in the region of 118,000 ly. This seems reasonable in comparison with our own Galaxy,

Fig. 4.43 Image of Messier 51, a spiral galaxy in Canes Venatici

which has a size of about 100,000 ly. It must be an interesting "Milky Way" in M51 for the residents of a planet in NGC 5195 looking towards NGC 5194. Two stars of similar magnitude have been identified on the negative image. The first is magnitude 11.37 UCAC4 686:54910 (the upper star in the pair), and the second is the magnitude 11.01 UCAC4 686:54909 (the lower of the pair). One can just barely distinguish the slight difference in magnitudes from the image.

The chart in Figure 4.45 shows the position of Messier 51 in the constellation of Canes Venatici. The star η Ursa Major lies not too far from M51. This star is better known as Alkaid. The star to the right of Alkaid in the Big Dipper asterism is the well-known multiple star, Mizar. There are four other Messier objects in Canes Venatici. These are M3, a fine globular cluster, M63, the spiral "Sunflower Galaxy" and M94 and M106, both spiral galaxies. The small black rectangle is the actual image superimposed on the chart. The chart is drawn with the image horizontal. As a result, north is not at the top. This is because the camera was not aligned with north when it was positioned on the telescope.

The image was taken on January 8 at 06h 46m 42s local time in Spain. This corresponded to a Universal Time of 05h 46m 42s. The Right Ascension is 13h 30m 26s, and the Local Sidereal Time was 12h 50m 52s. Subtracting one from the other gives a difference of 0h 39m 34s, indicating the time that had to go before M13 crossed the meridian. Adding this time to the local time the image was taken, the meridian transit time of M51 would be 07h 26m 16s. The telescope used to take the

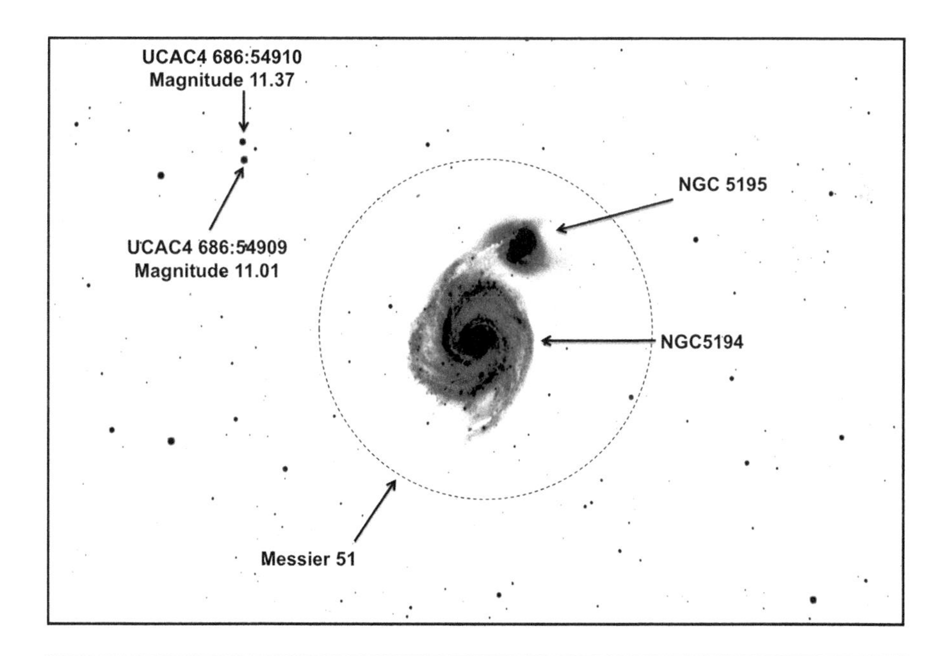

Fig. 4.44 Negative image of Messier 51

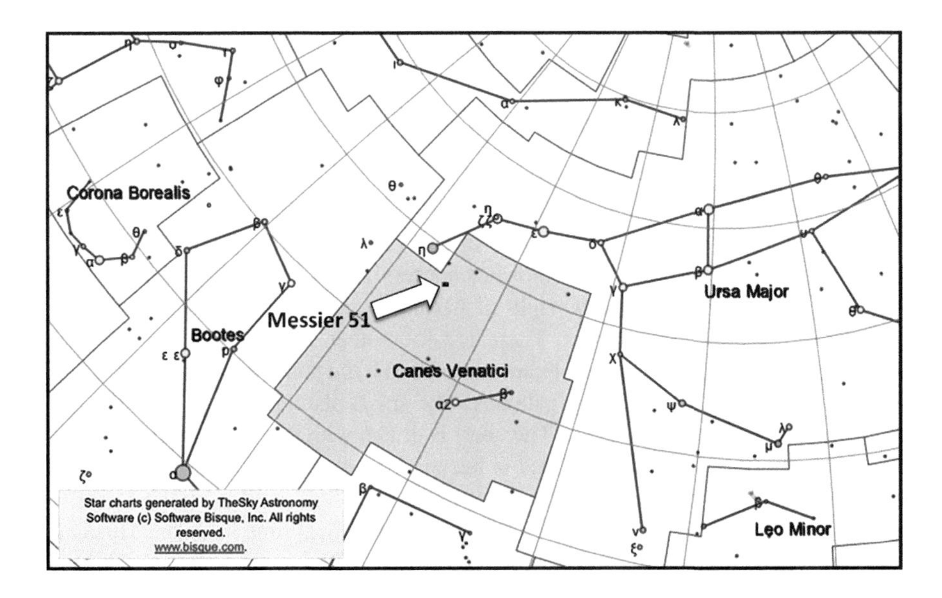

Fig. 4.45 Chart showing the location of M51

image of M51 was a Corrected Dall-Kirkham Astrograph with an aperture of 318 mm mounted on a Paramount ME.

Messier 51: Data relating to the image in Figure 4.43

REMOTELY IMAGED FROM SPAIN
NGC 5194 (WHIRLPOOL GALAXY)
OBJECT TYPE: Spiral Galaxy

RA: 13h 30m 26s
DEC: +47° 07' 23"
ALTITUDE: +77° 36' 49"
AZIMUTH: 33° 00' 48"
FIELD OF VIEW: 0° 37' 16" x 0° 24' 51"
OBJECT SIZE: 11' X 7'
POSITION ANGLE: 220° 19' from North
EXPOSURE TIME: 300 s
DATE: 8th January
LOCAL TIME: 06m 46m 42s
UNIVERSAL TIME: 05h 46m 42s
SCALE: 0.73 arcsec/pixel
MOON PHASE: 17.78% (waning)

Telescope Optics
OTA: Planewave CDK
Optical Design: Corrected Dall-Kirkham Astrograph

Aperture: 318 mm
Focal Length: 2541 mm
F/Ratio: f/7.9
Guiding: External
Mount: Paramount ME

Instrument Package
CCD: SBIG-STXL-6303E
A/D Gain: 1.47 e-/ADU
Pixel Size: 9μm
Resolution: 0.73 arcsec/pixel
Sensor: Front Illuminated
Cooling: -20°C default
Array: 3072 x 2048 (6.3 Megapixels)
FOV: 37.41 x 24.94 arcmin
Location
Observatory: AstroCamp – MPC Code -189
UTC +1.00 (Daylight savings time is observed)
Minimum Target Elevation: Approx 30 – 40 Degrees
N 38° 09'
W 002° 19'
Elevation: 1650 meters (5413 ft)

Messier 52

Messier 52, shown in Figure 4.46, is an outstanding open cluster in the constellation of Cassiopeia. It has an apparent diameter of 13' of arc and is believed to lie at a distance of 5,000 ly. This would give M52 an actual diameter of 19 ly. The image gives a good idea of the range of star colors in the frame. Just out of the frame at the bottom right of the image is the nebula known as the "Bubble Nebula". This is part of Caldwell C11, otherwise known as NGC 7635. Charles Messier discovered the M52 cluster on September 7, 1774. W.H. Smyth commented that M52 was "An irregular cluster of stars between the head of Cepheus and his daughter's throne; it lies north-west-by-west of Beta Cassiopeiae, and one third of the way towards Alpha Cephei. This object assumes somewhat of a triangular form, with an orange-tinted 8th-magnitude star at its vertex, giving it the resemblance of a bird with outspread wings. It is preceded by two stars of 7th and 8th magnitudes, and followed by another of similar brightness; and the field is one of singular beauty under a moderate magnifying power."

The negative image in Figure 4.47 identifies a couple of stars in the frame. The bright yellow/orange star on the right is HIP 115542 and is an F7 star with an approximate temperature of 6,400 K. The Hipparcos Catalog places it at 3,397.47 ly, in which case it is a foreground star not part of the cluster. The other star to the

Fig. 4.46 Image of Messier 52, an open cluster in Cassiopeia

southeast of M52 is HIP 115661, which is magnitude 7.82 and is an A5 star having a temperature in the region of 8,100 K. Robert Burnham Jr. described M52 as "one of the richer and more compressed clusters with a computed density of somewhat over 3 stars per cubic parsec, rising to more than 50 stars per cubic parsec near the centre" (1 Parsec = 3.26 ly).

The chart in Figure 4.48 shows the position of M52 in the constellation of Cassiopeia. It lies in a rich part of the Milky Way about 6° from the magnitude 2.27 star Caph (HIP 746), or 11-Beta Cassiopeiae, and only 38 arcmin from the H II region, Caldwell 11, as mentioned above. There is only one other Messier object in Cassiopeia, which is about 15° away from M52. There are 5 additional Caldwell objects in Cassiopeia as well as C11. These are open clusters C8, C10 and C13, and elliptical galaxies C17 and C18.

The image was taken on October 3 at 23h 36m 59s local time in New Mexico. This corresponded to a Universal Time of 05h 36m 59s (October 4). The Right Ascension of M52 is 23h 25m 38s, and the Local Sidereal Time was 23h 27m 06s. Clearly, the cluster was close to the meridian, as the RA and LST are very similar values. The LST is higher than the RA, so M52 must have already crossed the

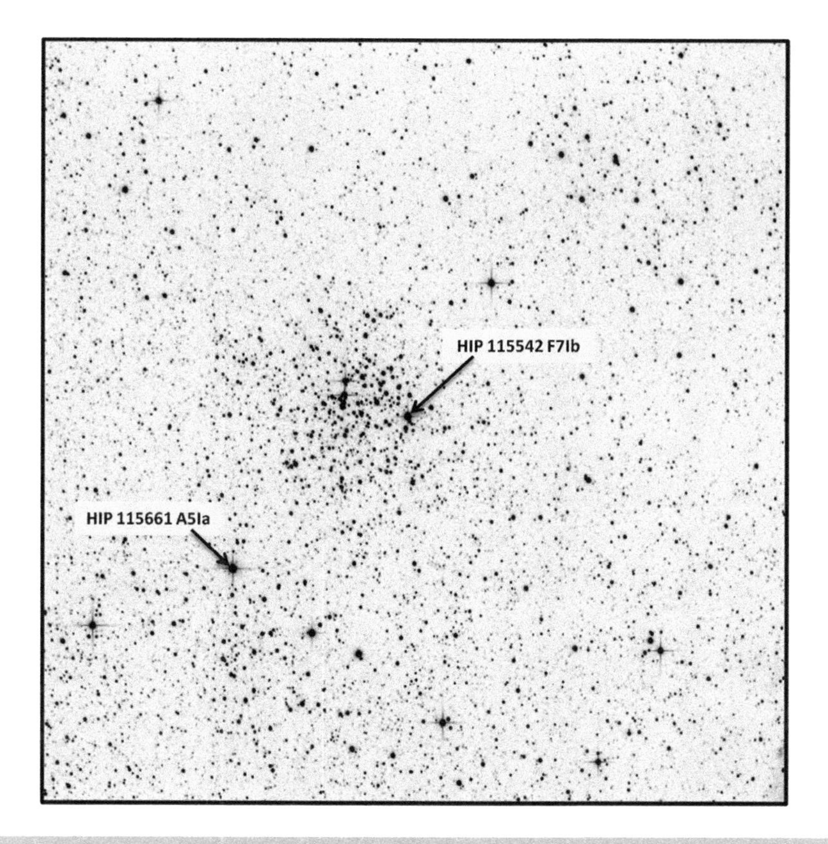

Fig. 4.47 Negative image of M52

meridian from east to west. This can be seen clearly on the chart. Subtracting one from the other gives a difference of 0h 1m 28s. Subtracting this from the local time when the image was taken gives a meridian transit time of 23h 35m 21s. The telescope used to take the image of M52 was a Takahashi TOA-150 apochromatic refractor with an aperture of 150 mm.

Messier 52: Data relating to the image in Figure 4.46

REMOTELY IMAGED FROM MAYHILL, NEW MEXICO
NGC 7654
OBJECT TYPE: Open Cluster

RA: 23h 25m 38s
DEC: +61° 40' 37"
ALTITUDE: +61° 06' 20"
AZIMUTH: 359° 38' 29"
FIELD OF VIEW: 47' 04"' X 47' 04"

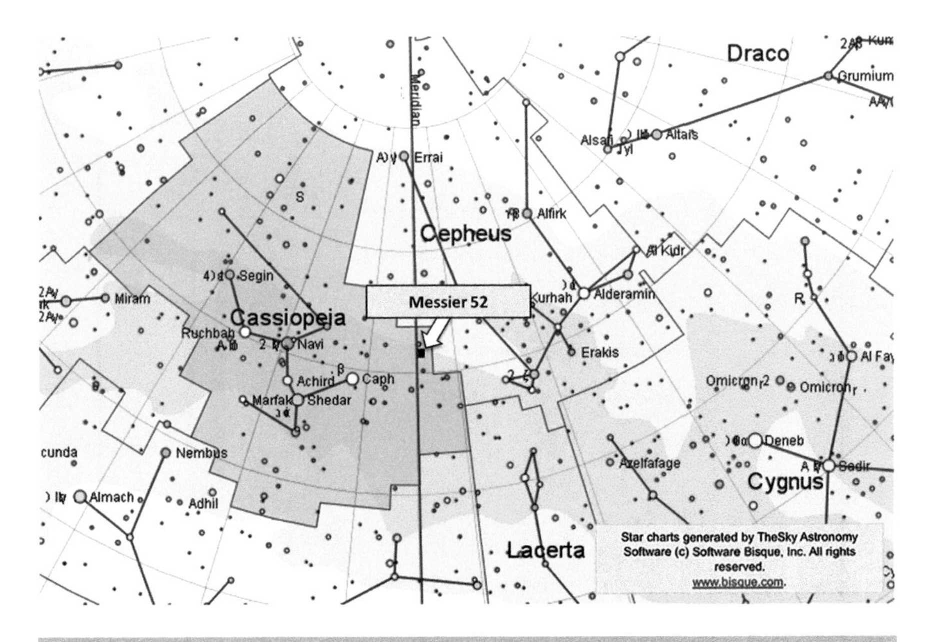

Fig. 4.48 Chart showing the location of M52

OBJECT SIZE: 13'
POSITION ANGLE: 358° 29' from North
EXPOSURE TIME: 600 s
DATE: 3rd October
LOCAL TIME: 23h 36m 59s
UNIVERSAL TIME: 05h 36m 59s
SCALE: 1.38 arcsec/pixel
MOON PHASE: 9.39% (waxing)

Telescope Optics
OTA: Takahashi TOA-150
Optical Design: Apochromatic Refractor
Aperture: 150 mm
Focal Length: 1095 mm
F/Ratio: f/7.3
Guiding: Internal
Mount: Paramount GTS

Instrument Package
SBIG ST-4000M One-Shot Color CCD
A/D Gain: 0.6e-/ADU
Pixel Size: 7.4um square
Resolution: 1.45 arcsec/pixel

Sensor: Frontlit
Cooling: -20°C Winter (-10°C Summer)
Array: 2048 x 2048 (8.3 Megapixels)
FOV: 49.6 x49.6 arcmin

Location
Observatory: New Mexico Skies
UTC Minus 7.00 (Daylight savings time is observed)
Minimum Target Elevation: Approx 25 – 45 Degrees
(N or S) 32° 54’ Decimal: 32.9 North
(W or E) 105° 31’ Decimal: 105.5 West
Elevation: 2225 meters (7298 ft)

Messier 53

Messier 53, shown in Figure 4.49, is a fine globular cluster in the constellation of Coma Berenices. It has an apparent diameter of 13’ of arc and is at a distance of 58,000 ly. With these values, it has a true diameter of 219 ly. Messier observed it on February 26, 1777: “Nebula without stars discovered below & near Coma Berenices, a little distant from the star 42 in that constellation, according to Flamsteed. This nebula is round and conspicuous.” The cluster had been discovered two years earlier by Bode who commented: “On February 3, 1775, early in the morning, I discovered a nebula north of the star Epsilon or *Vindemiatrix* at the northern wing of Virgo, about 1 degree east of the 42nd star of Coma Berenices, which appears through the telescope rather vivid and of round shape.” It has a smaller apparent size to M3, which is 25’, in comparison with M53’s size of 13’.

The negative image in Figure 4.50 shows that north is to the left and east is down. To get an idea of the scale of the globular cluster, the distance in arcmin is shown between the center of the cluster and two stars. The distance to Tycho 1454:308 is 9’ and the distance to HIP 64386 is 20.5’. The stars in the globular cluster itself appear to be magnitude 14 or fainter, as evidenced by the star UCAC4 542:52543 arrowed on the negative image. This star has a magnitude of 14.41. M53 has a nearby, non-identical, twin cluster NGC 5053. This globular cluster is only 1° away from M53 but could easily be mistaken for a galactic open cluster, as it does not have the usual appearance of increasing stellar concentration towards the center.

The chart in Figure 4.51 shows the location of M53 in the constellation of Coma Berenices. It is to be found less than 1° from the star 42-Alpha Comae Berenices, (Diadem or HIP 64241), as pointed out by both Messier and Bode. Bode also mentioned the star Vindemiatrix (47-Epsilon Virginis or HIP 63608) to the south of M53. The angular separation between them is just less than 8°. Coma Berenices contains seven more Messier objects – all of them spiral galaxies. They are M64,

Fig. 4.49 Image of Messier 53, a globular cluster in Coma Berenices

M85, M88, M91, M98, M99 and M100. There are three Caldwell objects in the constellation. C36 and C38 are spiral galaxies, and C35 is an elliptical galaxy.

The image was taken on February 21 at 04h 38m 37s local time in New Mexico. This corresponded to a Universal Time of 11h 38m 37s. The Right Ascension of M53 is 13h 13m 46s, and the Local Sidereal Time was 14h 41m 41s. So, subtracting one from the other gives a difference of 1h 27m 55s. Thus, the meridian transit time works out to be 03h 10m 42s. The telescope used to take the image of M53 was a Corrected Dall-Kirkham Astrograph with an aperture of 510 mm, mounted on a Plane Wave Ascension 200HR mount.

Messier 53: Data relating to the image in Figure 4.49

REMOTELY IMAGED FROM: New Mexico
NGC 5024
OBJECT TYPE: Globular Cluster
RA: 13h 13m 46s
DEC: +18° 04' 38"
ALTITUDE: +65° 23' 22"
AZIMUTH: 238° 41' 04"
FIELD OF VIEW: 0° 54' 47" x 0° 36' 31"
OBJECT SIZE: 13'
POSITION ANGLE: 271° 09' from North
EXPOSURE TIME: 300 s

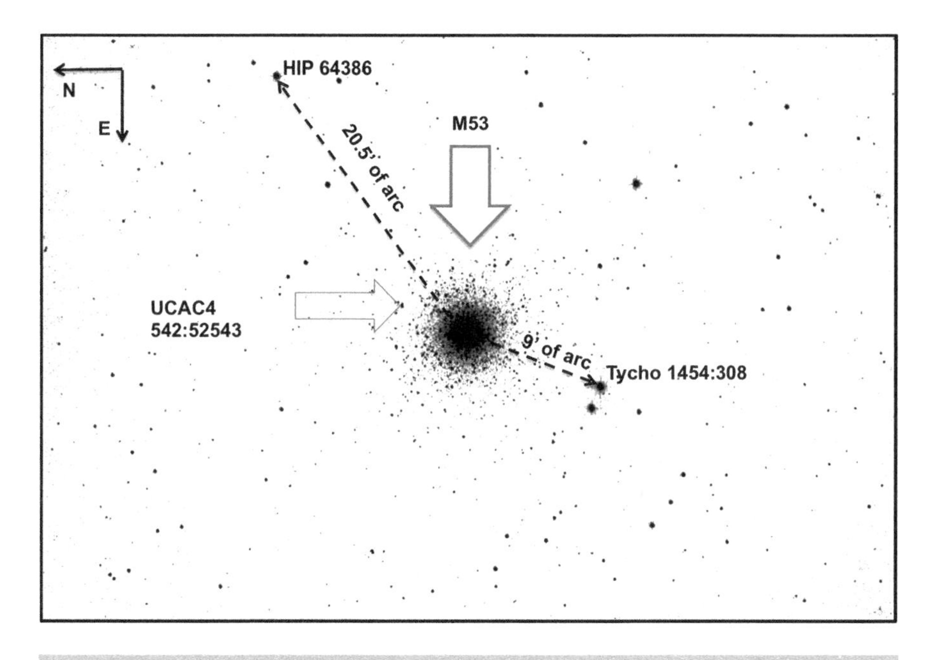

Fig. 4.50 Negative image of Messier 53

DATE: 21st February
LOCAL TIME: 04h 38m 37s
UNIVERSAL TIME: 11h 38m 37s
MOON PHASE: 25.76% (waning)

Telescope Optics
OTA: Planewave 20" CDK
Optical Design: Corrected Dall-Kirkham Astrograph
Aperture: 510 mm
Focal Length: 2280 mm (0.66 Focal Reducer Fitted)
F/Ratio: f/4.5
Guiding: Active Guiding Disabled
Mount: Planewave Ascension 200HR

Instrument Package
FLI Proline PL11002M CCD
Camera Pixel Size: 9 µm square
Resolution: 0.81 arcsec/pixel
Cooling: -30°C default
Array: 4008 x 2672 (10.7 Megapixels)

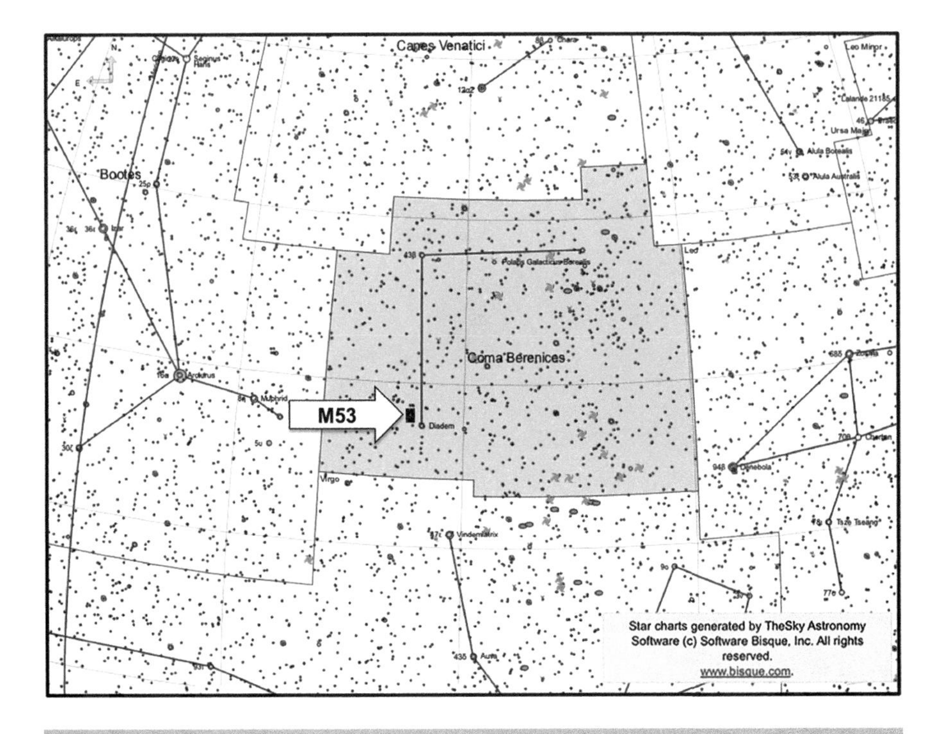

Fig. 4.51 Chart showing the location of M53

Location
Observatory: New Mexico Skies
UTC Minus 7.00 (Daylight savings time is observed)
Minimum Target Elevation: Approx 25 – 45 Degrees
(N or S) 32° 54’ Decimal: 32.9 North
(W or E) 105° 31’ Decimal: 105.5 West
Elevation: 2225 meters (7298 ft)

Messier 54

Messier 54, shown in Figure 4.52, is a globular cluster in the constellation of Sagittarius. It may appear to be just another globular in the Galaxy, but that is not the case. It certainly is a globular cluster, but in reality, it belongs to another galaxy altogether, the “Sagittarius Dwarf Elliptical Galaxy,” which is being absorbed into our own Milky Way Galaxy. Messier referred to his observation of M54 as follows: “Very faint nebula, discovered in Sagittarius; its center is brilliant & it contains no star, seen with an achromatic telescope of 3.5 feet. Its position has been determined from Zeta Sagittarii, of 3rd magnitude.” John Herschel noted: “Globular cluster;

Fig. 4.52 Image of Messier 54, a globular cluster in Sagittarius

very bright; large; round; gradually, then suddenly much brighter toward the middle; well resolved; stars of 15th magnitude." The image was taken with a 17-inch corrected Dall-Kirkham Telescope. This very telescope has been used to obtain images of the most distant object ever taken by amateur astronomers. The distant quasar was imaged on this telescope after 16 total hours of exposure. This is a telescope that you can use if you want to examine a distant, faint target, but I would not recommend any exposures adding up to that number of hours unless you are quite rich.

The negative image in Figure 4.53 shows M54 on a frame that is only about 16 arcmin wide. M54 sits in a busy region of the Milky Way, so there is a dense concentration of stars. The overexposed star at bottom right, Tycho 7409:106, is only magnitude 10.14, but bright enough to saturate the sensitive telescope and camera set up. Most of the images in this book have been taken with cameras that have an Anti Blooming Gate (ABG) that prevents oversaturation, or "blooming." The camera on this telescope does not (Non-ABG or NABG). It is mainly used as a scientific instrument, so the brightness of a star on the image is proportional to the amount of light being received in the camera. If you have an ABG camera, indi-

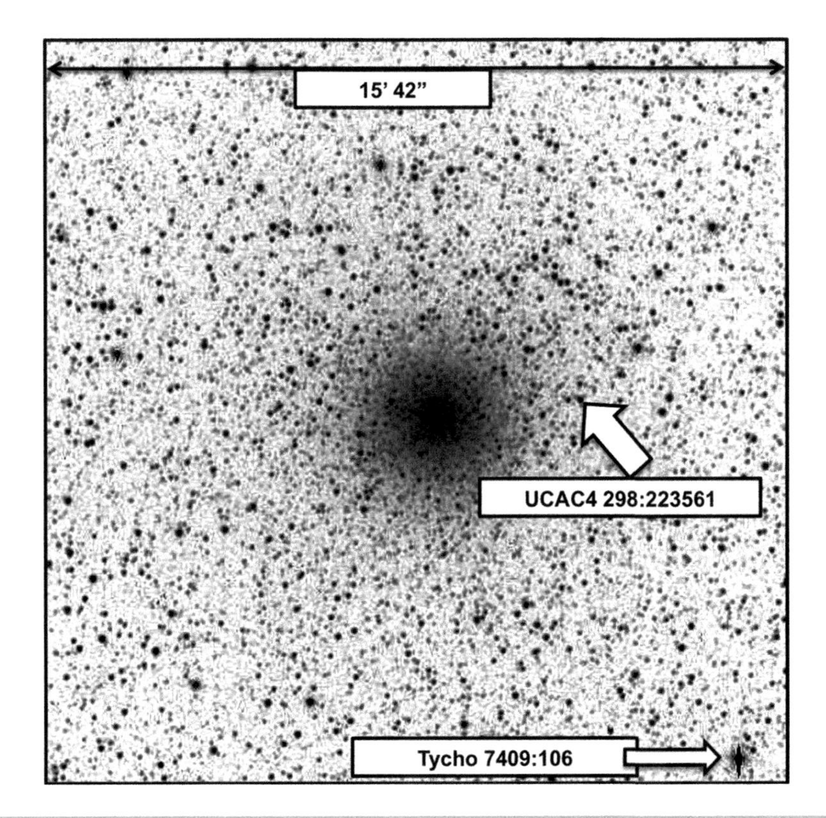

Fig. 4.53 Negative image of Messier 54

vidual stars may be saturated, giving a false brightness level. The magnitude of the arrowed star, UCAC4 298:223561, is 13.97.

The chart in Figure 4.54 shows the position of M54 in the constellation of Sagittarius. Perhaps we should call this the "Tea Leaf Globular," as it sits on the bottom of the Sagittarius teapot asterism. Of course, this could be applied to a number of objects in this area! The size of the image is quite small in comparison with many of the objects imaged in this book. The roughly 16' square image is tiny in comparison with the image of M24, which has a frame size of almost 4° x 4°, or 240' square. The enlarged inset to the chart was necessary to show the location of M54. There are other globulars in Sagittarius. There are 19 NGC objects, all globular clusters.

The image was taken at the local time 22h 33m 12s on the morning of October 6 from New South Wales. The corresponding Universal Time was 11h 33m 12s. Messier 54 has a Right Ascension of 18h 56m 07s and the local Sidereal Time was 22h 31m 27s, giving a difference of 3h 35m 20s. This means that M54 has already crossed the meridian from east to west at 18h 57m 52s. The telescope/camera setup is extremely sensitive, so exposures should be short. This image was a 5-minute exposure. Higher exposures are likely to be saturated.

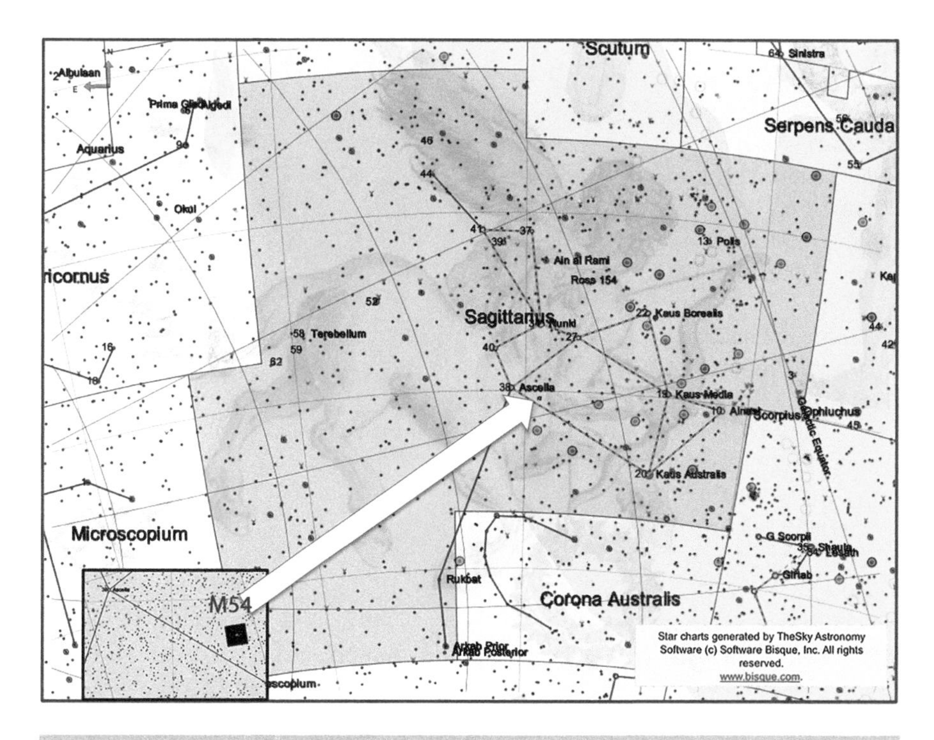

Fig. 4.54 Chart showing the location of M54

Messier 54: Data relating to the image in Figure 4.52

REMOTELY IMAGED FROM NEW SOUTH WALES, AUSTRALIA
NGC 6715
OBJECT TYPE: Globular Cluster

RA: 18h 56m 07s
DEC: -30° 27' 15"
ALTITUDE: +44° 15' 14"
AZIMUTH: 256° 18' 37"
FIELD OF VIEW: 0° 16' 08" x 0° 15' 42"
OBJECT SIZE: 12’
POSITION ANGLE: 188° 00' from North
EXPOSURE TIME: 300 s
DATE: 6th October
LOCAL TIME: 22h 33m 12s
UNIVERSAL TIME: 11h 33m 12s
SCALE: 0.92 arcsec/pixel
MOON PHASE: 25.03% (waxing)

Telescope Optics
OTA: Planewave 17" CDK
Optical Design: Corrected Dall – Kirkham Astrograph
Aperture: 431 mm
Focal Length: 2912 mm
F/Ratio: f/6.8
Guiding: External
Mount: Paramount PME

Instrument Package
CCD: FLI Proline PL4710
Pixel Size: 13µm square
Sensor: E2V CCD47-10-1-109 Deep Depletion Fused Silica
Cooling: -35°C default
Array: 1024 x 1024 pixels
FOV: 15.5 x 15.5 arcmin

Location
Observatory: Siding Spring
UTC +10.00 (Australia Daylight savings time is observed)
31° 16' 24" South
149° 03' 52" East
Elevation: 1122 meters (3681 ft)

Messier 55

Messier 55, shown in Figure 4.55, is a globular cluster in the constellation of Sagittarius. It has an apparent diameter of 19' of arc and is believed to lie at a distance of 17,300 ly. With those figures, I estimate an actual diameter of just less than 96 ly. It is smaller than the globular cluster M13, which has a diameter of 150 ly, but larger than M12 at 75 ly. Messier observed M55 on July 28 1778, noting, "A nebula which is a whitish spot, of about 6' extension, its light is even and does not appear to contain any star. Its position has been determined from Zeta Sagittarii, with the use of an intermediate star of 7th magnitude. This nebula has been discovered by M. l'Abbe de LaCaille." This French astronomer had in fact discovered it 26 years earlier on June 26, 1752. He saw it as "an obscure nucleus of a big comet."

The image had a position angle of 92° 17', so it has been rotated to place north at the top, as shown in the negative image in Figure 4.56. The angular size of the frame is shown as 43' 54", which can be used to get an idea of the size of the globular cluster. At the distance of M55 the width of the frame is equivalent to about 205 ly, as shown. To get an idea of stellar magnitudes, I have identified the star UCAC4 296:225403, which has a magnitude of 14.37. M55 is in Sagittarius, which is difficult for more northern observers but was at a good altitude from Siding Spring.

Fig. 4.55 Image of Messier 55, a globular cluster in Sagittarius

The chart in Figure 4.57 shows the position of M55 in the constellation of Sagittarius, with the square image superimposed on the chart. Messier had used the magnitude 2.6 star Ascella, or 38-Zeta Sagittarii as a reference, which is given the designation HIP 93506 in the Hipparcos Catalog. The distance from the center of the image to Astella is 8°10'. The star Rukbat is Alpha Sagittarii. There is an angular distance of 10° 10' from M55 to this magnitude 3.96 star. This star is HIP 95347 in the Hipparcos Catalog.

The image was taken at local time 22h 12m 44s on October 6 in Siding Spring. The corresponding Universal Time was 11h 12m 44s. Messier 55 has a Right Ascension of 19h 41m 03s, and the Local Sidereal Time was 22h 10m 55s. This means that M55 had already crossed the meridian from east to west some hours earlier. The difference between RA and LST is 2h 29m 52s, making the meridian crossing time 19h 42m 52s. The telescope used was a 431-mm Planewave instrument.

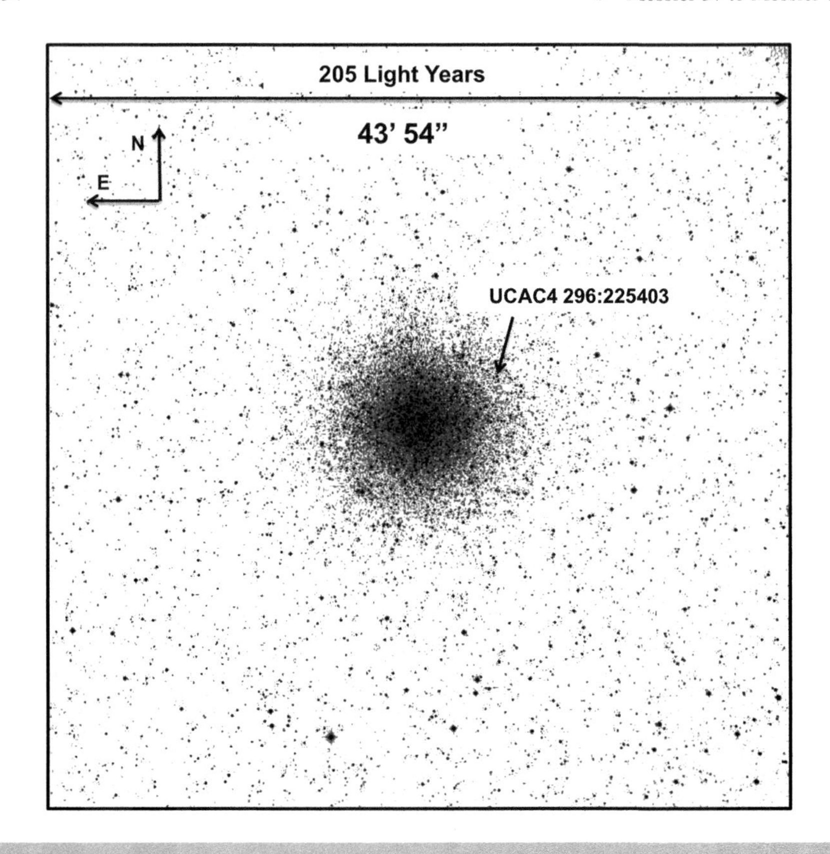

Fig. 4.56 Negative image of Messier 55

Messier 55: Data relating to the image in Figure 4.55

REMOTELY IMAGED FROM NEW SOUTH WALES, AUSTRALIA
NGC 6809
OBJECT TYPE: Globular Cluster

RA: 19h 41m 03s
DEC: -30° 55' 16"
ALTITUDE: +58° 03' 57"
AZIMUTH: 260° 37' 05"
FIELD OF VIEW: 0° 43' 54" x 0° 43' 54"
OBJECT SIZE: 19'
POSITION ANGLE: 92° 17' from North
EXPOSURE TIME: 300 s
DATE: 6th October
LOCAL TIME: 22h 12m 44s
UNIVERSAL TIME: 11h 12m 44s
SCALE: 0.64 arcsec/pixel
MOON PHASE: 24.91% (waxing)

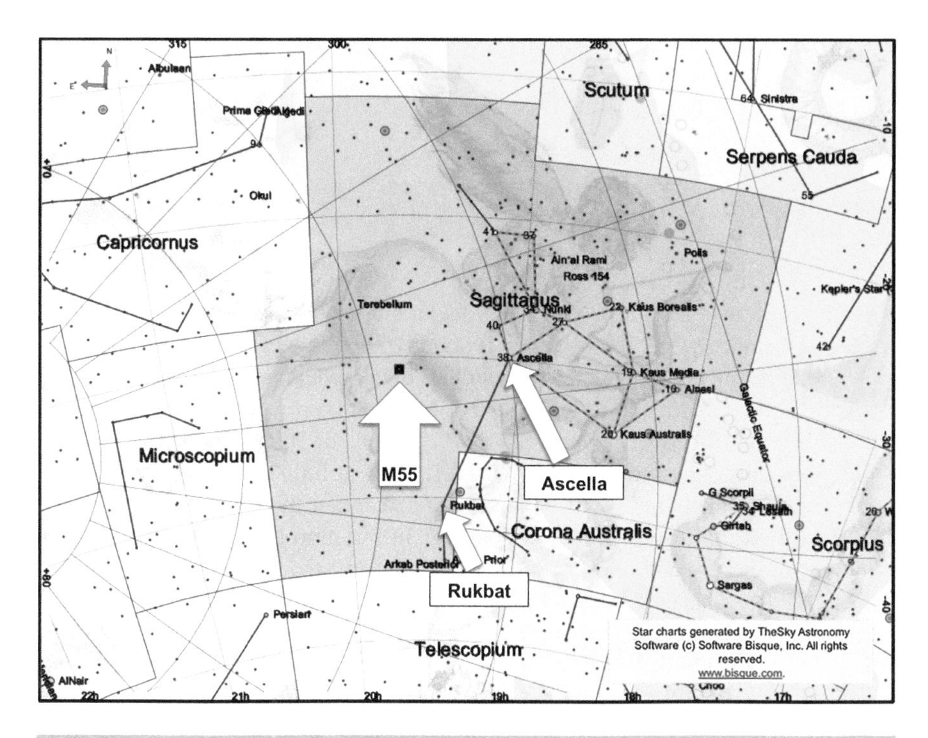

Fig. 4.57 Chart showing the location of M55

Telescope Optics
OTA: Planewave 17" CDK
Optical Design: Corrected Dall-Kirkham Astrograph
Aperture: 431 mm
Focal Length: 2912 mm
F/Ratio: f/6.8
Guiding: Active Guiding Disabled
Mount: Planewave Ascension 200HR

Instrument Package
CCD: FLI Proline 16803
Pixel Size: 9μm square
Sensor: KAF – 16803
Cooling: -35°C default
Array: 4096 x 4096 pixels
FOV: 43.2 x 43.2 arcmin

Location
Observatory: Siding Spring, New South Wales, Australia
UTC +10.00 (Australia Daylight savings time is observed)

31° 16' 24" South
149° 03' 52" East
Elevation: 1122 meters

Messier 56

Messier 56, shown in Figure 4.58, is a globular cluster in the constellation of Lyra. It lies at a distance of 32,900 ly, which is about 1/3 the size of the Galaxy and quite a way out. The apparent diameter is about 8.8' of arc, which would give it an actual diameter of 84.2 ly for the distance stated. M56 is one of Charles Messier's own discoveries, confirmed on January 23 1779. He noted, "On the 19th of January, 1779, when observing the Comet C/1779 A1, I saw at little distance of it, & on its parallel, a very faint nebula, which one cannot perceive without a refractor, the dusk then prevented me to determine its position: In the morning of the 23rd, I have compared it directly with the second star of Cygnus of the fifth magnitude."

The negative image of M56 in Figure 4.59 identifies two stars to give an indication of stellar magnitudes. They are UCAC4 601:74662, a magnitude 10.25 star and HIP 94790, a magnitude 8.97 star. In the image, north is to the left and east is down. This is because the position angle is approximately 272° from north. To get north to the top, the image needs to be rotated by roughly 88° clockwise. Stars in the cluster are much fainter, with the brightest stars being in the region of magnitude

Fig. 4.58 Image of Messier 56, a globular cluster in Lyra

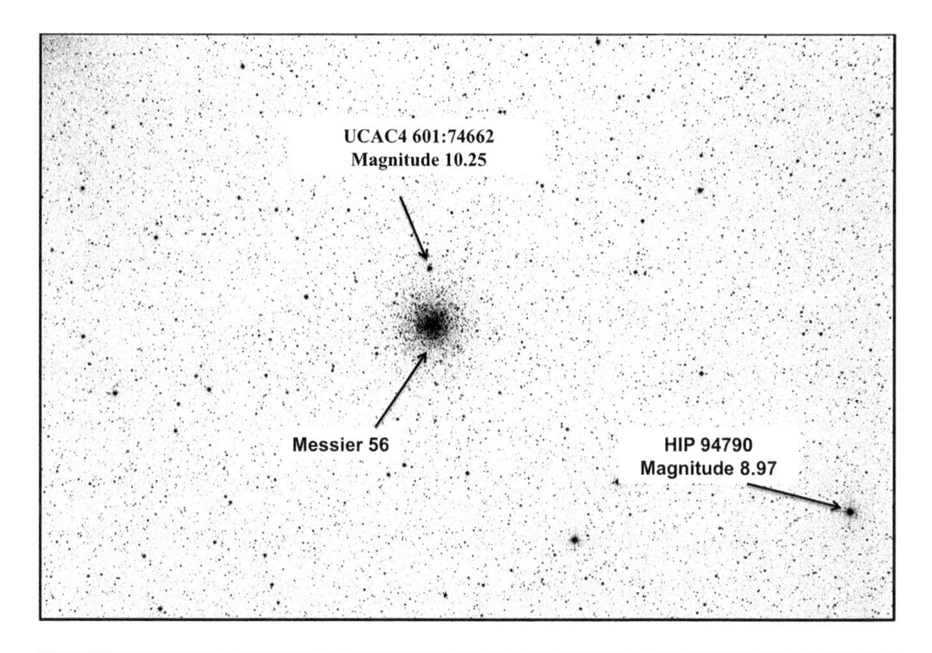

Fig. 4.59 Negative image of Messier 56

13. There are no other globular clusters in Lyra. The only other Messier object is M57, the Ring Nebula. There are no Caldwell objects in Lyra.

The chart in Figure 4.60 shows the location of M56 in the constellation of Lyra. There is an angular separation of 5° 40' between M56 and the Ring Nebula M57, 4° 30' to the star Sulafat (14-Gamma Lyrae) and almost 12° to Vega (3-Alpha Lyrae). You will see on the chart that it has been rotated to bring north to the top. Lyra has Cygnus to the east and Hercules to the west. Vulpecula is to the southeast and Draco is to the north.

The image was taken on September 25 at the local time of 22h 49m 48s. The corresponding Universal Time was 20h 49m 48s. The Local Sidereal Time was 21h 02m 34s, and the Right Ascension of M56 is 19h 17m 15s. This tells us that M56 has already crossed the meridian. The difference between LST and RA is 1h 44m 45s, meaning that M56 had a transit time at 21h 5m 3s. The telescope used to take the image from Spain was a 431-mm Corrected Dall-Kirkham Astrograph.

Messier 56: Data relating to the image in Figure 4.58

REMOTELY IMAGED FROM SPAIN
NGC 6779
OBJECT TYPE: Globular Cluster

RA: 19h 17m 15s
DEC: +30° 13' 21"

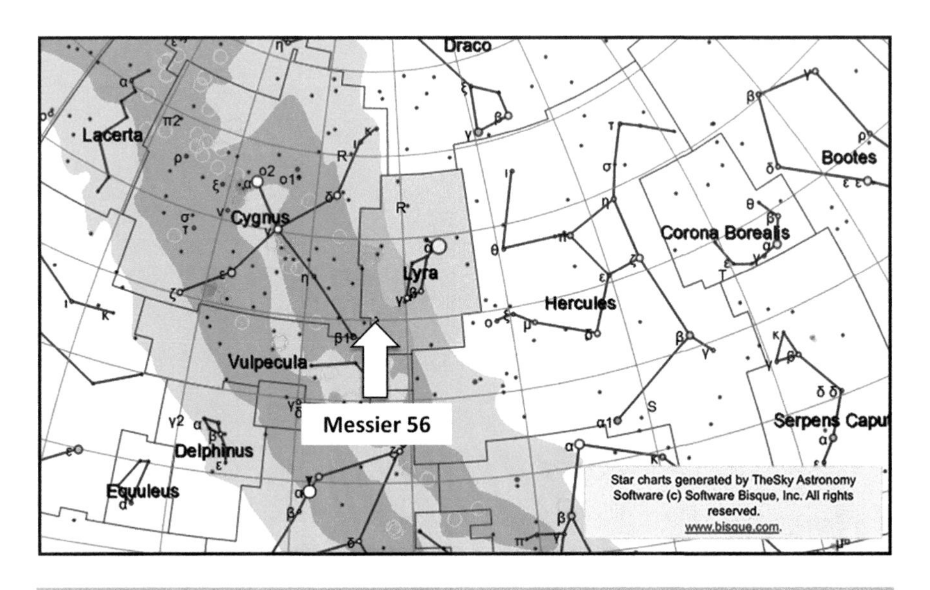

Fig. 4.60 Chart showing the location of M56

ALTITUDE: +67° 06' 53"
AZIMUTH: 260° 14' 21"
FIELD OF VIEW: 0° 42' 13" x 0° 28' 09"
OBJECT SIZE: 8.8'
POSITION ANGLE: 272° 12' from North
EXPOSURE TIME: 300 s
DATE: 25th September
LOCAL TIME: 22h 49m 48s
UNIVERSAL TIME: 20h 49m 48s
SCALE: 0.63 arcsec/pixel
MOON PHASE: 24.66% (waning)

Telescope Optics
OTA: Planewave 17" CDK
Optical Design: Corrected Dall – Kirkham Astrograph
Aperture: 431 mm
Focal Length: 2929 mm
F/Ratio: f/6.8
Guiding: External
Mount: Paramount ME

Instrument Package
SBIG STL - 1100M
A/D Gain: 2.2 e-/ADU

Pixel Size: 9µm square
Resolution: 0.63 arcsec/pixel
Sensor: Frontlit
Cooling: -20°C default
Array: 4008 x 2672 (10.7 Megapixels)
FOV: 28.2 x 42.3 arcmin

Location
Observatory: AstroCamp – MPC Code – I89
UTC +1.00 (Daylight savings time is observed)
Minimum Target Elevation: Approx 35 – 45 Degrees
North: 38° 09'
West 0002° 19'
Elevation: 1650 meters (5413 ft)

Messier 57

Messier 57 is a planetary nebula in the constellation of Lyra. At the center of the "ring" in the image shown in Figure 4.61, you can see a small white dot – the remains of a star that ran out of fuel to support fusion reactions. As the internal pressure reduces, gravity takes over and the star collapses, resulting in the outer parts of the star being blown off in gaseous form. The remains of the star have

Fig. 4.61 Image of Messier 57, a planetary nebula in Lyra

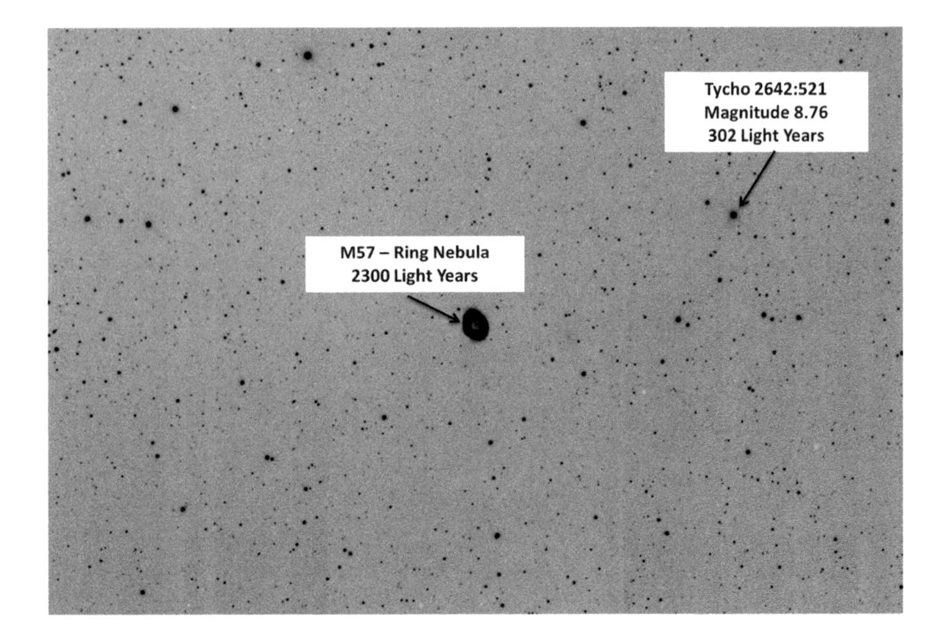

Fig. 4.62 Negative image of Messier 57

heated up in the process and provided energy to cause the expelled gas to glow. The Ring Nebula is an excellent example of this process. Messier observed the object on January 31, 1779, noting, "When comparing, in this morning, the Comet [C/1779 A1] with Beta in Lyra, I have perceived with the refractor a small cluster of light, which has appeared to me to be composed of very small Stars, which one could not distinguish with the instrument: This rounded cluster of light was placed between Gamma & Beta Lyrae."

The negative image in Figure 4.62 shows that M57 is at a distance of 2,300 ly. Its apparent dimensions are 1.4' x 1.0' of arc, which corresponds to an actual size of 0.94 X 0.67 ly. It is believed to expanding at about 1" per century, so the measurement of 84" (1.4') corresponds to a time of 8,400 years, and the measurement of 60" (1') corresponds to 6,000 years, to give a rough idea of its age. The central star, GSC 2642:1202, has a magnitude of 15.31 and is a white dwarf, which has a temperature of about 100,000 K. The star will cool down and fade into a black dwarf over the next few billion years. A field star is identified on the negative image as Tycho 2642:521 of magnitude 8.76, lying at a distance of 302 ly (Tycho Catalog) so is a foreground star.

The chart in Figure 4.63 shows the position of the Ring Nebula in the constellation of Lyra. It lies between the stars Sulafat and Sheliak at the base of the Lyre asterism in Lyra. Sulafat (HIP 93194) is a magnitude 3.25 star and Sheliak (HIP 92420) is magnitude 3.52. The angle between Sulafat and Sheliak is slightly less than 2°. The distance between Sulafat and M57 is 1° 11' and the distance between Sheliak and M57 is 48' 40" by my measurements. The angular distance from M57

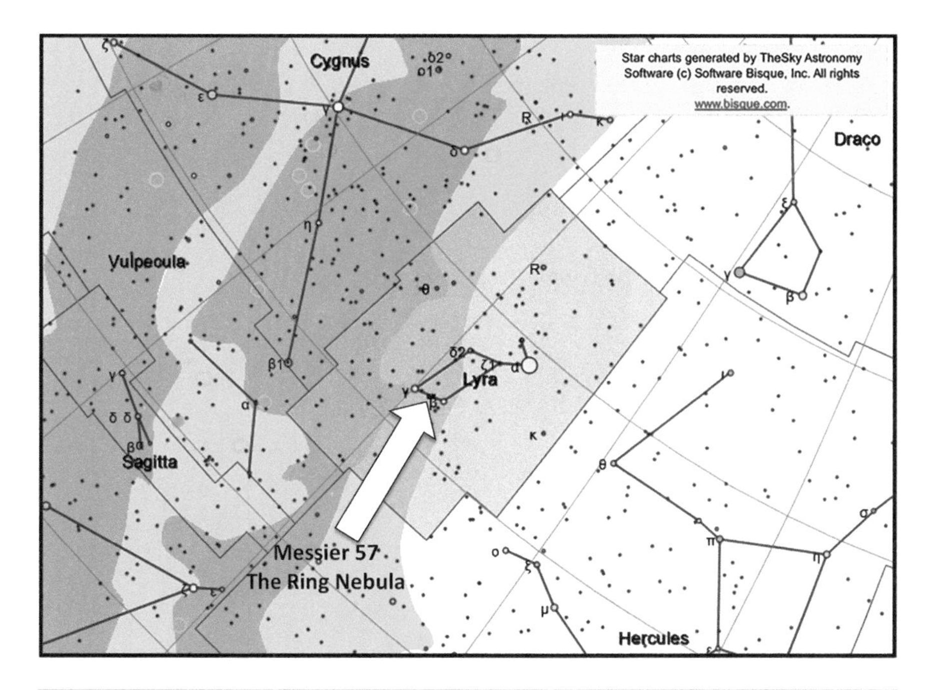

Fig. 4.63 Chart showing the location of M57

to 3-Alpha Lyrae (Vega) is 6° 38'. The only other Messier object in Lyra is M56, a globular cluster. With no Caldwell objects and apparently no other planetary nebulae, the only other brighter object of interest is NGC 6791, an open cluster of 10' x 10'. M57 is quite easy to find between Sulafat and Sheliak.

The image was taken at 01h 34m 37s on September 13. The corresponding Universal Time was 23h 34m 37s. The Right Ascension of the Ring Nebula is 18h 54m 13s, and the Local Sidereal Time was 22h 54m 57s. The difference between the LST and RA is 4h 0m 44s. This is the time that had elapsed since M57 was on the meridian. The telescope used to take the image of M13 was a Corrected Dall-Kirkham Astrograph with an aperture of 318 mm mounted on a Paramount ME.

Messier 57: Data relating to the image in Figure 4.61

REMOTELY IMAGED FROM SPAIN
NGC 6720
OBJECT TYPE: Planetary Nebula

RA: 18h 54m 13s
DEC: +33° 03' 29"
ALTITUDE: +41° 05' 46"
AZIMUTH: 284° 20' 24"
FIELD OF VIEW: 0° 37' 16" x 0° 24' 51"

OBJECT SIZE: 1.4' X 1.0'
POSITION ANGLE: 220° 33' from North
EXPOSURE TIME: 300 s
DATE: 13th September
LOCAL TIME: 01h 34m 37s
UNIVERSAL TIME: 23m 34m 37s
SCALE: 0.73 arcsec/pixel
MOON PHASE: 82.50% (waxing)

Telescope Optics
OTA: Planewave CDK
Optical Design: Corrected Dall-Kirkham Astrograph
Aperture: 318 mm
Focal Length: 2541 mm
F/Ratio: f/7.9
Guiding: External
Mount: Paramount ME

Instrument Package
CCD: SBIG-STXL-6303E
A/D Gain: 1.47 e-/ADU
Pixel Size: 9μm
Resolution: 0.73 arcsec/pixel
Sensor: Front Illuminated
Cooling: -20°C default
Array: 3072 x 2048 (6.3 Megapixels)
FOV: 37.41 x 24.94 arcmin

Location
Observatory: AstroCamp – MPC Code -189
UTC +1.00 (Daylight savings time is observed)
Minimum Target Elevation: Approx 30 – 40 Degrees
N 38° 09'
W 002° 19'
Elevation: 1650 meters (5413 ft)

Messier 58

Messier 58 (NGC 4579), shown in Figure 4.64, is a barred spiral galaxy in the constellation of Virgo and is part of the Virgo Cluster of galaxies. There are around 2,000 galaxies in this cluster, and 16 of them are Messier objects. It was discovered by Messier himself on April 15, 1779. He noted: "Very faint nebula discovered in Virgo, almost on the same parallel as epsilon, 3rd mag. The slightest light for illu-

Fig. 4.64 Image of Messier 58, a spiral galaxy in Virgo

minating the micrometer wires makes it disappear." I actually tried to obtain images of M58 with smaller aperture telescopes, but the 20-inch astrograph in New Mexico gave a better image. M58 is probably the brightest galaxy in the Virgo cluster. It lies at a distance of 60,000,000 ly and has an apparent size of 4.5' X 5.5' of arc. With these figures, the actual size is calculated to be 78,500 X 96,000 ly. That would make it slightly smaller than our Galaxy in its longest dimension.

The negative image in Figure 4.65 shows M58 in relation to stars in the frame. In this image, north is to the left and east is down. Part of the elliptical galaxy NGC 4564 is just caught in the frame as shown. It is located 29' away from M58 – only a full Moon's width. The bright star Tycho 878:506, to the west of Messier 58, is located about 7.5' from the galaxy. It is magnitude 8.01 and is only 459.38 ly away. It might seem far to us but M58 is 130 times further away! To the north is the magnitude 9.06 star Tycho 878:664. Further west towards the top of the image is Tycho 878:569, fainter still with a magnitude of 9.99.

The chart in Figure 4.66 shows the location of M58 in Virgo. Note that north is at the top and east is to the left, thus the portrait view of the superimposed M58 image. The nearest bright star is Vindemiatrix to the east of M58. This is the magnitude 2.85 spectral type G8 star 47-Epsilon Virginis. It has a separation of 6° from Messier 58. The brightest star in Virgo is 67-Alpha Virginis (HIP 65474 in the Hipparcos Catalog) and is 26° away from M58 to the southeast. There are 3 other Messier spiral galaxies in Virgo and 7 Messier elliptical galaxies. Quite a busy area for Charles Messier! There is only 1 Caldwell object in Virgo -- C52, an elliptical galaxy.

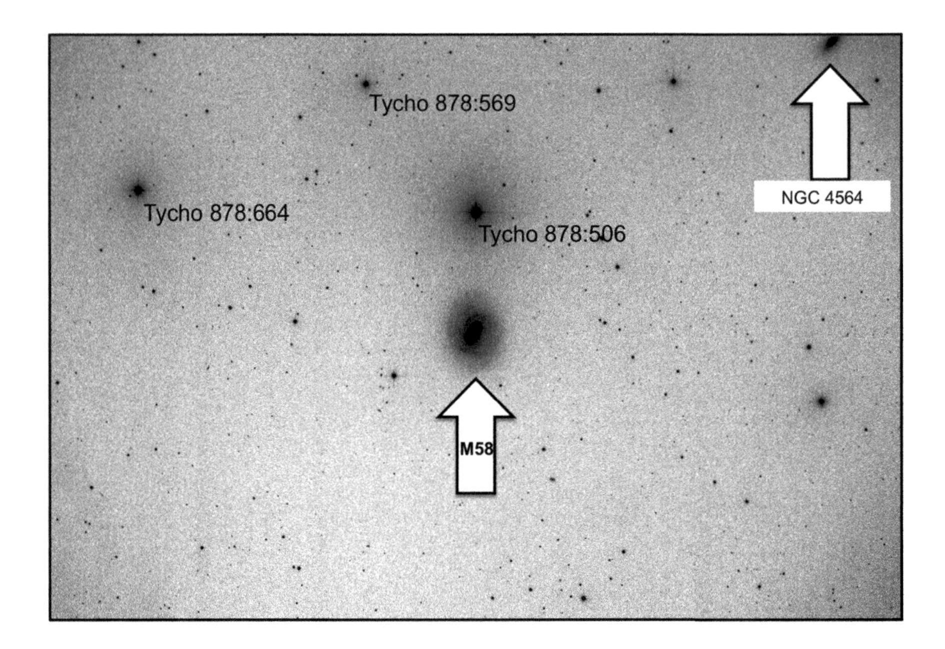

Fig. 4.65 Negative image of Messier 58

The image was taken at 00h 10m 13s on March 3. The Universal Time was 07h 10m 13s. M58 has a Right Ascension of 12h 38m 36s, and the Local Sidereal Time when the image was taken was 10h 51m 58s. This means that M58 had yet to cross the meridian. The difference between the RA and LST is 1h 46m 38s, so the M58 meridian transit would take place at 01h 56m 51s. The altitude of M58 when the image was taken was +57° 44' 46", and the azimuth angle was 124° 35' 39". The telescope used is the 508-mm Corrected Dall Kirkham astrograph mounted on a Paramount ME.

Messier 58: Data relating to the image in Figure 4.64

REMOTELY IMAGED FROM: New Mexico
NGC 4579
OBJECT TYPE: Spiral Galaxy

RA: 12h 38m 36s
DEC: +11° 43' 21"
ALTITUDE: +57° 44' 46"
AZIMUTH: 124° 35' 39"
FIELD OF VIEW: 0° 54' 47" x 0° 36' 31"
OBJECT SIZE: 5.5’ x 4.5’
POSITION ANGLE: 271° 07' from North
EXPOSURE TIME: 300 s

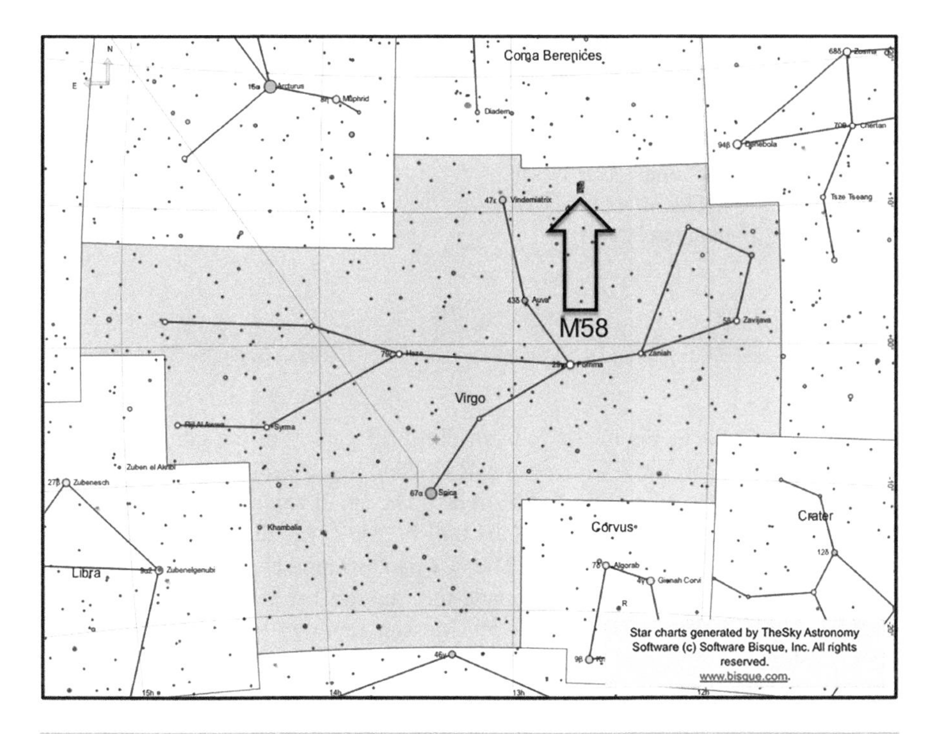

Fig. 4.66 Chart showing the location of M58

DATE: 3rd March
LOCAL TIME: 00h 10m 13s
UNIVERSAL TIME: 07h 10m 13s
MOON PHASE: 25.99% (waxing)

Telescope Optics
OTA: Planewave 20" CDK
Optical Design: Corrected Dall-Kirkham Astrograph
Aperture: 510 mm
Focal Length: 2280 mm (0.66 Focal Reducer Fitted)
F/Ratio: f/4.5
Guiding: Active Guiding Disabled
Mount: Planewave Ascension 200HR

Instrument Package
FLI Proline PL11002M CCD
Camera Pixel Size: 9 μm square
Resolution: 0.81 arcsec/pixel
Cooling: -30°C default
Array: 4008 x 2672 (10.7 Megapixels)

Location
Observatory: New Mexico Skies
UTC Minus 7.00 (Daylight savings time is observed)
Minimum Target Elevation: Approx 25 – 45 Degrees
(N or S) 32° 54' Decimal: 32.9 North
(W or E) 105° 31' Decimal: 105.5 West
Elevation: 2225 meters (7298 ft)

Messier 59

Messier 59, shown in Figure 4.67, is an elliptical galaxy in the constellation of Virgo. It is part of the Virgo Cluster of galaxies and lies at a distance of 60,000,000 ly. It has an apparent size of 5' X 3.5' of arc. The small angle formula shows that this is equivalent to about 87,000 X 61,000 ly. M59 was discovered by Johann Gottfried Koehler on April 11, 1779. The German astronomer discovered M60 that same night. Another discovery that made the Messier list was M67. Four nights after Koehler's discovery of Messier 59, Messier reported his own discovery of M59: "Nebula in Virgo & in the neighborhood of the preceding [M58] on the parallel of Epsilon Virginis, which has served for its determination: it is of the same light

Fig. 4.67 Image of Messier 59, an elliptical galaxy in Virgo

as the above [M58], equally faint." M59 possesses a very large number of globular clusters. It is estimated that there are roughly 2,000 of these.

The negative image in Figure 4.68 identifies M59 and some of the field objects. Note that north is to the left and east is down. NGC 4638 is an elliptical galaxy with dimensions 2.2' X 1.4". It lies 16' 30" to the southeast of M59. Just below NGC 4638 on the image is the much fainter spiral galaxy NGC 4637. It is about 17' 45" away from M59. Roughly to the north of Messier 59 at an angular separation of 6' 37" is the elliptical galaxy IC 3672. This has equal major and minor axes of 1'. At the top left (northwest) of the image at a distance of 18' 52" from M59 is the spiral galaxy NGC 4607 with dimensions of 2.9' X 0.7'. To the southwest of M59 at a distance of 10' 16" is the irregular galaxy IC 3665. This is only just visible as a smudge on the image.

The chart in Figure 4.69 shows the location of M59 in the constellation of Virgo. Galaxy symbols are included on the chart, and you can see that Virgo and the surrounding areas are swarming with galaxies. Just over 1° to the west of M59 is its predecessor in Messier's list, M58. Messier 59 is just over 1.5° from the southern border of Coma Berenices, which itself is brimming with galaxies. The position angle of the M59 image was just over 270°, so the embedded image has been rotated from its landscape view by just less than 90° to put north at the top of the chart. The field of view of the image is only 0° 42' 13" x 0° 28' 09", so it the embedded image appears very small on the chart.

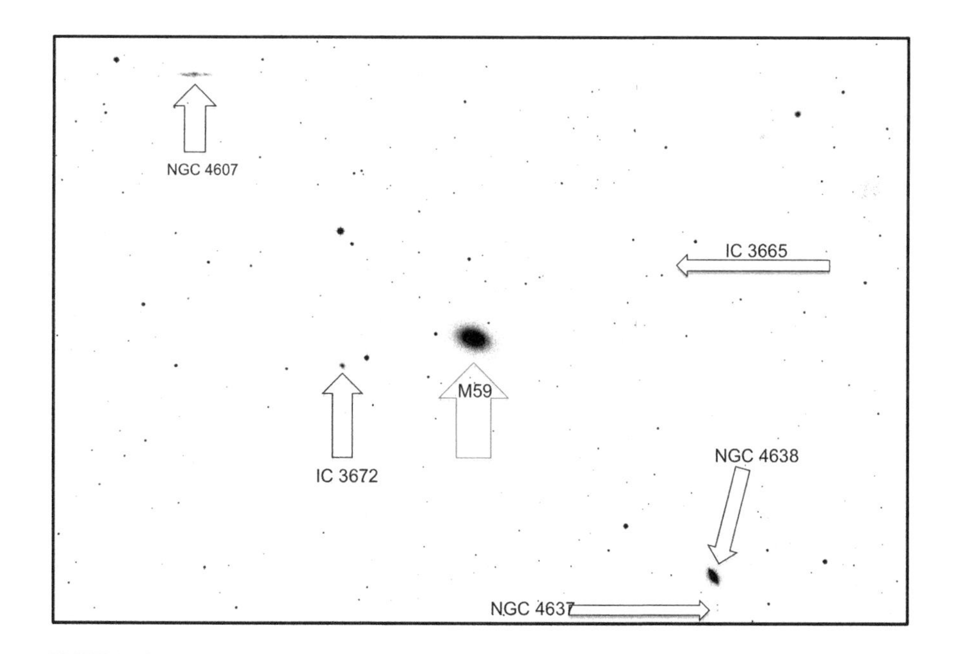

Fig. 4.68 Negative image of Messier 59

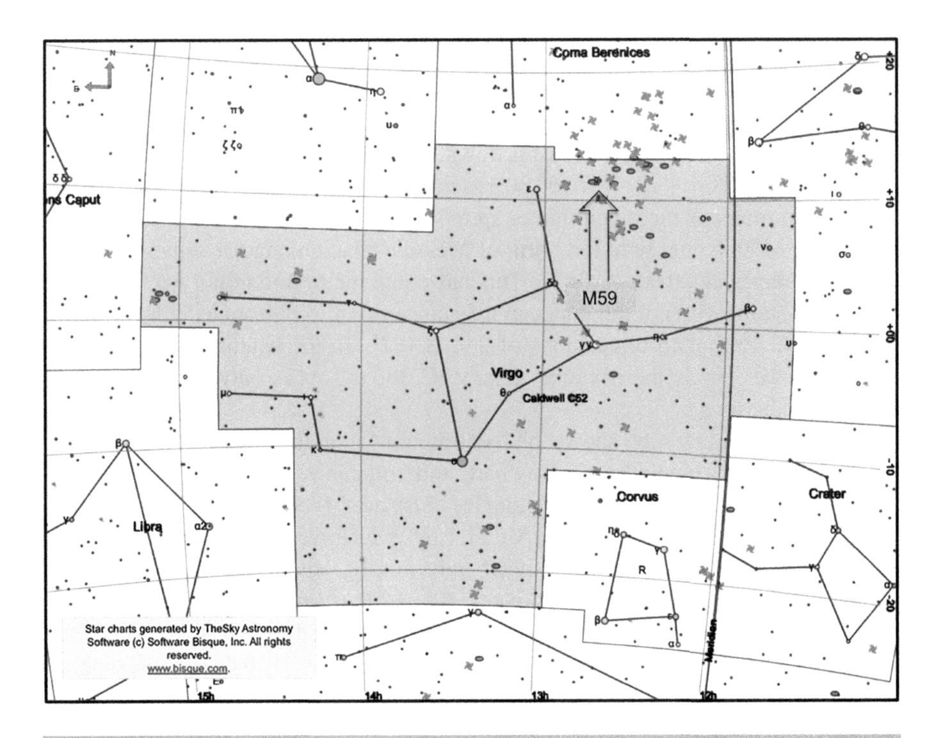

Fig. 4.69 Chart showing the location of M59

The image was taken at 05h 51m 23s on January 9 from Nerpio, Spain. The Universal Time was 04h 51m 23s. The Right Ascension of M59 is 12h 42m 53s, and the Local Sidereal Time was 11h 57m 48s. From this, you can tell that M59 was not too far from the meridian, moving towards it from east to west. The difference between RA and LST is 45m 5s, so the meridian transit would take place at 06h 36m 28s. You can see on the chart that the star 5-Beta Virginis had just crossed the meridian line at the time the image was taken. The telescope used at the site in Southern Spain was a corrected Dall-Kirkham astrograph with an aperture of 431 mm.

Messier 59: Data relating to the image in Figure 4.67

REMOTELY IMAGED FROM SPAIN
NGC4621
OBJECT TYPE: Elliptical Galaxy

RA: 12h 42m 53s
DEC: +11° 33' 12"
ALTITUDE: +44° 20' 59"
AZIMUTH: 161° 52' 53"

FIELD OF VIEW: 0° 42' 13" x 0° 28' 09"
OBJECT SIZE: 5' X 3.5'
POSITION ANGLE: 272° 03' from North
EXPOSURE TIME: 300 s
DATE: 9th January
LOCAL TIME: 05h 51m 23s
UNIVERSAL TIME: 04h 51m 23s
SCALE: 0.63 arcsec/pixel
MOON PHASE: 85.28% (waxing)

Telescope Optics
OTA: Planewave 17" CDK
Optical Design: Corrected Dall – Kirkham Astrograph
Aperture: 431 mm
Focal Length: 2929 mm
F/Ratio: f/6.8 Guiding: External
Mount: Paramount ME

Instrument Package
SBIG STL - 1100M
A/D Gain: 2.2 e-/ADU
Pixel Size: 9μm square
Resolution: 0.63 arcsec/pixel
Sensor: Frontlit
Cooling: -20°C default
Array: 4008 x 2672 (10.7 Megapixels)
FOV: 28.2 x 42.3 arcmin

Location
Observatory: AstroCamp Observatory – MPC Code – I89
UTC +1.00 (Daylight savings time is observed)
Minimum Target Elevation: Approx 35 – 45 Degrees
North: 38° 09'
West 0002° 19'
Elevation: 1650 meters (5413 ft)

Messier 60

Messier 60 (NGC 4649), shown in Figure 4.70, is an elliptical galaxy in the constellation of Virgo. It is unusual in that it forms a close pair with the spiral galaxy NGC 4647. M60 was discovered by Johann Gottfried Koehler on April 11, 1779, on the same night that he discovered Messier 59. M60 lies at a distance of 60,000,000 ly and has an apparent size of 7' X 6' of arc. This would correspond to an actual size

Fig. 4.70 Image of Messier 60, an elliptical galaxy in Virgo

of 122 X 105 million ly approximately. This makes it one of the largest elliptical galaxies. M60 has an estimate of over 5,000 globular clusters in its halo. Messier reported on April 15, 1779: "Nebula in Virgo, a little more distinct than the two preceding [M58 and M59] on the same parallel as Epsilon Virginis, which has served for its determination."

The negative image in Figure 4.71 shows Messier 60 and its "companion" galaxy NGC 4647. NGC 4647 is a spiral galaxy with a major axis of 2.9' and a minor axis of 2.3'. Note that north is to the left and east is down on this image. There is an angular separation of 2' 33" between M60 and NGC 4647. Two more galaxies are identified on the negative image. The first is NGC 4660, an elliptical galaxy that is 25' 10" to the southeast of M60. This galaxy has a listed size of 2.1' X 1.7'. The other galaxy pointed out is the elliptical galaxy NGC 4638 that is 14' 22" to the southwest of Messier 60. This galaxy has a listed size of 2.2' X 1.4'.

The chart in Figure 4.72 that shows the location of M60 in Virgo with the embedded image securely locked to the chart. Note that the image has been rotated to situate north at the top and east to the left. M59 and M58 are identified to show just how close these galaxies are to one another. There is 25' between M60 and M59 and 1' 30" between M60 and M58. They all lie fairly close to the border between Virgo and Coma Berenices. Note the large number of galaxies in this area. You may have noticed the other two marked Messier objects. These are M89 and M90.

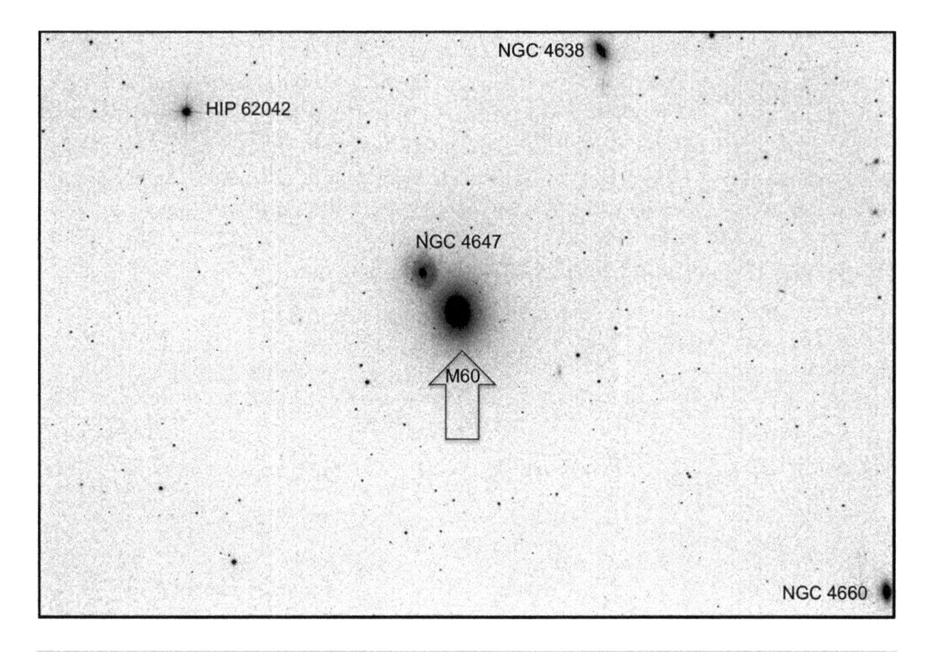

Fig. 4.71 Negative image of Messier 60

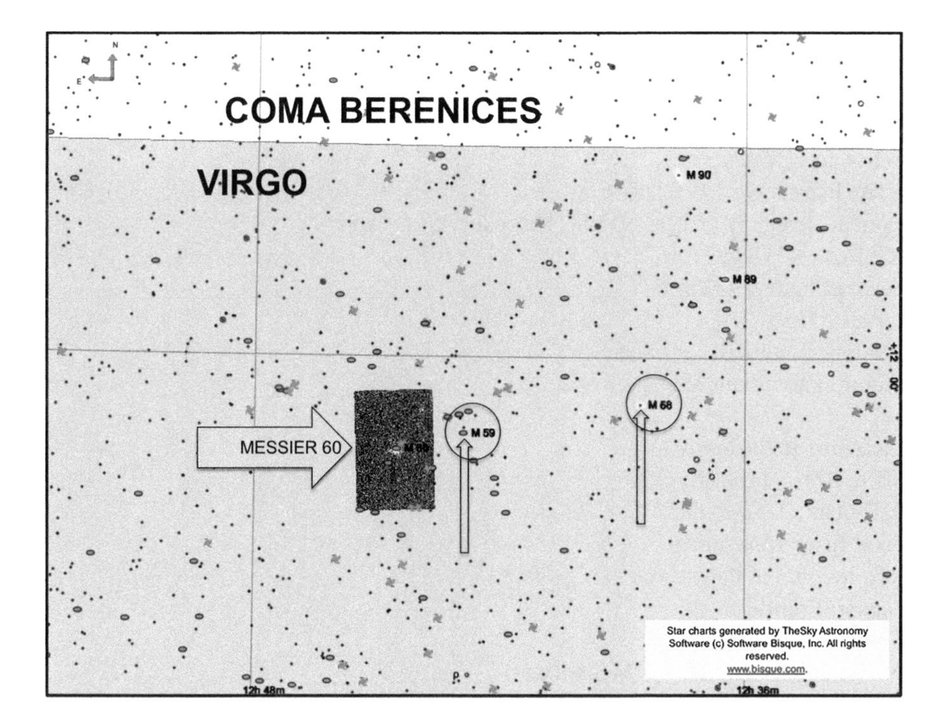

Fig. 4.72 Chart showing the location of M60

The image was taken at 06h 3m 34s on January 9 from Nerpio, Southern Spain. The Universal Time was 05h 03m 43s. The Right Ascension of M60 is 12h 44m 31s, and the Local Sidereal Time was 12h 10m 10s. The difference between RA and LST is 34m 21s, so M60 would cross the meridian from east to west in just over half an hour at 6h 37m 55s. The telescope used was a corrected Dall-Kirkham astrograph with an aperture of 431 mm, mounted on a Paramount ME.

Messier 60: Data relating to the image in Figure 4.70

REMOTELY IMAGED FROM SPAIN
NGC4649
OBJECT TYPE: Elliptical Galaxy

RA: 12h 44m 31s
DEC: +11° 27' 34"
ALTITUDE: +62° 13' 39"
AZIMUTH: 162° 29' 57"
FIELD OF VIEW: 0° 42' 13" x 0° 28' 09"
OBJECT SIZE: 7' X 6'
POSITION ANGLE: 272° 03' from North
EXPOSURE TIME: 300 s
DATE: 9th January
LOCAL TIME: 06h 03m 43s
UNIVERSAL TIME: 05h 03m 43s
SCALE: 0.63 arcsec/pixel
MOON PHASE: 85.35% (waxing)

Telescope Optics
OTA: Planewave 17" CDK
Optical Design: Corrected Dall – Kirkham Astrograph
Aperture: 431 mm
Focal Length: 2929 mm
F/Ratio: f/6.8
Guiding: External
Mount: Paramount ME

Instrument Package
SBIG STL - 1100M
A/D Gain: 2.2 e-/ADU
Pixel Size: 9μm square
Resolution: 0.63 arcsec/pixel
Sensor: Frontlit
Cooling: -20°C default
Array: 4008 x 2672 (10.7 Megapixels)
FOV: 28.2 x 42.3 arcmin

Location

Observatory: AstroCamp – MPC Code – I89
UTC +1.00 (Daylight savings time is observed)
Minimum Target Elevation: Approx 35 – 45 Degrees
North: 38° 09'
West 0002° 19'
Elevation: 1650 meters (5413 ft)

Messier 61

Messier 61, shown in Figure 4.73, is a spiral galaxy in Virgo. It forms part of the Virgo cluster of galaxies, lying at a distance of 60 million ly. It has an apparent size of 6 X 5.5 arcmin, which corresponds to an actual size of 105,000 X 96,000 ly. On May 5, 1779, Italian astronomer Barnabus Oriani was observing Comet C/1779 A1 when he discovered a "nebulous star" that was in fact M61. Messier was observing the same night and spotted M61 but thought it was comet C/1779 A1, so the discovery is credited to Oriani. It took Messier until May 11 to realize his mistake.

Fig. 4.73 Image of Messier 61, a spiral galaxy in Virgo

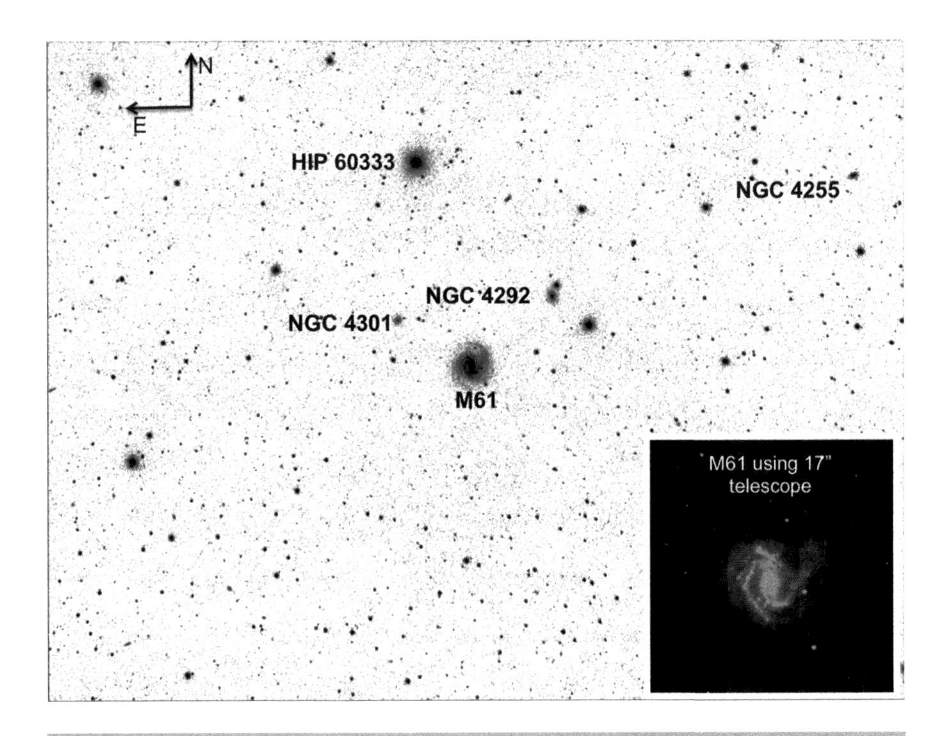

Fig. 4.74 Negative image of M61

In Figure 4.74 the negative of the desaturated color image shows the galaxy and several other nearby objects. The inset is the positive image taken with the 17" instrument. To the northwest of M61 is the spiral galaxy NGC 4292. This has a major axis of 1.6 arcmin and a minor axis of 1.2 arcmin. The angular distance between M61 and NGC 4292 is just over 12'. To the northeast of M61 is the spiral galaxy NGC 4301. The dimensions of this galaxy are 1.6' X 1.3'. This is separated from M61 by 10'. Further west is the spiral galaxy NGC 4255 with major and minor axes of 1.3' and 0.2', respectively. There is 48' 21" between M61 and NGC 4255. The magnitude 7.64 foreground star HIP 60333 is labeled. This is a cool M-type star that lies at a distance of 595.18 ly. To put things into perspective, M61 is 100,000 times further away than HIP 60333. At the speed of light, it would take us 8 lifetimes to reach HIP 60333, so M61 really does seem to be unattainable!

The chart in Figure 4.75 shows the position of M61 in the constellation of Virgo. The position angle of the camera was 357°, only 3° away from having north at the top of the image. You can see from the embedded image on the chart that this is the case. M61 lies between the stars 43-Delta Virginis (Auva or HIP 63090) and 5-Beta Virginis (Zavijava or HIP 57757), which are both in the Virgo asterism. Auva has a magnitude of 3.39, and Zavijava has a magnitude of 3.59. Angular separation between M61 and Auva is 8° 28', and between M61 and Zavijava, 8° 14'.

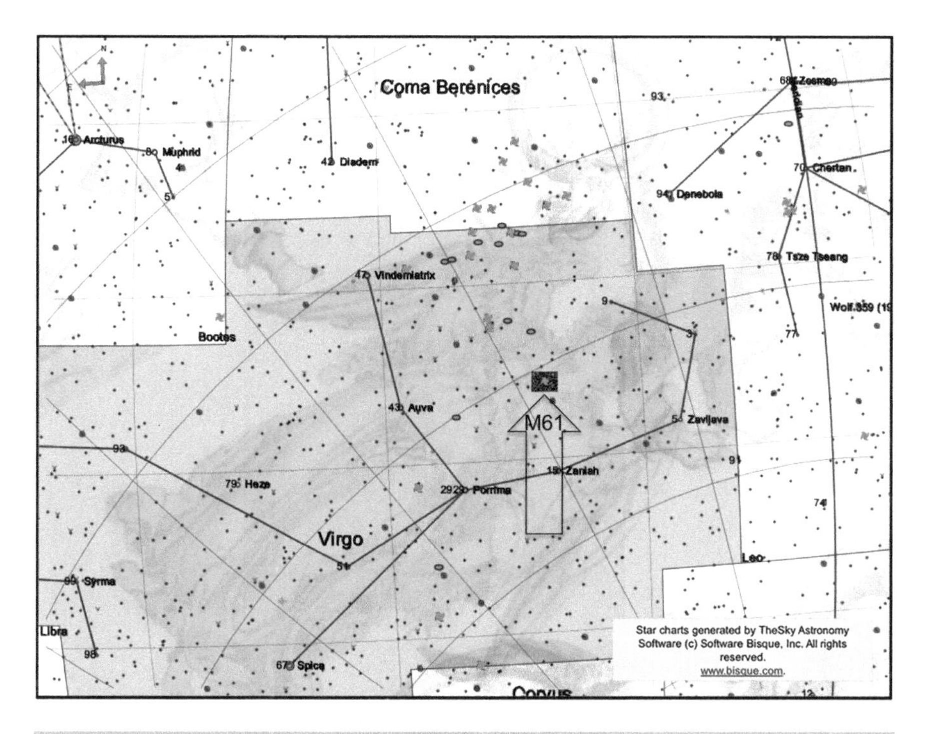

Fig. 4.75 Chart showing the location of M61

The image was taken at 02h 15m 05s on February 21. The corresponding Universal Time was 15h 15m 05s. The Local Sidereal Time was 11h 14 49s, and the Right Ascension of M61 is 11h 51m 35s. As the RA of M61 is greater than the LST, the difference between these two values is the amount of time to go before M61 reaches the meridian. That means there is 36m 46s before transit. This puts the transit time of M61 at 02h 51m 51s. Two telescopes based in New South Wales were used to image Messier 61, the first image taken with a single-shot color camera on a small, 3-inch Takahashi Sky90 and the second with a monochrome camera on a 17-inch Planewave CDK.

Messier 61: Data relating to the image in Figure 4.73

REMOTELY IMAGED FROM NEW SOUTH WALES, AUSTRALIA
NGC4303
OBJECT TYPE: Spiral Galaxy
RA: 12h 22m 48s
DEC: +04° 22' 35"
ALTITUDE: +50° 52' 08"
AZIMUTH: 27° 30' 11"

FIELD OF VIEW: 1° 37' 42" x 1° 13' 17"
OBJECT SIZE: 6’ X 5.5’
POSITION ANGLE: 357° 00' from North
EXPOSURE TIME: 300 s
DATE: 21st February
LOCAL TIME: 02h 15m 50s
UNIVERSAL TIME: 15h 15m 50s (20th February)
SCALE: 3.66 arcsec/pixel
MOON PHASE: 33.21% (waning)

Telescope Optics
OTA: Takahashi Sky 90
Optical Design: Apochromatic Refractor
Aperture: 90 mm
Focal Length: 417 mm
F/Ratio: f/5.6
Guiding: None
Mount: Paramount PME

Instrument Package
CCD: SBIG ST2000 XMC
Colour CMOS
Pixel Size: 7.4µm square
Sensor: Frontlit
Cooling: CNA
Array: 1600 x 1200 pixels
FOV: 60.5 x 80.7 arcmin

Location
Observatory: Siding Spring
UTC +10.00 (Australia Daylight savings time is observed)
31° 16’ 24” South
149 03’ 52” East
Elevation: 1122 meters (3681 ft)

Messier 62

M62, shown in Figure 4.76, is a globular cluster in the constellation of Ophiuchus. It does not seem to be particularly exciting on the image taken with the 3-inch refractor. A larger aperture would have revealed more detail in the globular. It certainly does not compare with the spectacular M13. It does seem to be an odd shape, which could be caused by its proximity to the galactic center and the resultant

Fig. 4.76 Image of Messier 62, a globular cluster in Ophiuchus

gravitational effects. It lies 22,500 ly from us and only 6,500 ly from the galactic center. It has an apparent size of 15' of arc, which corresponds to about 98 ly. Messier first observed it on June 4, 1779 but only confirmed its position on June 7: "Very beautiful nebula, discovered in Scorpio, it resembles a little Comet, the center is brilliant & surrounded by a faint glow. Its position was determined, by comparing it with the star Tau of Scorpius." You will have noticed the reference to M62's location in "Scorpio." Constellation boundaries were redefined in 1930, which repositioned M62 into Ophiuchus.

The negative image in Figure 4.77 shows the position of M62 in Ophiuchus in relation to a star that is in Scorpius. A line has been drawn that approximates the constellation boundary. Robert Burnham Jr. commented, "The cluster is seen against, and is probably embedded in, a rich Milky Way star field, so that the area, for many degrees around the group, is sprinkled with multitudes of tiny star sparks. This is one of the globulars which appears to be actually immersed in the starry hub of the Galaxy." The negative image certainly does bring out the sheer number of stars in the area with its multitude of black dots!

The chart in Figure 4.78 shows Messier 62 "sitting on the fence" between the two constellations. If you look to the west (right) of M62, you will see the magnitude 2.83 star, marked 23, on the chart. Its Flamsteed-Bayer designation is 23-Tau Scorpii. It has the Hipparcos Catalog designation HIP 81266. This is the star that

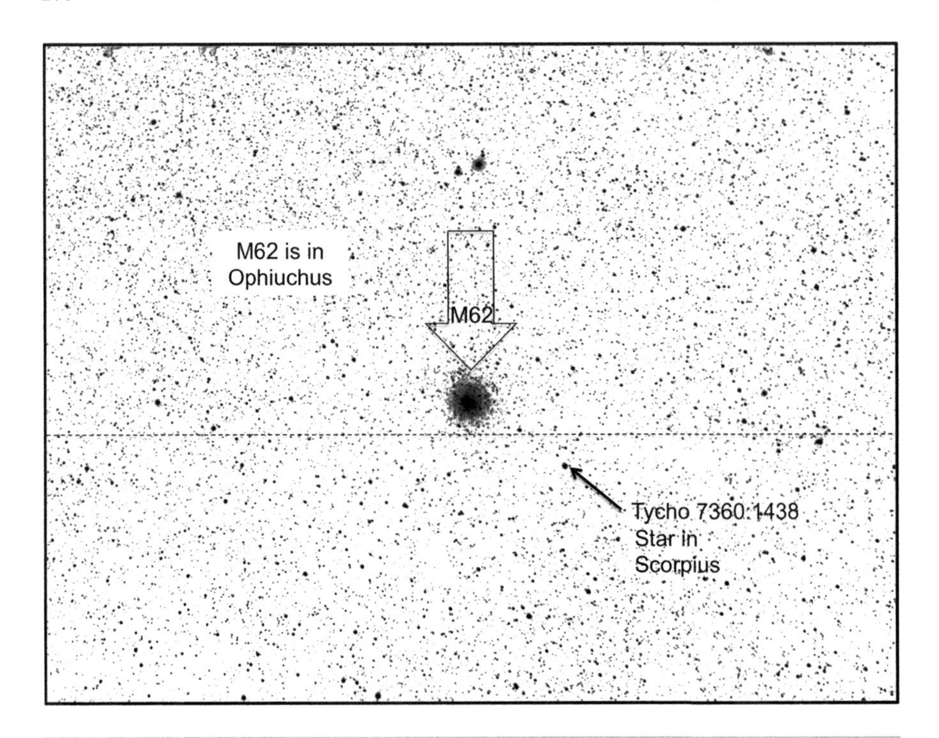

Fig. 4.77 Negative image of Messier 62

Messier used to determine the position of M62, as mentioned previously. The angular separation between Tau and M62 is 5° 52'. Further south you will notice the star 26, which is 26-Epsilon Scorpii. This magnitude 2.29 star is Wei or HIP 82396. It has an angular distance of 4° 47' from M62. Notice also that the bright star Antares is just above 23-Tau Scorpii, so the bright red star would be a good place to start if you are hunting for M62 visually.

The color image was taken at 23h 02m 25s on the evening of April 14. The corresponding Universal Time was 13h 02m 25s. The Right Ascension of M62 is 17h 02m 19s and the Local Sidereal Time was 12h 30m 00s. The difference between RA and LST is 4h 32 19s, so adding this to the local time gives a transit time for M62 of 03h 34m 44s. The telescope used was the 90-mm refractor with a single-shot color camera.

Messier 62: Data relating to the image in Figure 4.76

REMOTELY IMAGED FROM NEW SOUTH WALES, AUSTRALIA
NGC 6266
OBJECT TYPE: Globular Cluster

RA: 17h 02m 19s
DEC: -30° 08' 00"

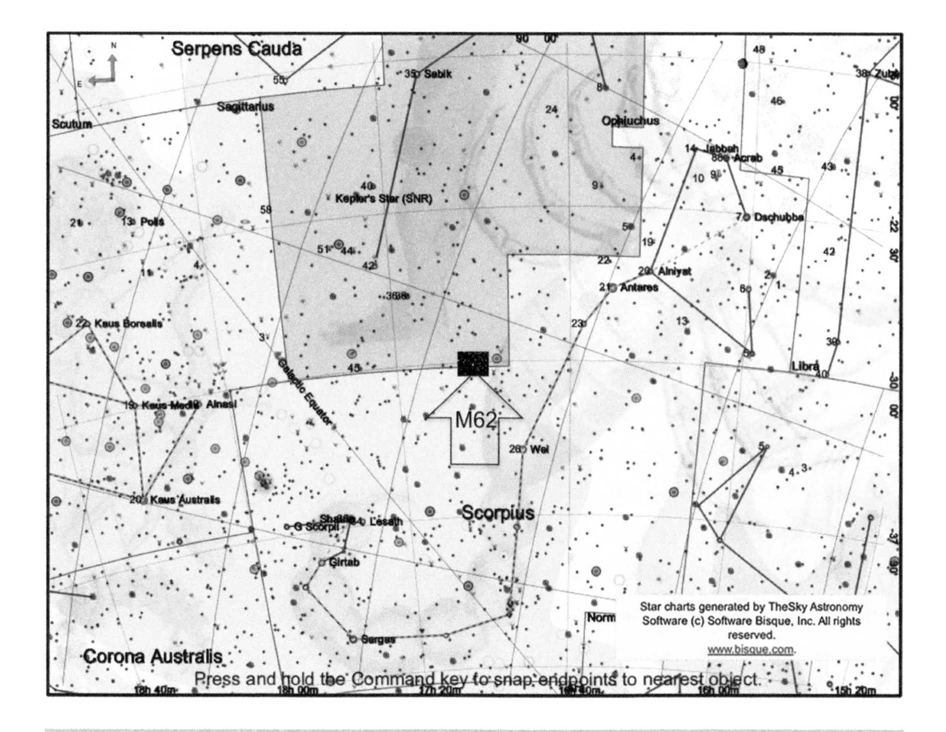

Fig. 4.78 Chart showing the location of M62

ALTITUDE: +32° 26' 18"
AZIMUTH: 108° 04' 13"
FIELD OF VIEW: 1° 37' 46" x 1° 13' 19"
OBJECT SIZE: 15'
POSITION ANGLE: 357° 22' from North
EXPOSURE TIME: 300 s
DATE: 14th April
LOCAL TIME: 23h 02m 25s
UNIVERSAL TIME: 13h 02m 25s
SCALE: 3.67 arcsec/pixel
MOON PHASE: 90.08% (waning)

Telescope Optics
OTA: Takahashi Sky 90
Optical Design: Apochromatic Refractor
Aperture: 90 mm
Focal Length: 417 mm
F/Ratio: f/5.6
Guiding: None
Mount: Paramount PME

Instrument Package
CCD: SBIG ST2000 XMC
Colour CMOS
Pixel Size: 7.4μm square
Sensor: Frontlit
Cooling: CNA
Array: 1600 x 1200 pixels
FOV: 60.5 x 80.7 arcmin

Location
Observatory: Siding Spring
UTC +10.00 (Australia Daylight savings time is observed)
31° 16' 24" South
149 03' 52" East
Elevation: 1122 meters (3681 ft)

Messier 63

Messier 63, shown in Figure 4.79, is a spiral galaxy located in the constellation of Canes Venatici. It has an apparent size of 10' X 6' of arc. It lies at a distance of 37 million ly, which would give it an actual size of approximately 108,000 X 65,000

Fig. 4.79 Image of Messier 63, a spiral galaxy in Canes Venatici

ly. Messier's colleague and close friend Pierre Méchain discovered the galaxy on June 14, 1779. Messier later commented, "…It is faint, it has nearly the same light as the nebula reported previously -- it contains no star, & the slightest illumination of the micrometer wires makes it disappear: it is close to a star of 8th magnitude, which precedes the nebula on the hour wire."

It is often surprising how much enlargement can be achieved on a wide-angle high-resolution image. M63 is called the Sunflower Galaxy, and this becomes evident when the galaxy from the original image is cropped and enlarged, as you can see. The star arrowed near the galaxy is Tycho 3024:814. This is a magnitude 9.3 star that lies in front of M63 at 3,262 ly. On the negative image in Figure 4.80, I have identified the spiral galaxy NGC 5103 at upper right and show an enlarged view from the original. I have done the same with PGC 46386. In fact, from the original image, I found a large number of faint galaxies that I was able to identify using the solved plate image superimposed on the chart. There were too many to show on the negative image without it becoming even more cluttered.

The chart in Figure 4.81 shows the location of M63 in the constellation of Canes Venatici. There are four other Messier objects in this constellation. M51 is the best in my eyes. It lies at an angular separation of about 5° 45' from M63. Having said that, however, the constellation also contains the spectacular globular cluster M3, which is roughly 14° 30' away. Then there are spiral galaxies M94 and M106 at 4° 40' and 11° 20' from M63, respectively. M63 is 5° 13' from the star 12-Alpha1 Canum Venaticorum, alternatively known as HIP 63121 in the Hipparcos Catalog.

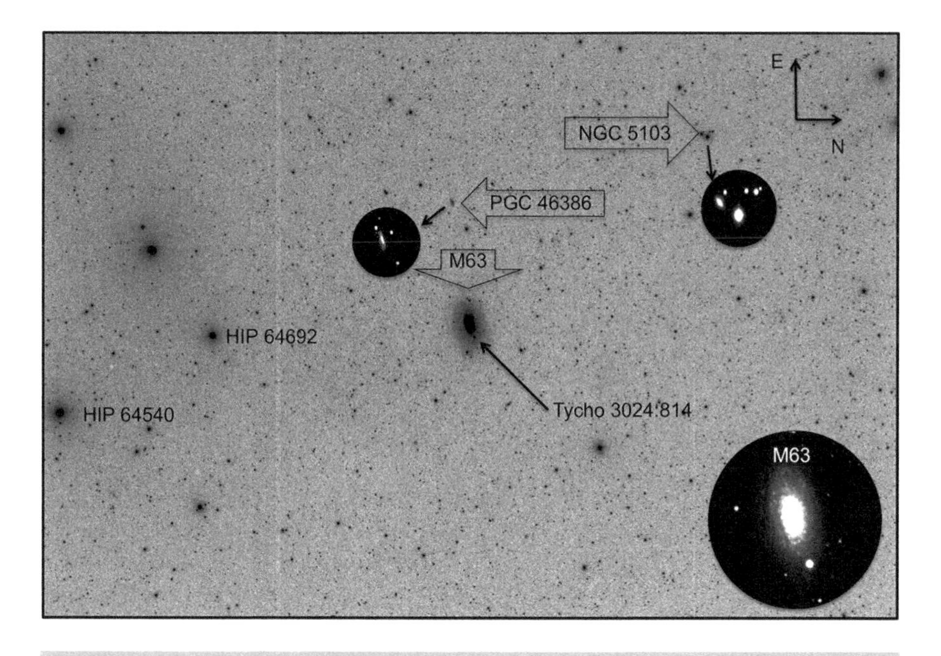

Fig. 4.80 Negative image of M63

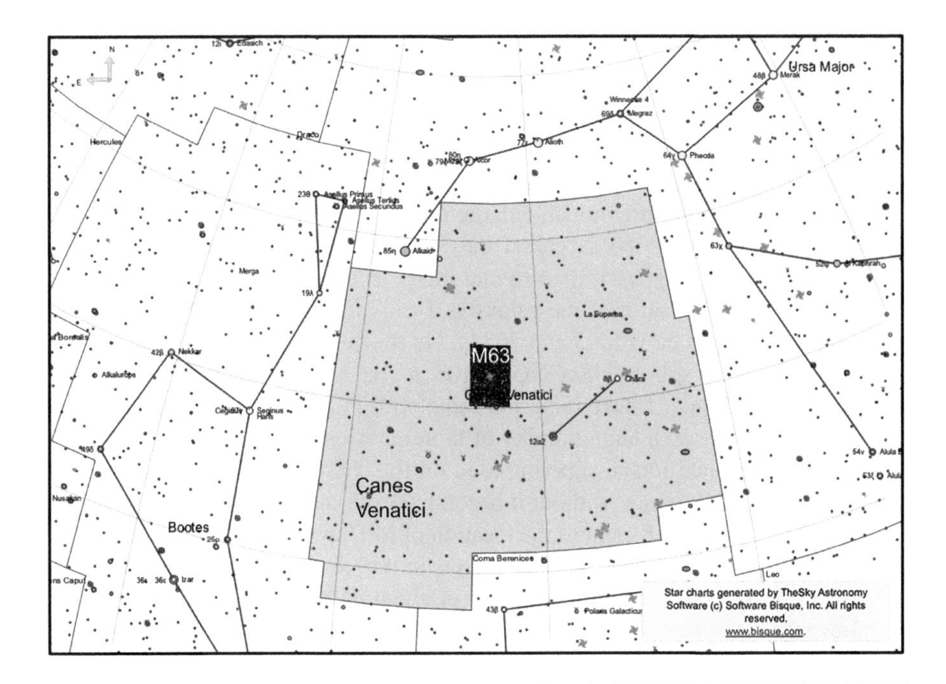

Fig. 4.81 Chart showing the location of M63

The image was taken at 01h 02m 23s on February 2. The corresponding Universal Time was 08h 02m 23s. The Right Ascension of M63 is 13h 16m 35s, and the Local Sidereal Time was 09h 49m 57s. If the difference between RA and LST is added to the local time the image was taken, it gives a meridian transit time of about 04h 29m. The telescope used to capture Messier 63 was a wide-angle instrument with an aperture of 106 mm on a Paramount ME mount.

Messier 63: Data relating to the image in Figure 4.79

REMOTELY IMAGED FROM MAYHILL, NEW MEXICO
NGC 5055 (SUNFLOWER GALAXY)
OBJECT TYPE: Spiral Galaxy

RA: 13h 16m 35s
DEC: +41° 56' 24"
ALTITUDE: +48° 34' 34"
AZIMUTH: 61° 51' 35"
FIELD OF VIEW: 3° 54' 08" x 2° 36' 05"
OBJECT SIZE: 10' X 6'
POSITION ANGLE: 90° 07' from North
EXPOSURE TIME: 300 s
DATE: 2nd February
LOCAL TIME: 01h 02m 23s

UNIVERSAL TIME: 08h 02m 23s
SCALE: 3.5 arcsec/pixel
MOON PHASE: 29.79% (waxing)

Telescope Optics
OTA: Takahashi FSQ - ED
Optical Design: Petzval Apochromatic Astrograph
Aperture: 106 mm
Focal Length: 530 mm
F/Ratio: f/5.0
Guiding: External
Mount: Paramount PME

Instrument Package
CCD: SBIG STL-11000M
Pixel Size: 9μm square
Sensor: Frontlit
Cooling: -15°C
Array: 4008 x 2672 pixels
FOV: 188.8 x 233.7 arcmin
Position Angle: 092.7°

Location
Observatory: New Mexico Skies
UTC Minus 7.00 (Daylight savings time is observed)
Minimum Target Elevation: Approx 25 – 45 Degrees
(N or S) 32° 54' Decimal: 32.9 North
(W or E) 105° 31' Decimal: 105.5 West
Elevation: 2225 meters (7298 ft)

Messier 64

Messier 64 (NGC 4826), shown in Figure 4.82, is a spiral galaxy in the constellation of Coma Berenices. It is known as the Black Eye Galaxy. The dark feature that earns it that name is a dust cloud that blocks the lights from stars in the background. You may see that the discovery of M64 is credited to different people in the literature. Famous German astronomer Johann Bode discovered it on April 4, 1779, and then Messier discovered it independently on March 1, 1780. However, more recent research credits Edward Pigott with its discovery on March 23, 1779. Edward was the son of astronomer Nathaniel Pigott and discovered M64 from Frampton House in Glamorganshire. It was William Herschel who first observed the dark dust lane and compared it with a Black Eye. He commented: "A very remarkable object, much elongated, about 12' long, 4' or 5' broad, contains one lucid spot like a star

Fig. 4.82 Image of Messier 64, a spiral galaxy in Coma Berenices

with a small black arch under it, so that it gives one the idea of what is called a black eye, arising from fighting."

The negative image in Figure 4.83 shows the dimensions of M64, which are 9.3' x 5.4' of arc. The distance is usually quoted as 24 million ly, and if this distance is adopted, then the actual dimensions are 65,000 x 38,000 ly. However, the Space Telescope Science Institute has published a figure of 19 Million ly. If this figure is adopted, then the actual dimensions of M64 are 51,000 x 30,000 ly. Some magnitudes of field stars are marked on the negative image. Many of them are around magnitude 10, going down to around magnitude 15.

The chart in Figure 4.84 shows the location of Messier 64 in the constellation of Coma Berenices. The angular separation between M64 and the nearby star 35 Comae Berenices is about 55 arcmin. This G8 star is also known as HIP 62886 and is magnitude 4.88, lying at a distance of 324.21 ly -- a fraction of the distance to M64. The star 36 Comae Berenices (HIP 63355) is about 4° 16' from M64. It is a cool M-type star with a magnitude of 4.76 and is 298.95 ly distant. A nearby star of real interest is 31 Comae Berenices, or Polaris Galacticus Borealis (HIP 62763). This star is effectively in the position of the North Galactic Pole, so you can think of it as the Polaris of the Galaxy. It is roughly 6° from Messier 64. The star is a magnitude 4.93 G-type at a distance of 307.12 ly.

The local time in New Mexico when the image of M64 was taken is 23h 02m 08s. The corresponding Universal Time was 05h 02m 08s. The Right Ascension of M64 is 12h 57m 35s, and the Local Sidereal Time was 12h 47m 58s. The difference between these indicates the time to go to before M64 crossed the meridian (09m

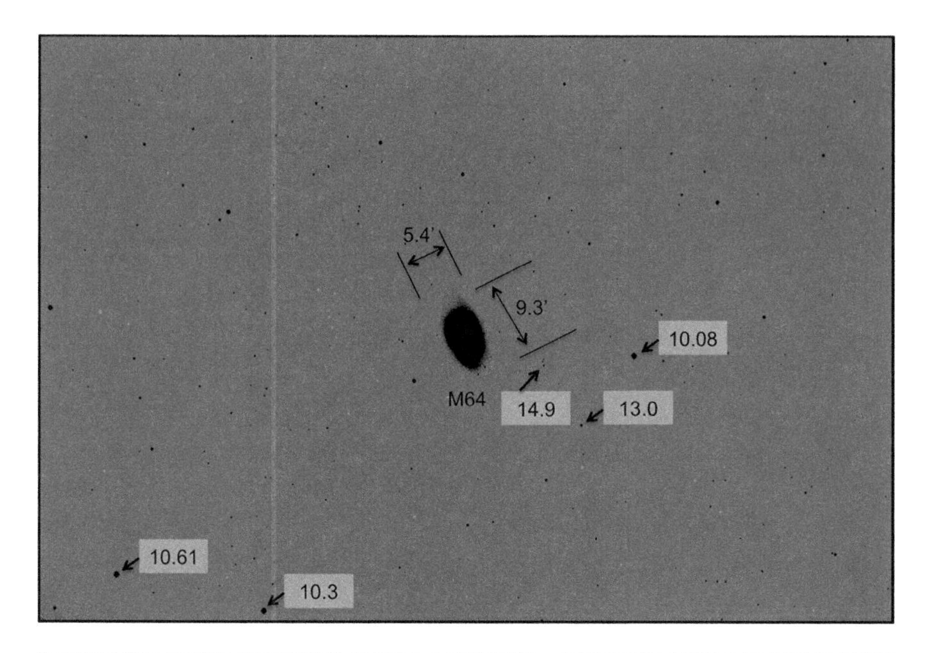

Fig. 4.83 Negative image of Messier 64

37s). M64 would transit the meridian at about 23h 12m. If you look back to the chart, you can see M64 approaching the meridian line, which is just to the west. The telescope used to take the image of M53 was a Corrected Dall-Kirkham Astrograph with an aperture of 510 mm mounted on a Plane Wave Ascension 200HR.

Messier 64: Data relating to the image in Figure 4.82

REMOTELY IMAGED FROM MAYHILL, NEW MEXICO
NGC 4826 (Black Eye Galaxy)
OBJECT TYPE: Spiral Galaxy

RA: 12h 57m 35s
DEC: +21° 35' 26"
ALTITUDE: +78° 36' 22"
AZIMUTH: 168° 36' 59"
FIELD OF VIEW: 0° 54' 47" x 0° 36' 31"
OBJECT SIZE: 9.3' X 5.4'
POSITION ANGLE: 271° 08' from North
EXPOSURE TIME: 300 s
DATE: 3rd May
LOCAL TIME: 23h 02m 08s
UNIVERSAL TIME: 05h 02m 08s

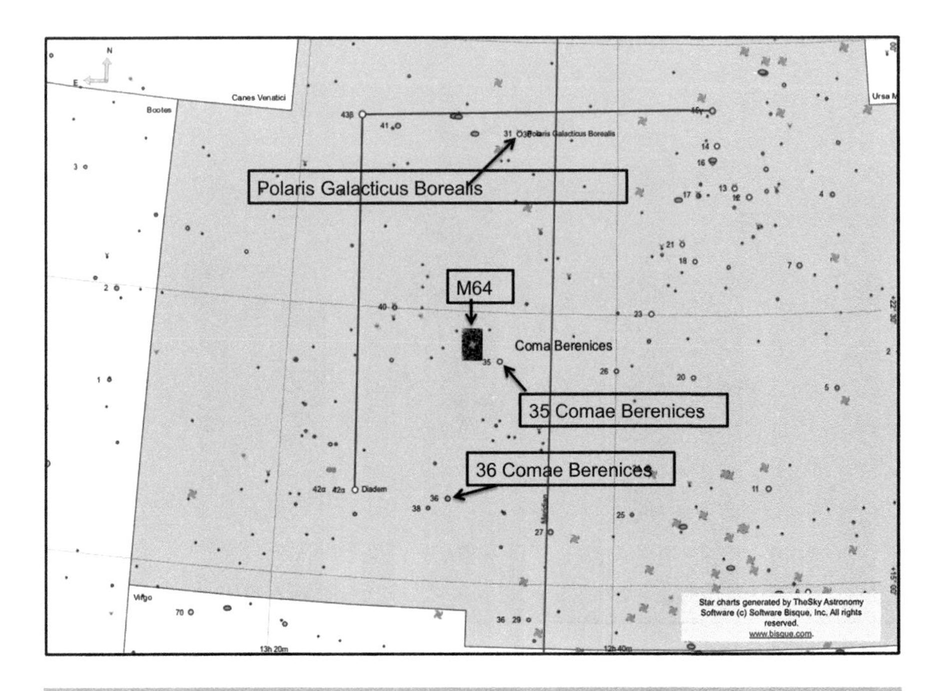

Fig. 4.84 Chart showing the location of M64

SCALE: 0.82 arcsec/pixel
MOON PHASE: 61.76% (waxing)

Telescope Optics
OTA: Planewave 20" CDK
Optical Design: Corrected Dall-Kirkham Astrograph
Aperture: 510 mm
Focal Length: 2280 mm (0.66 Focal Reducer Fitted)
F/Ratio: f/4.5
Guiding: Active Guiding Disabled
Mount: Planewave Ascension 200HR

Instrument Package
FLI Proline PL11002M CCD
Camera Pixel Size: 9 µm square
Resolution: 0.81 arcsec/pixel
Cooling: -30°C default
Array: 4008 x 2672 (10.7 Megapixels)

Location
Observatory: New Mexico Skies
UTC Minus 7.00 (Daylight savings time is observed)

Minimum Target Elevation: Approx 25 – 45 Degrees
(N or S) 32° 54' Decimal: 32.9 North
(W or E) 105° 31' Decimal: 105.5 West
Elevation: 2225 meters (7298 ft)

Messier 65

Messier 65 (NGC 3523), shown in Figure 4.85, is a spiral galaxy that sits in the constellation of Leo. It is part of a triplet of galaxies that include M66 and NGC 3628. The apparent size of M65 is 8' x 1.5' of arc. The distance to this object is 35 million ly, so the actual size of M65 is about 81,000 x 15,000 ly. Messier observed this galaxy on March 1, 1780, commenting, "Nebula discovered in Leo: It is very faint and contains no star." There has been disagreement on who actually discovered the galaxy and its companion, M66. French astronomer Pierre Méchain is quoted as the discoverer by some, Messier by others. Robert Burnham Jr. had no doubt: "M65 and M66 were discovered by P. Méchain in March 1780; it seemed that Messier's great comet of 1773 passed directly through the field on November 2nd 1773, but the two galaxies were not noted by that diligent observer."

The negative image in Figure 4.86 identifies M65 and its two companions in what is called the "M66 Group." Note that north is to the right and east is up. The

Fig. 4.85 Image of Messier 65, a spiral galaxy in Leo

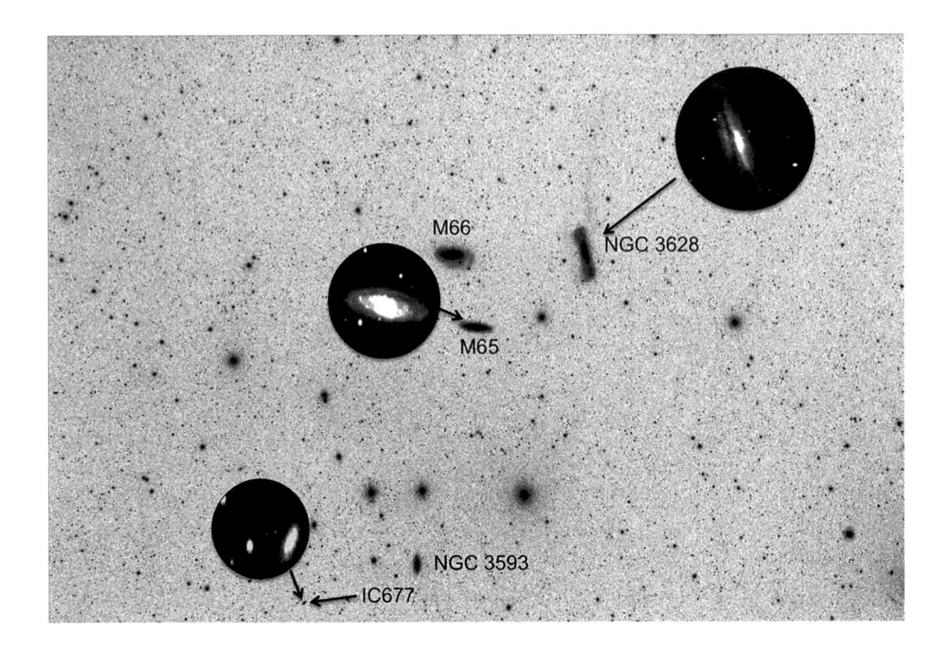

Fig. 4.86 Negative image of Messier 65

image of M65 has been enlarged and is shown next to the object. The same has been done for the galaxy NGC 3628. The latter galaxy has a major axis of 13.1' and a minor axis of 3.1'. A small galaxy, NGC 3593, is also pointed out, having a major axis of 5' 20" and a minor axis of just over 2'. The tiny galaxy IC 677 is shown enlarged. It was too small to measure using my software, but the values given in the Index Catalog are 1.4' x 0.6'.

The chart in Figure 4.87 shows the location of M65 in Leo with north at the top. The image had a position angle of 89° 52' from north, so the image was rotated counterclockwise through almost 90° to get north at the top. The image sits directly between the stars 78-Iota Leonis and 70-Theta Leonis, otherwise known as Tsze Tseang (HIP 55642) and Chertan (HIP 54879), respectively. Along with spiral M66, Leo has three more Messier galaxies. M95 and M96 are spirals, and M105 is an elliptical galaxy. M96 is about 9 arcmin from M65, and M96 is about 8' 18". M105 is about 8' away from M65.

The image was taken at 01h 11m 28s on February 2. The corresponding Universal Time was 08h 11m 28s. M65 has a Right Ascension of 11h 19m 50s, and the Local Sidereal Time when the image was taken was 9h 59m 03s. This means that in 1h 20m 47s, M65 would transit the meridian at about 02h 30m. The meridian line can be seen on the chart passing close to the head of Leo. The telescope used to capture Messier 63 was a wide-angle instrument with an aperture of 106 mm on a Paramount ME mount.

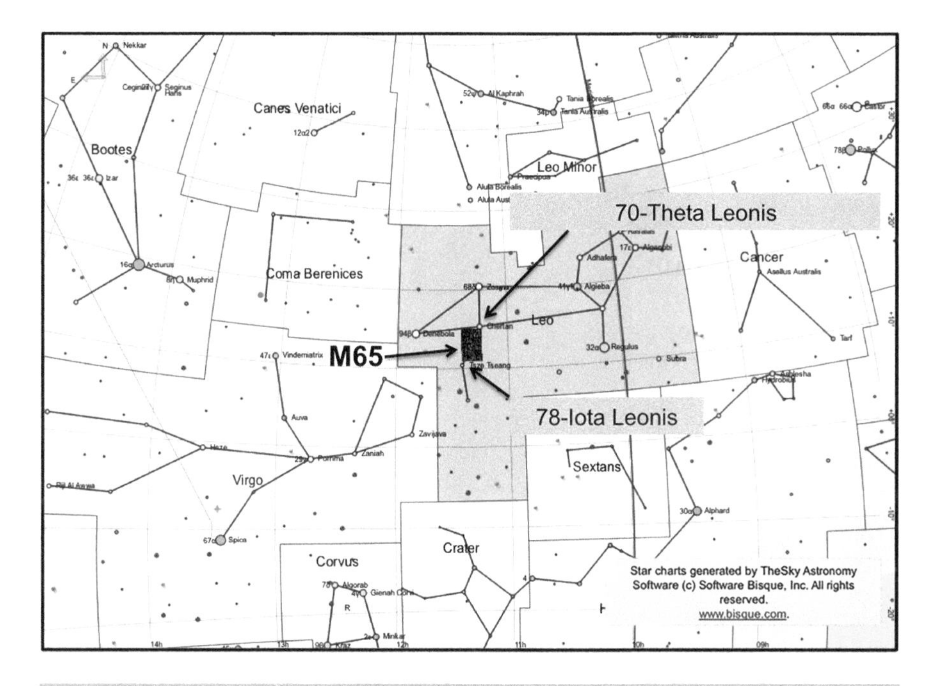

Fig. 4.87 Chart showing the location of M65

Messier 65: Data relating to the image in Figure 4.85

REMOTELY IMAGED FROM MAYHILL, NEW MEXICO
NGC 3523
OBJECT TYPE: Spiral Galaxy

RA: 11h 19m 50s
DEC: +12° 59' 42"
ALTITUDE: +62° 56' 54"
AZIMUTH: 132° 18' 21"
FIELD OF VIEW: 3° 54' 12" x 2° 36' 08"
OBJECT SIZE: 8' X 1.5'
POSITION ANGLE: 89° 52' from North
EXPOSURE TIME: 300 s
DATE: 2nd February
LOCAL TIME: 01h 11m 28s
UNIVERSAL TIME: 08h 11m 28s
SCALE: 3.51 arcsec/pixel
MOON PHASE: 29.86% (waxing)

Telescope Optics
OTA: Takahashi FSQ - ED
Optical Design: Petzval Apochromatic Astrograph
Aperture: 106 mm
Focal Length: 530 mm
F/Ratio: f/5.0
Guiding: External
Mount: Paramount PME

Instrument Package
CCD: SBIG STL-11000M
Pixel Size: 9μm square
Sensor: Frontlit
Cooling: -15°C
Array: 4008 x 2672 pixels
FOV: 188.8 x 233.7 arcmin
Position Angle: 092.7°

Location
Observatory: New Mexico Skies
UTC Minus 7.00 (Daylight savings time is observed)
Minimum Target Elevation: Approx. 25 – 45 Degrees
(N or S) 32° 54' Decimal: 32.9 North
(W or E) 105° 31' Decimal: 105.5 West
Elevation: 2225 meters (7298 ft)

Messier 66

Messier 66, shown in Figure 4.88, is a spiral galaxy located in the constellation of Leo. It forms part of the trio of galaxies consisting of M66, M65 and NGC 3628, collectively known as the M66 Group. M66 is at a distance of 35 million ly, the same distance as M65 and NGC 3628. The apparent dimensions are 8' x 2.5' of arc, which makes the real size 81,000 x 25,000 ly using the small angle formula. This means it is larger than M65. Messier observed M66 on the same night as M65. He saw it as a starless nebula through his limited equipment. M66 is also known as NGC 3627. Detailed images of M66 from the Hubble NASA/ESA space telescope show that the galaxy has been gravitationally distorted by the other galaxies in the trio.

M66 is identified on the negative image in Figure 4.89, and an enlargement from the original image is also shown. On this image, north is to the right and east is up. The small galaxy at top left is NGC 3666, or PGC 35043 in the Principal Galaxy Catalog. It is shown in the enlarged view and has a major axis of 3.5' and a minor

Fig. 4.88 Image of Messier 66, a spiral galaxyin Leo

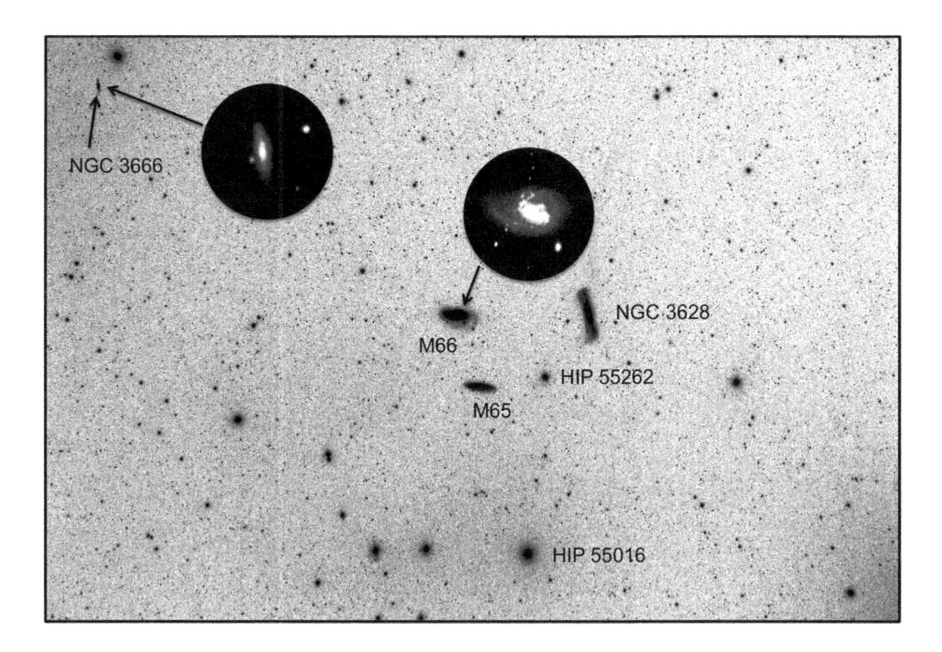

Fig. 4.89 Negative image of Messier 66

axis of 1'. The main sequence F8 star HIP 55262, close to the M66 trio, is magnitude 7.12 and is at a distance of 155.61 ly – somewhat closer than M66 at 35 million ly! The brighter star at the bottom of the image (west) is HIP 55016. This is a magnitude 5.31 K3 star at a distance of 478.24 ly in the Hipparcos Catalog and is the naked-eye star 73n Leonis.

The image taken is shown on the chart in Figure 4.90 between Iota and Theta Leonis. The chart is aligned with north at the top and east to the left. The star Regulus (HIP 49669) in Leo is 32-Alpha Leonis and is magnitude 1.36. It is a main sequence B7 star at a distance of 77.49 ly. It lies at an angular distance of 17° 41' from M66. The star Denebola (HIP 57632) in Leo is 94-Beta Leonis and is magnitude 2.14. It is a main sequence variable A3 star at a distance of 36.18 ly. It has an angular separation of 7° 6' from Messier 66.

The image of M66 was taken at local time 01h 31m 38s on February 3 from New Mexico. The Universal Time was 08h 31m 38s. Messier 66 has a Right Ascension of 11h 21m 09s, and the Local Sidereal Time was 10h 23m 13s. The difference between these two is 1h 31m 38s, which was the time to go before M66 reached the meridian. Adding this to the current local time gives a meridian transit time of 02h 29m 34s. M66 was at an altitude of +66° 06' 28" when the image was taken and was at azimuth 142° 59' 31". The telescope used to capture Messier 63 was a wide-angle instrument with an aperture of 106 mm on a Paramount ME mount.

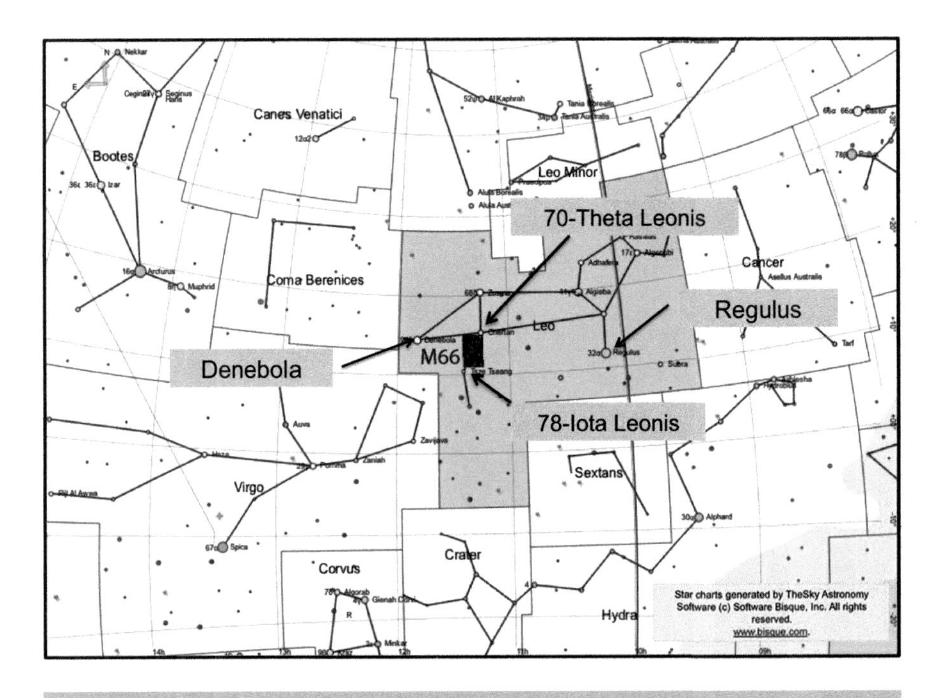

Fig. 4.90 Chart showing the location of M66

Messier 66: Data relating to the image in Figure 4.88

REMOTELY IMAGED FROM MAYHILL, NEW MEXICO
NGC 3627
OBJECT TYPE: Spiral Galaxy

RA: 11h 21m 09s
DEC: +12° 53' 39"
ALTITUDE: +66° 06' 28"
AZIMUTH: 142° 59' 31"
FIELD OF VIEW: 3° 54' 04" x 2° 36' 03"
OBJECT SIZE: 8' X 2.5'
POSITION ANGLE: 89° 52' from North
EXPOSURE TIME: 300 s
DATE: 3rd February
LOCAL TIME: 01h 31m 38s
UNIVERSAL TIME: 08h 31m 38s
SCALE: 3.5 arcsec/pixel
MOON PHASE: 40.78% (waxing)

Telescope Optics
OTA: Takahashi FSQ - ED
Optical Design: Petzval Apochromatic Astrograph
Aperture: 106 mm
Focal Length: 530 mm
F/Ratio: f/5.0
Guiding: External
Mount: Paramount PME

Instrument Package
CCD: SBIG STL-11000M
Pixel Size: 9μm square
Sensor: Frontlit
Cooling: -15°C
Array: 4008 x 2672 pixels
FOV: 188.8 x 233.7 arcmin
Position Angle: 092.7°

Location
Observatory: New Mexico Skies
UTC Minus 7.00 (Daylight savings time is observed)
Minimum Target Elevation: Approx 25 – 45 Degrees
(N or S) 32° 54' Decimal: 32.9 North
(W or E) 105° 31' Decimal: 105.5 West
Elevation: 2225 meters (7298 ft)

Messier 67

Messier 67 (NGC 2682), shown in Figure 4.91, is an open cluster in the constellation of Cancer. It has an apparent size of 30' of arc, so the full moon would just cover it. The image size is just less than 55' on each side, taken with a 20" f/4.4 astrograph from New South Wales. The distance to M67 is about 2,700 ly, so, using the small angle formula, the actual size can be calculated to be in the region of 24 ly. M67 is unusual for a galactic cluster in that it is believed to be about 3 to 4 billion years old. If you compare this with M36, for example, which is a young cluster around 25 million years old, there is a large age gap. As a result, there would be no old, red giants in M36, but red giants are found in M67. M67 was originally discovered by Sicilian astronomer Giovanni Battista Hodierna in 1779. Messier discovered it independently on April 6, 1780. He noted it was a "Cluster of small stars with nebulosity, below the southern claw of Cancer. The position determined from the star Alpha [Acubens]."

Fig. 4.91 Image of Messier 67, an open cluster in Cancer

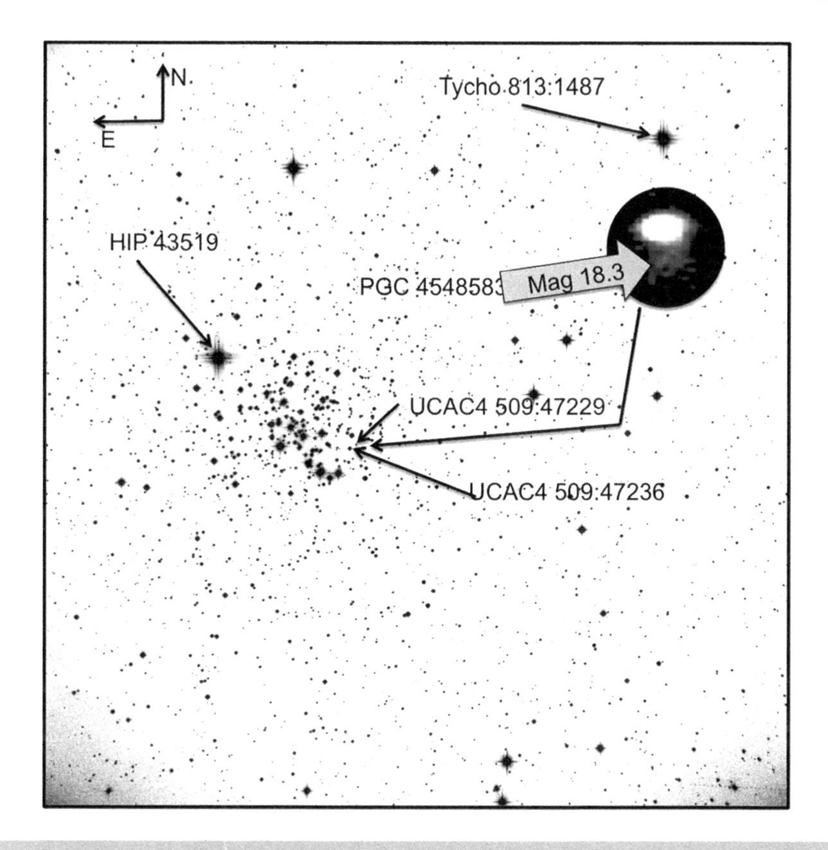

Fig. 4.92 Negative image of Messier 67

The negative image in Figure 4.92 shows M67 with north at the top and east to the left. The identified star HIP 43519 is a line-of-sight star at 737.91 ly distant (Hipparcos Catalog). This is a magnitude 7.83 K0 star. Even fainter is the star UCAC4 509:47236 at magnitude 12.16. So how much fainter can we go? The star UCAC4 509:47229, only just visible on the negative image, is much fainter at magnitude 16.04. This star was selected because immediately below it, one can just make out the faint galaxy PGC 4548583 (shown on the enlargement from the original positive image, below UCAC4 509:47229). This galaxy has a magnitude of 18.32.

The chart in Figure 4.93 shows the location of M67 in the constellation of Cancer. Acubens is the magnitude 4.26 star 65-Alpha Cancri and M67 lies directly to the west of this star. The angular separation between Acubens and Messier 67 is 1° 45'. Tegmine is the magnitude 4.67 Star 16-Zeta2 Cancri. The angular separation between Tegmine and M67 is 11° 4'. There is only one other Messier object in Cancer and that is M44, the Beehive Cluster. There is only one Caldwell object in Cancer -- Caldwell 48, a spiral galaxy.

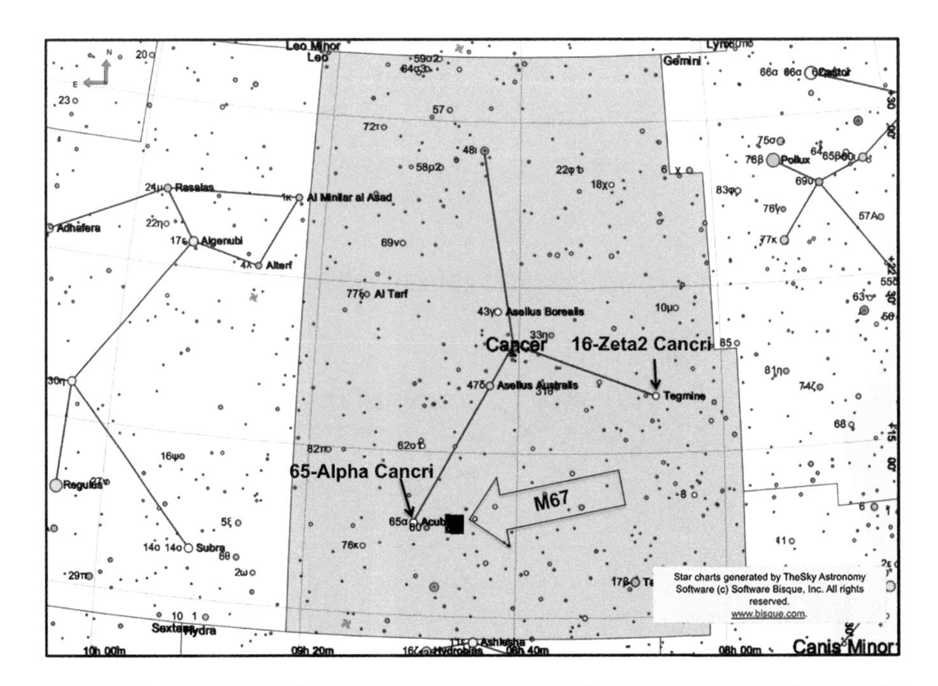

Fig. 4.93 Chart showing the location of M67

The image of Messier 67 was taken from Siding Spring at the local time of 23h 59m 28s, 32 seconds before midnight, on January 17. The Right Ascension is 08h 52m 21s and the Local Sidereal Time was 05h 44m 12s. This tells us that M67 still had a few hours to go before crossing the meridian. RA – LST gives us a difference of 3h 8m 9s. Adding this to the local time provides a meridian transit time of 03h 07m on January 18. The altitude was very low at +27° 41' 14" and the azimuth was only 54° 00' 33". The telescope used is shown in the photograph.

Messier 67: Data relating to the image in Figure 4.91

REMOTELY IMAGED FROM NEW SOUTH WALES, AUSTRALIA
NGC 2682
OBJECT TYPE: Open Cluster
RA: 08h 52m 21s
DEC: +11° 44' 57"
ALTITUDE: +27° 41' 14"
AZIMUTH: 54° 00' 33"
FIELD OF VIEW: 0° 55' 43" x 0° 55' 43"
OBJECT SIZE: 30'
POSITION ANGLE: 90° 51' from North
EXPOSURE TIME: 300 s

DATE: 17th January
LOCAL TIME: 23h 59m 28s
UNIVERSAL TIME: 11h 59m 28s
SCALE: 1.09 arcsec/pixel
MOON PHASE: 72.94% (waning)

Telescope Optics
OTA: Planewave 20" (0.51m) CDK
Optical Design: Corrected Dall-Kirkham Astrograph
Aperture: 510 mm
Focal Length: 2259 mm (focal reducer)
F/Ratio: f/4.4
Guiding: Active Guiding Disabled
Mount: Planewave Ascension 200HR

Instrument Package
CCD: PL09000 CCD Camera
Pixel Size: 12μm square
Sensor: Frontlit
Cooling: -30°C default
Array: 3056 x 3056 pixels
FOV: 55.9 x 55.9 arcmin

Location
Observatory: Siding Spring
UTC +10.00 (Australia Daylight savings time is observed)
31° 16' 24" South
149 03' 52" East
Elevation: 1122 meters (3681 ft)

Messier 68

Messier 68 (NGC 4590), shown in Figure 4.94, is a globular cluster in the constellation of Hydra. It has an apparent diameter of 11' of arc and lies at a distance of 33,000 ly. These values indicate an actual diameter of 106 ly. It was observed by Messier on April 8, 1780: "Nebula without stars below Corvus & Hydra; it is very faint, very difficult to see with the refractors; near it is star of sixth magnitude." William Herschel referred to M68 as, "A cluster of very compressed small stars, about 3 minutes broad and 4 minutes long. The stars are so compressed, that most of them are blended together."

The negative image in Figure 4.95 shows M68 with identified stars on the 1° 37' 39" x 1° 13' 14" frame. The magnitude 5.41 star HIP 61621 is the magnitude 6 star

Fig. 4.94 Image of Messier 68, a globular cluster in Hydra

identified by Charles Messier. This star is a main sequence F0 star at a distance of 112.82 ly. To the northeast of M68 is the variable star GCVS FI Hydra, which is identified on the negative image and incorporated as an enlargement. The enlargement shows the variable and an adjacent star, UCAC4 317:68770, for comparison. The magnitude range of the variable is 10.19 to 17.39 over a 324.1 day period. The brightness changes in the variable an be checked on the AAVSO (American Association of Variable Star Observers) website. AAVSO members track thousands of variable stars and the results are stored on the website. By searching for the variable star it is possible to see recent magnitude estimates and a graph showing the variation over a long period. The star close to the variable, UCAC4 317:68770, has a magnitude of 12.53, so sometimes the variable is brighter or fainter than this star.

The chart in Figure 4.96 shows the location of M68 in the constellation of Hydra, between the stars 46-Gamma Hydrae and Xi Hydrae. The angular separation between M68 and 46-Gamma Hydrae is 9° 38' and between M68 and Xi Hydrae is 15° 12'. There are two other Messier objects in Hydra, M48, an open cluster, and M83, a spiral galaxy. Hydra contains one Caldwell object, C59, which is a planetary nebula.

The image of M68 was taken on February 3 at 01h 00m 05s. The Universal Time was 14h 00m 05s on February 2. The Right Ascension of M68 is 12h 40m 23s, and

Fig. 4.95 Negative image of Messier 68

the Local Sidereal Time was 08h 57m 54s. This tells that there were a few hours to go before M68 reached the meridian: 3h 52m 29s, to be precise. Adding this to the local time gives a meridian transit time of 04h 52m. The image was taken using a 90-mm Takahashi Sky90 telescope.

Messier 68: Data relating to the image in Figure 4.94

REMOTELY IMAGED FROM NEW SOUTH WALES, AUSTRALIA
NGC 4590
OBJECT TYPE: Globular Cluster

RA: 12h 40m 23s
DEC: -26° 50' 03"
ALTITUDE: +39° 34' 12"
AZIMUTH: 100° 34' 55"
FIELD OF VIEW: 1° 37' 39" x 1° 13' 14"
OBJECT SIZE: 11'
POSITION ANGLE: 357° 16' from North
EXPOSURE TIME: 300 s
DATE: 3rd February

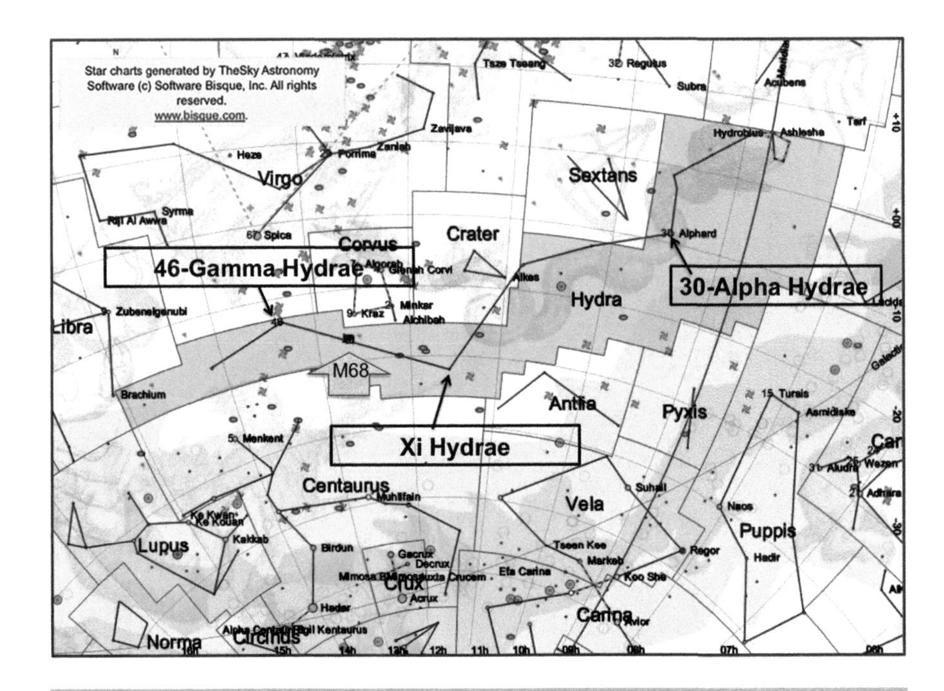

Fig. 4.96 Chart showing the location of M68

LOCAL TIME: 01h 00m 05s
UNIVERSAL TIME: 14h 00m 05s
SCALE: 3.66 arcsec/pixel
MOON PHASE: 32.38% (waxing)

Telescope Optics
OTA: Takahashi Sky 90
Optical Design: Apochromatic Refractor
Aperture: 90 mm
Focal Length: 417 mm
F/Ratio: f/5.6
Guiding: None
Mount: Paramount PME

Instrument Package
CCD: SBIG ST2000 XMC
Colour CMOS
Pixel Size: 7.4μm square
Sensor: Frontlit
Cooling: CNA
Array: 1600 x 1200 pixels
FOV: 60.5 x 80.7 arcmin

Location
Observatory: Siding Spring
UTC +10.00 (Australia Daylight savings time is observed)
31° 16' 24" South
149 03' 52" East
Elevation: 1122 meters (3681 ft)

Messier 69

Messier 69, shown in Figure 4.97, is a relatively faint cluster when compared to some of its fellow Messier globulars. It is found in the constellation of Sagittarius. It has an apparent size of 9.8' of arc and is at a distance of 29,700 ly. This implies an actual diameter of about 85 ly. Messier discovered this cluster on August 31, 1780 and commented, "Nebula without star, in Sagittarius, below his left arm & near the arc; near it is a star of 9th magnitude; its light is very faint, one can only see it under good weather, & the least light employed to illuminate the micrometer wires makes it disappear: its position has been determined from Epsilon Sagittarii." M69 is believed to be only 6,200 ly from the Galactic center.

Fig. 4.97 Image of Messier 69, a globular cluster in Sagittarius

Fig. 4.98 Negative image of Messier 69

Note that the plate solution gives a position angle of 86° 31', so east is up (approximately) and north is to the right in the image. It needs to be rotated counterclockwise by that angle to put north at the top. Messier referred to a magnitude 9 star near M69. This is probably the star, HIP 90772, which is given a magnitude 8.02 in the Hipparcos Catalog. It is labeled on the negative image in Figure 4.98. The angular separation between this star and M69 is roughly 4.5'. HIP 90772 is a main sequence star of spectral type A8 that lies at a distance of 629.65 ly. (Hipparcos Data). There is a bright star in the image field to the south (left) of M69, remembering that east is up in this image. This is the star HIP 90763, which has a magnitude of 5.37 and is a main sequence star of spectral type A5, at a distance of 349.95 ly. There are a number of variables in the field of view of the image. One of these has been identified as GCVS V3480 Sgr. This is very difficult to see on the main image without enlargement. The area in question has been enlarged to pinpoint the variable on the negative image. The maximum magnitude is 15.5 and the minimum magnitude is 18.2. It has a period of 277 days, going through a full cycle of brightness variation in that time.

The chart in Figure 4.99 shows the location of M69 in the constellation of Sagittarius. M69 is another tealeaf at the bottom of the Sagittarius teapot. Messier derived the position of M69 from the known location of the star Epsilon Sagittarii, pointed out on the chart. The angular separation between M69 and the magnitude 1.79 Epsilon is 2.5°. This star is spectral type B9 and is also known as Kaus Australis. It lies at a distance of 144.64 ly. The star that is the tip of the teapot spout

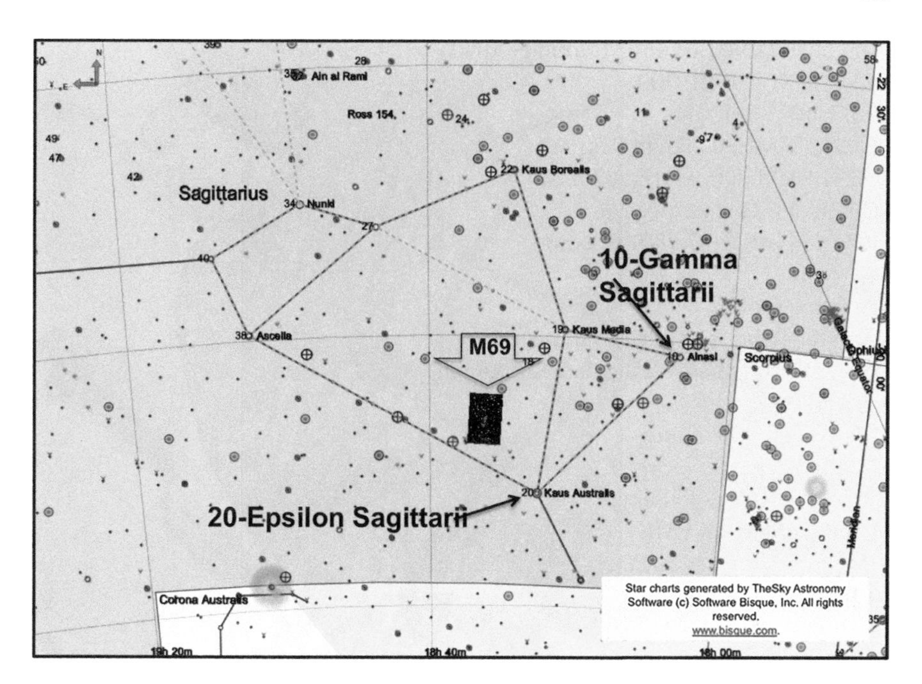

Fig. 4.99 Chart showing the location of M69

is Alnasl. This is 10-Gamma Sagittarii or HIP 88635, a magnitude 2.98 star of spectral type K0 some 96.1 ly away. There is an angular separation of 5° 47' between Alnasl and M69.

The image was taken at 04h 32m 01s on April 10. The corresponding Universal Time was 18h 32m 01s (April 9). Messier 69 has a Right Ascension of 18h 32m 30s, and the Local Sidereal Time was 17h 40m 47s. That indicates less than an hour to go before M69 crossed the meridian at almost 24m past 5 am. The telescope used was a Fast Newtonian Astrograph with an aperture of 400 mm and a focal length of 1,425 mm. The camera is an Apogee Aspen CG16070 with a Class 1 CCD.

Messier 69: Data relating to the image in Figure 4.97

REMOTELY IMAGED FROM NEW SOUTH WALES, AUSTRALIA
NGC 6637
OBJECT TYPE: Globular Cluster

RA: 18h 32m 30s
DEC: -32° 19' 53"
ALTITUDE: +78° 57' 40"
AZIMUTH: 99° 03' 43"
FIELD OF VIEW: 1° 26' 35" x 0° 57' 32"
OBJECT SIZE: 9.8'

POSITION ANGLE: 86° 31' from North
EXPOSURE TIME: 300 s
DATE: 10th April
LOCAL TIME: 04h 32m 01s
UNIVERSAL TIME: 18h 32m 01s (9th April)
SCALE: 1.07 arcsec/pixel
MOON PHASE: 97.78% (waxing)

Telescope Optics
OTA: 16” 0.4m Astro Systeme Austria
Optical Design: Fast Newtonian Astrograph
Aperture: 400 mm
Focal Length: 1425 mm
F/Ratio: f/3.5
Guiding: Off-Axis Guiding
Mount: Paramount PME

Instrument Package
CCD: Apogee Aspen CG16070 Class 1 CCD
Pixel Size: 7.4μm square
Sensor: KAI – 16070 ABG
Cooling: -30°C default
Array: 4864 x 3232 pixels
FOV: 88.3 x 58.7 arcmin

Location
Observatory: Siding Spring
UTC +10.00 (Australia Daylight savings time is observed)
31° 16’ 24” South
149 03’ 52” East
Elevation: 1122 meters (3681 ft)

Messier 70

Messier 70, shown in Figure 4.100, is a globular cluster that is relatively small in comparison with other Messier globulars like M3, M4, M5 or M13. If you compare the color image of M70, taken with a modest 90-mm telescope from Siding Spring in Australia, with the image of M4 taken with the same equipment and settings, you can see the difference. M70 has an apparent size of 8’ of arc, whereas M4 is 36’. M70 lies at a distance of 29,300 ly, which would indicate an actual size of about 68 ly. In reality this is only slightly smaller than the size of M4, which is 75 ly, however, M70 is almost 4 times further away than M4, which is 7,200 ly distant. Messier discovered M70 on August 31, 1780, the same night that he discovered the

Fig. 4.100 Image of Messier 70, a globular cluster in Sagittarius

nearby M69. He commented, "Nebula without star, near the preceding (M69) & on the same parallel: near it is a star of the ninth magnitude & four small telescopic stars, almost on the same straight line, very close to one another, & are situated above the nebula, as seen in a reversing telescope; the nebula [location] was determined from the same star Epsilon Sagittarii."

The negative image in Figure 4.101 shows M70 with north at the top and east to the left. Messier made reference to a nearby magnitude 9 star. He could have been referring to the brightest nearby star, which is HIP 91713, but is magnitude 7.78 in the Hipparcos Catalog. Otherwise, there are a number of fainter magnitude 9 stars to choose from. If he was looking much closer in to M70, it could be the star Tycho 7411:25, but this is magnitude 9.89. Messier commented on 4 stars in a straight line above the nebula. He was using an inverting telescope, so when seen with north at the top in the negative image, they are below M70. The four stars are identified in the negative image. The star on the right is HIP 91866 with a magnitude of 9.21. The next stars are Tycho 7411:192 with a magnitude of 9.09, Tycho 7411:87 with a magnitude of 9.64, and the leftmost star Tycho 7411:684 with a magnitude of 9.18.

M70 is not far from M69, which you can see by comparing the M70 chart in Figure 4.102 with that of M69 in Figure 4.99. Messier referred to them being on the same parallel. M70 has an RA of 18h 44m and M69 an RA of 18h 33m. There

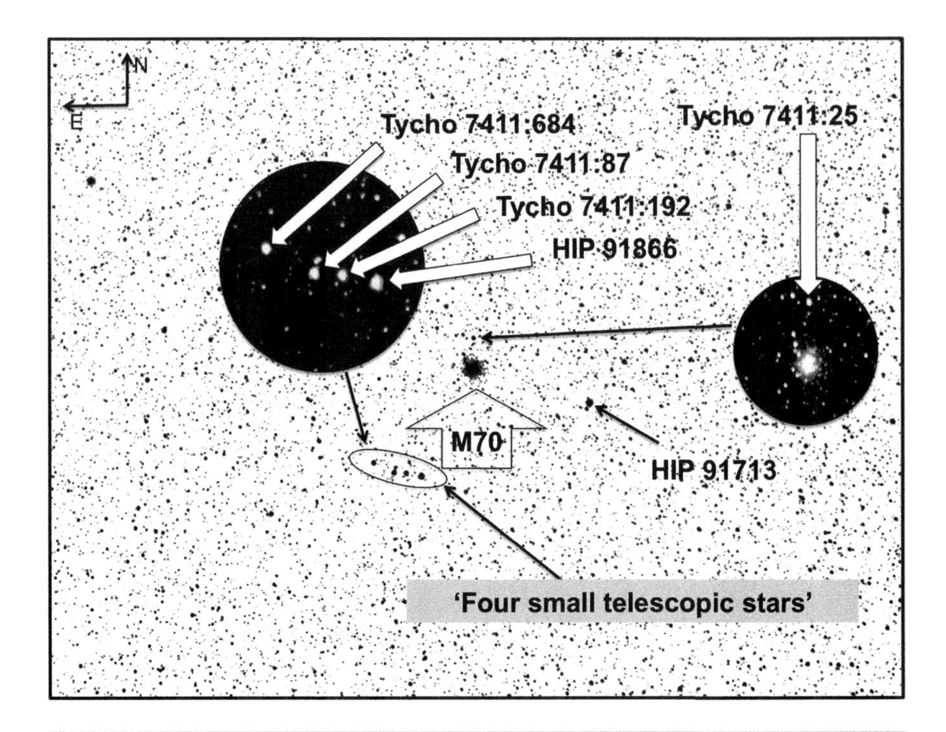

Fig. 4.101 Negative image of Messier 70

is roughly 2.5° angular separation between the two Messier globulars. M70 is another tealeaf at the bottom of the teapot, situated between the star 22-Lambda Sagittarii or Kaus Borealis, and the star Alpha Sagittarii or Rukbat. The angular separation between Kaus Borealis and M70 is roughly 7° 34'. Kaus Borealis is magnitude 2.82 of spectral type K1 and is at a distance of 77.29 ly. The angular separation between Rukbat and M70 is just less than 12°. Rukbat is magnitude 3.96 of spectral type B8 and is at a distance of 169.87 ly.

The image was taken on April 1 at 04h 01m 53s in Siding Spring. The corresponding Universal Time was 17h 01m 53s. (March 31). The Right Ascension of M70 is 18h 44m 19s, and the Local Sidereal Time was 15h 34m 55s. It would therefore take M70 over 3 hours to reach the meridian. The exact time to go was 03h 09m 24s, giving a meridian transit time for M70 of 07h 11m 17s. The equipment image shows the 90-mm telescope.

Messier 70: Data relating to the image in Figure 4.100

REMOTELY IMAGED FROM NEW SOUTH WALES, AUSTRALIA
NGC 6681
OBJECT TYPE: Globular Cluster

RA: 18h 44m 19s
DEC: -32° 16' 13"

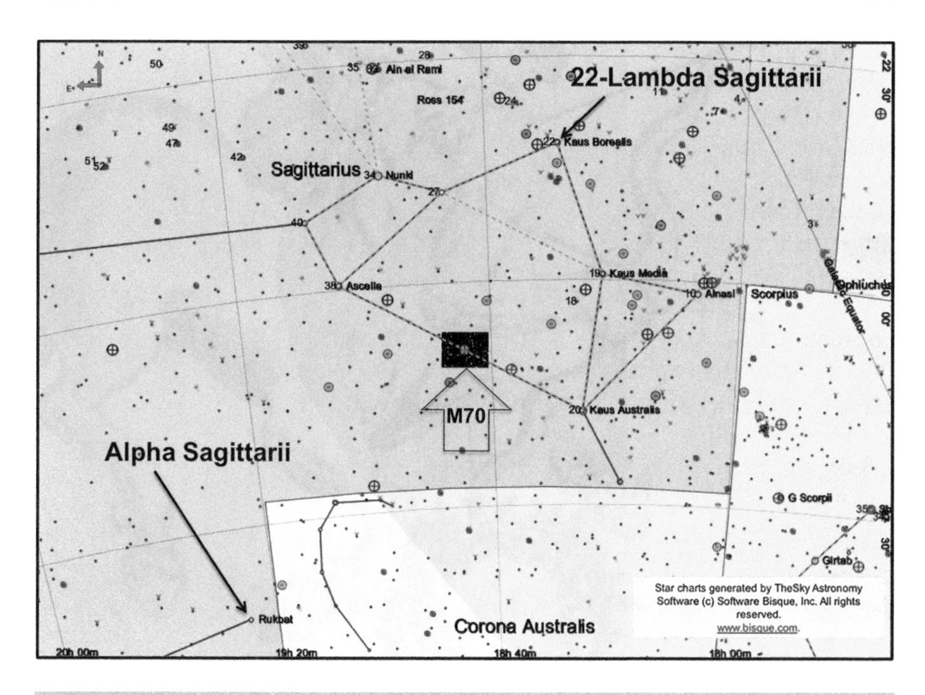

Fig. 4.102 Chart showing the location of M70

ALTITUDE: +50° 03' 30"
AZIMUTH: 104° 22' 33"
FIELD OF VIEW: 1° 37' 44" x 1° 13' 18"
OBJECT SIZE: 8’
POSITION ANGLE: 357° 24' from North
EXPOSURE TIME: 300 s
DATE: 1st April
LOCAL TIME: 04h 01m 53s
UNIVERSAL TIME: 17h 01m 53s
SCALE: 3.67 arcsec/pixel
MOON PHASE: 17.42% (waxing)

Telescope Optics
OTA: Takahashi Sky 90
Optical Design: Apochromatic Refractor
Aperture: 90 mm
Focal Length: 417 mm
F/Ratio: f/5.6
Guiding: None
Mount: Paramount PME

Instrument Package
CCD: SBIG ST2000 XMC
Colour CMOS
Pixel Size: 7.4μm square
Sensor: Frontlit
Cooling: CNA
Array: 1600 x 1200 pixels
FOV: 60.5 x 80.7 arcmin

Location
Observatory: Siding Spring
+10.00 (Australia Daylight savings time is observed)
31° 16' 24" South
149 03' 52" East
Elevation: 1122 meters (3681 ft)

Messier 71

Messier 71 in the constellation of Sagitta, shown in Figure 4.103, is marked as a globular cluster, but for a long time, there has been discussion about whether this should be classified as either a globular or open cluster. It is currently regarded as a globular. It has an apparent size of 7.2' of arc and is at a distance of 13,000 ly. With those values, the actual size would be about 27 ly. The field of view of the telescope and camera combination is 37' 16" x 24' 51", so M71 appears large in Figure 4.103 in comparison with the M70 image with its large field of 97' 44" x 73' 18". Messier 71 was originally discovered by Swiss astronomer Jean-Philippe de Chéseaux in 1745-46, then independently rediscovered by German astronomer Johann Koehler between 1772 and 1779. Again, it was discovered independently by Messier's friend Méchain on June 28, 1780. Messier noted, "Nebula discovered by M. Méchain on June 28, 1780, between the stars Gamma and Delta Sagittae… its light is very faint & it contains no star; the least light makes it disappear. It is situated about 4 degrees below that discovered in Vulpecula [M27]."

The negative image in Figure 4.104 shows M71 and a few objects in the field. Note that the camera was not oriented on the telescope with north at the top. The compass on the image shows the directions of north and east. The bright star at bottom right is HIP 97840, which is a magnitude 8.34 star of spectral class K lying at a distance of 821.55 ly. The fainter star above M71 is Tycho 1624:1944 with a magnitude of 10.1. A well-known variable star in M71, GCVS Z Sagittae, is identified in the image and varies between magnitude 13.5 and 15.7. There are not many variable stars in M71, and this one is only one of eight listed.

The chart in Figure 4.105 shows the location of M71 in the little constellation of Sagitta with the plate-solved image superimposed. The tilt of the camera is obvious

Fig. 4.103 Image of Messier 71, a globular cluster in Sagitta

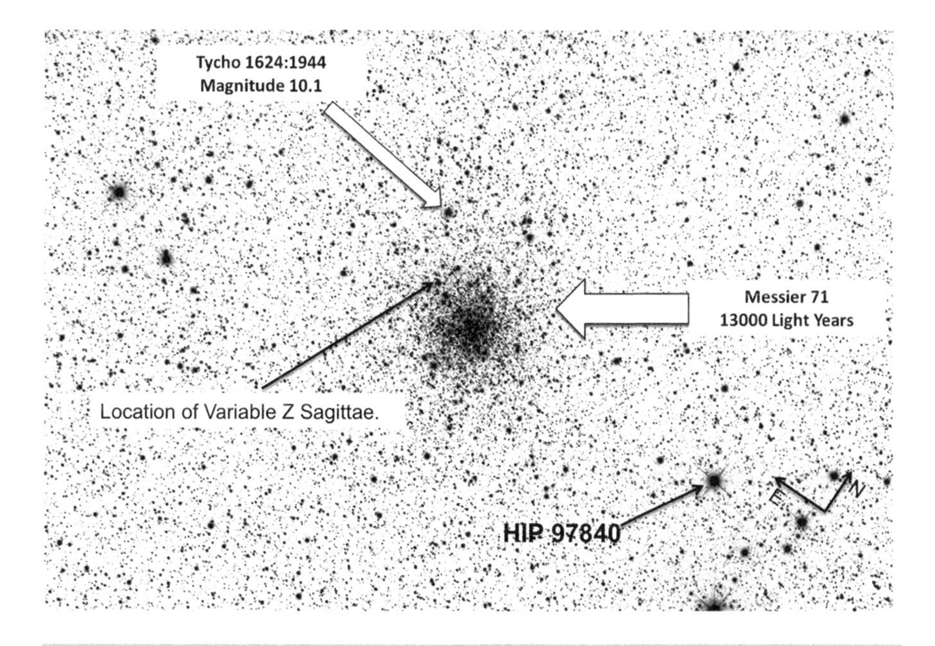

Fig. 4.104 Negative image of Messier 71

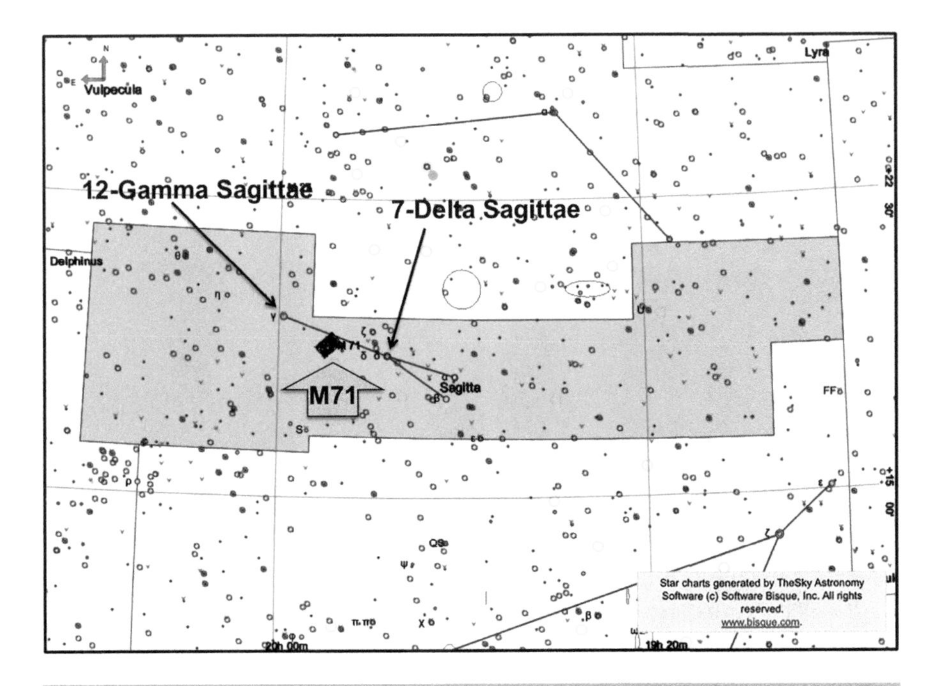

Fig. 4.105 Chart showing the location of M71

here as north is at the top of the chart. The cluster lies between two stars in the Sagitta asterism, just to the south (below) the line joining the two stars. To the east (left) is 12-Gamma Sagittae (HIP 98337). This is a magnitude 3.51 star of spectral type K5 and is given a distance of 274.08 ly in the Hipparcos Catalog. There is an angular separation of 1° 22' between this star and Messier 71. To the west (right) of M71 is the star 7-Delta Sagittae (HIP 97365). This is a magnitude 3.68 star. It is in fact a close double star that needs spectroscopy to separate the two stars. Professor Jim Kaler at the University of Illinois describes this in detail. He points out that the distance of 448 ly was found by direct parallax. I measure an angular separation of about 1° 32' between M71 and 7-Delta Sagittae.

The image was taken at local time 00h 07m 55s on September 22 in Spain. The corresponding Universal Time was 22h 07m 55s (September 21). The Right Ascension of M71 is 19h 54m 31s, and the Local Sidereal Time was 22h 03m 29s. This shows that M71 has already crossed the meridian from east to west at 21h 59m on September 21. The telescope used to take the image of M13 was a Corrected Dall-Kirkham Astrograph with an aperture of 318 mm mounted on a Paramount ME.

Messier 71: Data relating to the image in Figure 4.103

REMOTELY IMAGED FROM SPAIN
NGC 6838
OBJECT TYPE: Globular Cluster

RA: 19h 54m 31s
DEC: +18° 49' 43"
ALTITUDE: +55° 59' 12"
AZIMUTH: 244° 30' 44"
FIELD OF VIEW: 0° 37' 16" x 0° 24' 51"
OBJECT SIZE: 7.2'
POSITION ANGLE: 220° 32' from North
EXPOSURE TIME: 300 s
DATE: 22nd September
LOCAL TIME: 00h 07m 55s
UNIVERSAL TIME: 22h 07m 55s (21st September)
SCALE: 0.73 arcsec/pixel
MOON PHASE: 66.80% (waning)

Telescope Optics
OTA: Planewave CDK
Optical Design: Corrected Dall-Kirkham Astrograph
Aperture: 318 mm
Focal Length: 2541 mm
F/Ratio: f/7.9
Guiding: External
Mount: Paramount ME

Instrument Package
CCD: SBIG-STXL-6303E
A/D Gain: 1.47 e-/ADU
Pixel Size: 9μm
Resolution: 0.73 arcsec/pixel
Sensor: Front Illuminated
Cooling: -20°C default
Array: 3072 x 2048 (6.3 Megapixels)
FOV: 37.41 x 24.94 arcmin

Location
Observatory: AstroCamp – MPC Code -189
UTC +1.00 (Daylight savings time is observed)
Minimum Target Elevation: Approx 30 – 40 Degrees
N 38° 09'
W 002° 19'
Elevation: 1650 meters (5413 ft)

Messier 72

Messier 72 (NGC 6981), shown in Figure 4.106, is a globular cluster in the constellation of Aquarius. It is quite a long way from us, so it is a relatively difficult object to image compared to many of the other Messier globulars. Pierre Méchain originally discovered the cluster in 1780. Messier observed it on October 4 of the same year, noting, "Nebula seen by M. Méchain in the night of August 29-30, 1780, above the neck of Capricorn.... near it is a small telescopic star: the position was determined from the star Nu Aquarii, of fifth magnitude." In 1810, William Herschel used the 40-ft telescope at Slough to observe M72 and commented: "It is a cluster of stars of a round figure, but the very faint stars on the outside of these sorts of clusters are generally a little dispersed so as to deviate from a very perfect circular form."

Messier 72 lies at a distance of 55,400 ly and has an apparent diameter of 6.6' of arc. This gives it an actual diameter of 106 ly. Thus, it is a considerable size, but it is at a distance roughly twice of that to the center of the Galaxy and as a result appears much small. Two foreground stars are identified in the field on the negative image in Figure 4.107 to give an idea of the range of magnitudes visible. Note that

Fig. 4.106 Image of Messier 72, a globular cluster in Aquarius

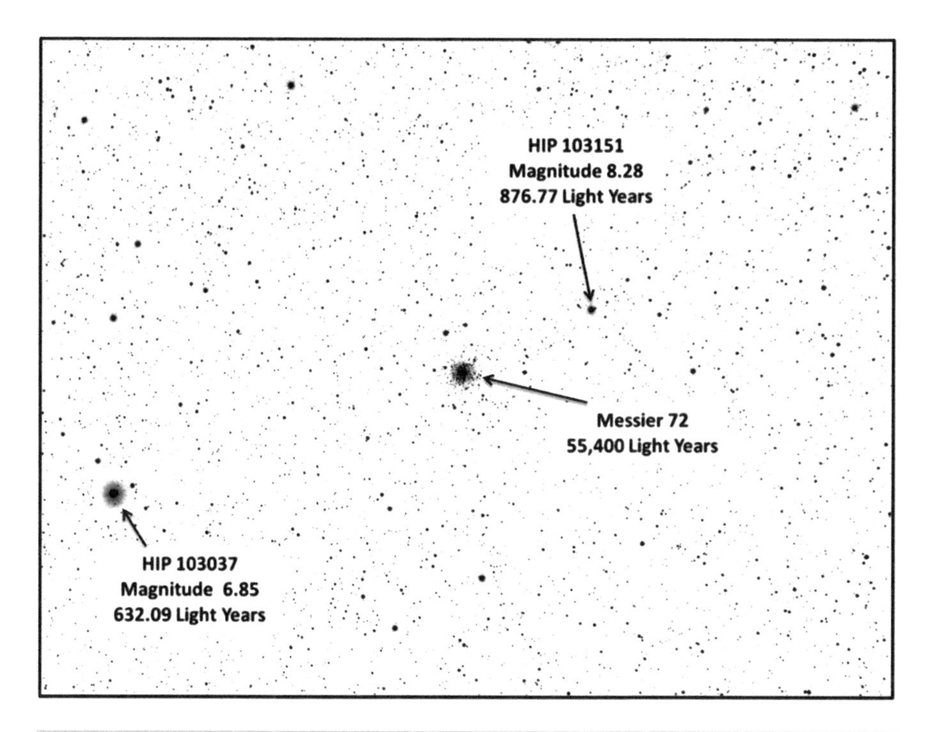

Fig. 4.107 Negative image of Messier 72

north is to the right and east is up on the image. HIP 103037 is magnitude 6.85 and is spectral type K4 at a distance of 632.09 ly. HIP 103151 is magnitude 8.28 and is spectral type F7 at a distance of 876.77 ly. The stars visible on the edge of the cluster in the image are in the region of magnitude 14 or 15.

Messier referred to the star Nu Aquarii, which he used to help calculate the position of M72. The star is 13-Nu Aquarii or HIP 104459 and is magnitude 4.5 in the Hipparcos Catalog. It is a G8 star at a distance of 163.65 ly. Measuring the angular separation between Nu Aquarii and Messier 72 gave me a value of roughly 4° 5'. This star is labeled on the chart in Figure 4.108 and also the star 2-Epsilon Aquarii or HIP 102618. This star is of spectral type A1 with a magnitude of 3.78 at a distance of 229.53 ly. The angular separation between Epsilon Aquarii and M72 is 3° 23'. There are two other Messier objects in Aquarius, M73 and the much more impressive globular cluster M2.

The 10-minute image was taken from Siding Spring, New South Wales on October at 22h 34m 17s. The corresponding Universal Time was 11h 34m 17s. The Right Ascension of M72 is 20h 54m 23s, and the Local Sidereal Time was 23h 55m 19s. Thus, M72 crossed the meridian 3h 0m 56s before. The meridian transit time of M72 was 19h 33 21s. The single-shot color image camera was used with a 90-mm refractor to obtain the image.

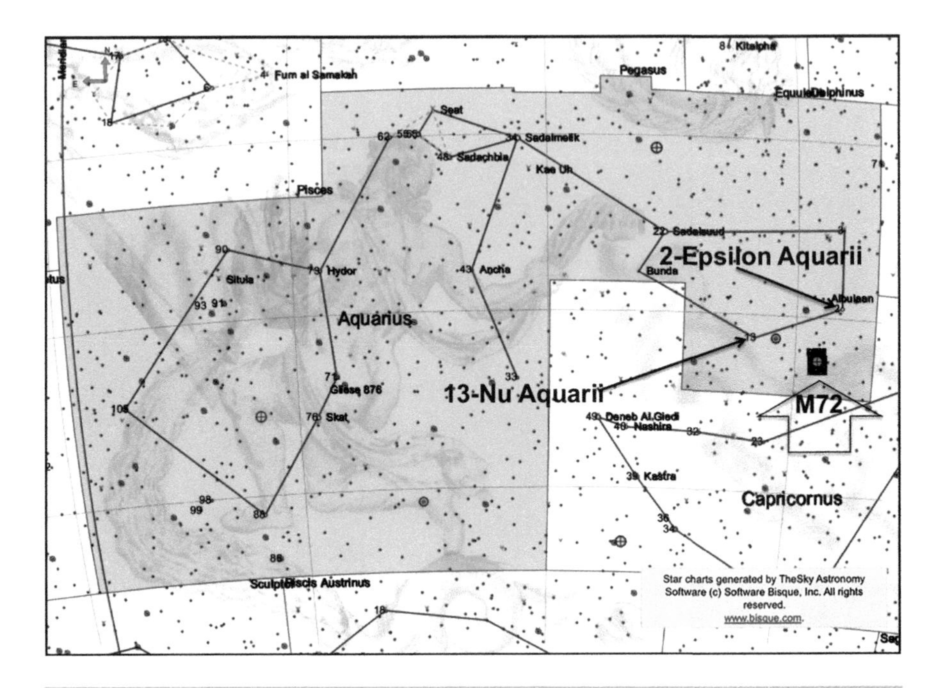

Fig. 4.108 Chart showing the location of M72

Messier 72: Data relating to the image in Figure 4.106

REMOTELY IMAGED FROM NEW SOUTH WALES, AUSTRALIA
NGC 6981
OBJECT TYPE: Globular Cluster

RA: 20h 54m 23s
DEC: -12° 28' 13"
ALTITUDE: +44° 25' 00"
AZIMUTH: 283° 55' 55"
FIELD OF VIEW: 1° 37' 41" x 1° 13' 16"
OBJECT SIZE: 6.6’
POSITION ANGLE: 90° 21' from North
EXPOSURE TIME: 600 s
DATE: 27th October
LOCAL TIME: 22h 34m 17s
UNIVERSAL TIME: 11h 34m 17s
SCALE: 3.66 arcsec/pixel
MOON PHASE: 9.43% (waning)

Telescope Optics
OTA: Takahashi Sky 90
Optical Design: Apochromatic Refractor

Aperture: 90 mm
Focal Length: 417 mm
F/Ratio: f/5.6
Guiding: None
Mount: Paramount PME

Instrument Package
CCD: SBIG ST2000 XMC
Colour CMOS
Pixel Size: 7.4μm square
Sensor: Frontlit
Cooling: CNA
Array: 1600 x 1200 pixels
FOV: 60.5 x 80.7 arcmin
Location
Observatory: Siding Spring
UTC +10.00 (Australia Daylight savings time is observed)
31° 16' 24" South
149 03' 52" East
Elevation: 1122 meters (3681 ft)

Messier 73

Messier 73, shown in Figure 4.109, is an unusual Messier object in that it is simply a small group of stars of no apparent particular significance. M73 is very close to M72, and Messier observed them both on the same evening. On October 4, 1780 Messier wrote that M73 was a "Cluster of three or four small stars, which resembles a nebula at first sight, containing a little nebulosity: this cluster is situated on the same parallel as the preceding nebula [M72] its position was determined from the same star Nu Aquarii." Looking at my image of the object, it is difficult to understand why Messier would say that it resembled a nebula, until one remembers that the quality of the telescopes available now far exceed the quality of Messier's instruments in the 18th Century. Almost two centuries later, Robert Burnham Jr. took issue with the presence of "nebulosity," commenting "On the last point [the presence of nebulosity] Messier was definitely in error, as the best of modern photographs show no signs of nebulosity in the group."

In Figure 4.110 the negative image of the original shows an enlarged view of M73 and identifies each of the four stars in the object. The brightest star is Tycho 5778:802, which has a magnitude of 10.31. Tycho 5778:509 is magnitude 11.1, UCAC4 387:141727 is magnitude 11.57 and the faintest of the four is UCAC4 387:141722 at a magnitude of 12.14. The diameter of the cluster is roughly only 1'

Fig. 4.109 Image of Messier 73, an asterism in Aquarius

20" of arc measured on my solved image, and evidently the likelihood of the four stars not being in a cluster is low as a result of this. SEDS gives a distance of 2,500 ly, which would give the cluster an estimated actual diameter of about 0.95 of a light-year – assuming it is a cluster. The bright star on the image is HIP 103640, which has a magnitude of 6.6.

The chart in Figure 4.111 shows the location of Messier 73 in the constellation of Aquarius. Notice how close M73 is to the globular cluster M72. The angular separation of M73 and M72 is 1° 20' from my chart measurement. There is an impressive Caldwell object close by M73. This is Caldwell 55 or NGC 7009, also known as the Saturn Nebula. There is 1° 48' of separation between M73 and the Saturn Nebula.

The image was taken on April 17 from Siding Spring at 04h 02m 57s. The corresponding Universal Time was 18h 02m 57s (April 16). The Right Ascension of M73 is 20h 59m 52s, and the Local Sidereal Time was 17h 39m 14s. Thus, M73 would cross the meridian in 03 h 20m 38s. Meridian transit of M73 would be at 07h 23m 19s. The single-shot color camera was used with a 90-mm refractor to obtain the image.

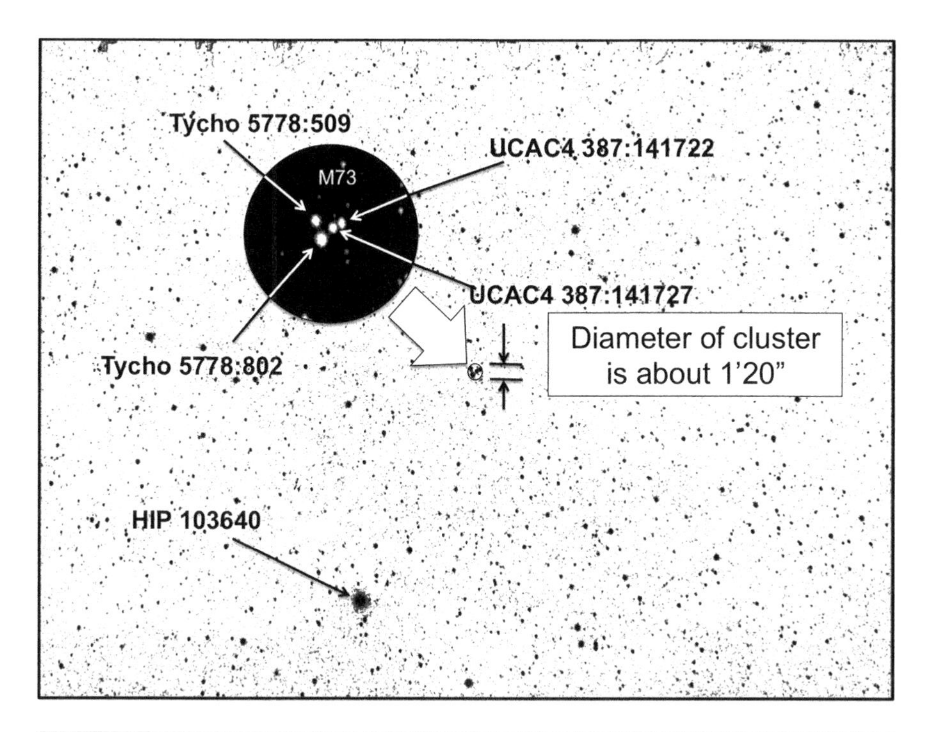

Fig. 4.110 Negative image of Messier 73

Messier 73: Data relating to the image in Figure 4.109

REMOTELY IMAGED FROM NEW SOUTH WALES, AUSTRALIA
NGC 6994
OBJECT TYPE: Open Cluster

RA: 20h 59m 52s
DEC: -12° 34' 03"
ALTITUDE: +40° 21' 14"
AZIMUTH: 79° 32' 45"
FIELD OF VIEW: 1° 37' 44" x 1° 13' 18"
OBJECT SIZE: 2.8'
POSITION ANGLE: 357° 12' from North
EXPOSURE TIME: 300 s
DATE: 17th April
LOCAL TIME: 04h 02m 57s
UNIVERSAL TIME: 18h 02m 57s (16th April)
SCALE: 3.67 arcsec/pixel
MOON PHASE: 74.60% (waning)

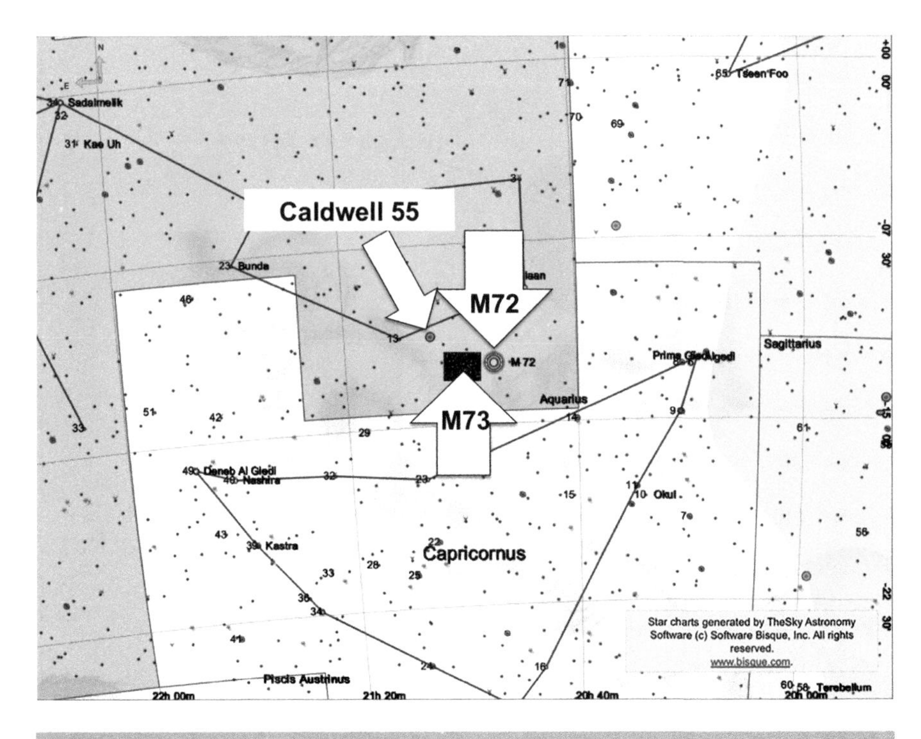

Fig. 4.111 Chart showing the location of M73

Telescope Optics
OTA: Takahashi Sky 90
Optical Design: Apochromatic Refractor
Aperture: 90 mm
Focal Length: 417 mm
F/Ratio: f/5.6
Guiding: None
Mount: Paramount PME

Instrument Package
CCD: SBIG ST2000 XMC
Colour CMOS
Pixel Size: 7.4μm square
Sensor: Frontlit
Cooling: CNA
Array: 1600 x 1200 pixels
FOV: 60.5 x 80.7 arcmin

Location
Observatory: Siding Spring
UTC +10.00 (Australia Daylight savings time is observed)
31° 16' 24" South
149 03' 52" East
Elevation: 1122 meters (3681 ft)

Messier 74

Messier 74 (NGC 628), shown in Figure 4.112, is a spiral galaxy in the constellation of Pisces. It is a face-on spiral that allows us to see the spiral structure with 2 spiral arms. It is not particularly bright, so is not the easiest galaxy to observe without a large aperture. Pierre Méchain discovered the galaxy in September 1780 and told his friend Messier, who added it to his catalog as number 74. Messier

Fig. 4.112 Image of Messier 74, a spiral galaxy in Pisces

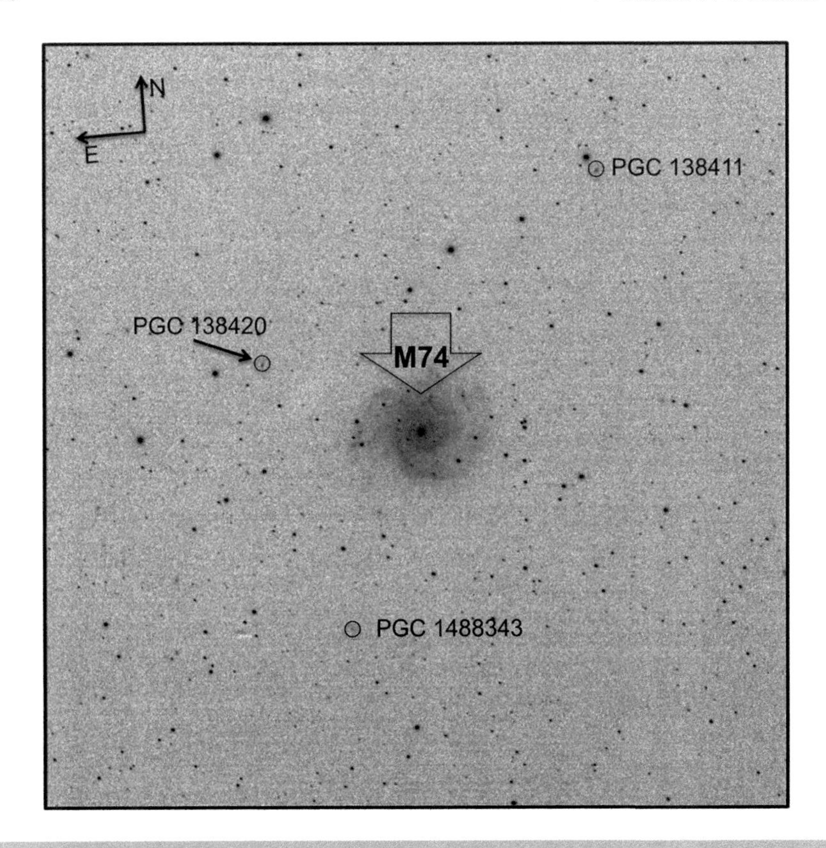

Fig. 4.113 Negative image of Messier 74

observed it on October 18, 1780, noting: "Nebula without stars, near the star Eta Piscium, seen by M. Méchain at the end of September 1780. This nebula doesn't contain any stars; it is fairly large, very obscure, and extremely difficult to observe; one can recognize it with more certainty in fine, frosty conditions." William Herschel used the 40-ft telescope in Slough to observe M74 on December 28,1799, commenting that it was "Very bright in the middle, but the brightness confined to a very small part, and is not round; about the bright middle is a very faint nebulosity to a considerable extent. The bright part seems to be of resolvable kind, but my mirror has been injured by condensed vapors."

M74 has dimensions of 10.2' x 9.5' of arc. It lies at a distance of 35 million ly, giving it actual dimensions of roughly 104,000 ly x 97,000 ly. While it has been noted that M74 is not particularly bright, there are more galaxies captured in the image frame that are considerably fainter. Their locations are circled on the negative image in Figure 4.113. These galaxies are listed in the Principal Galaxy Catalog. PGC 1488343 is located to the south east of M74 (north is at the top) and has a magnitude of 17.78 with dimensions of 0.4' x 0.2' listed in the PGC catalog. PGC 138420 is found slightly northeast of M74 with a magnitude of 17.16 and

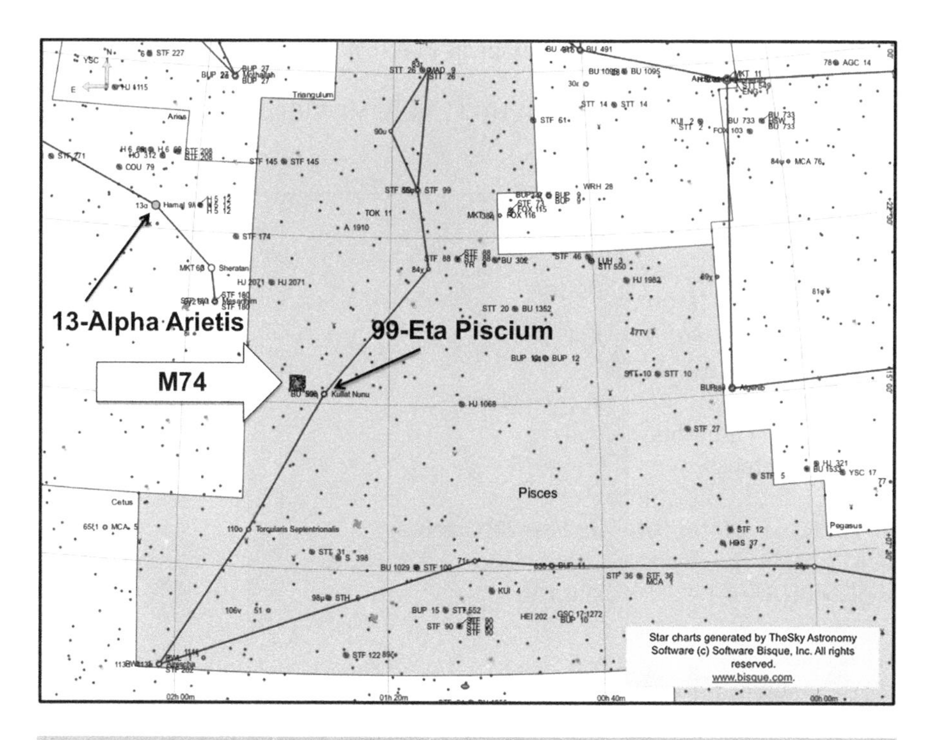

Fig. 4.114 Chart showing the location of M74

dimensions 0.4' x 0.2'. PGC 138411 is located to the northwest of M74 with a magnitude of 16.95 and dimensions 0.4' x 0.3'.

The chart in Figure 4.114 shows the location of Messier 74 in the constellation of Pisces. Messier commented that the galaxy was near the star Eta Piscium, and this is labeled on the chart as 99-Eta Piscium to give it its full Flamsteed- Bayer designation. This star, of spectral type G8, has the name "Kullat Nunu" and is HIP 7097 in the Hipparcos Catalog. It has a magnitude of 3.62 and is 294.10 ly away. There is an angular separation of roughly 1° 20' between Eta Piscium and M74. The star 13-Alpha Arietis is also labeled on the chart. This is the star Hamal in Aries also designated HIP 9884. This is a magnitude 2 spectral type K2 star at a distance of 65.92 ly. There is an angular separation of about 10° 28' between Hamal and Messier 74.

The image was taken on October 1 at 23h 24m 55s local time in New Mexico. This corresponded to a Universal Time of 05h 24m 55s (October 2). The Right Ascension of M74 is 01h 37m 37s, and the Local Sidereal Time was 23h 07m 07s. Subtracting one from the other gives a difference of 02h 30m 30s. Adding this to the local time when the image was taken gives a meridian transit time of 01h 55m 25s. The telescope used to take the image of M74 was a Takahashi TOA-150 apochromatic refractor with an aperture of 150 mm.

Messier 74: Data relating to the image in Figure 4.112

REMOTELY IMAGED FROM MAYHILL, NEW MEXICO
NGC 628
OBJECT TYPE: Spiral Galaxy

RA: 01h 37m 37s
DEC: +15° 52' 07"
ALTITUDE: +52° 02' 58"
AZIMUTH: 107° 16' 29"
FIELD OF VIEW: 47' 04"' X 47' 04"
OBJECT SIZE: 10.2' X 9.5'
POSITION ANGLE: 177° 59' from North
EXPOSURE TIME: 600 seconds
DATE: 1st October
LOCAL TIME: 23h 24m 55s
UNIVERSAL TIME: 05h 24m 55s (2nd October)
SCALE: 1.38 arcsec/pixel
MOON PHASE: 1.46% (waxing)

Telescope Optics
OTA: Takahashi TOA-150
Optical Design: Apochromatic Refractor
Aperture: 150 mm
Focal Length: 1095 mm
F/Ratio: f/7.3
Guiding: Internal
Mount: Paramount GTS

Instrument Package
SBIG ST-4000M One-Shot Color CCD
A/D Gain: 0.6e-/ADU
Pixel Size: 7.4um square
Resolution: 1.45 arcsec/pixel
Sensor: Frontlit
Cooling: -20°C Winter (-10°C Summer)
Array: 2048 x 2048 (8.3 Megapixels)
FOV: 49.6 x49.6 arcmin

Location
Observatory: New Mexico Skies
UTC Minus 7.00 (Daylight savings time is observed)
Minimum Target Elevation: Approx 25 – 45 Degrees
(N or S) 32° 54' Decimal: 32.9 North
(W or E) 105° 31' Decimal: 105.5 West
Elevation: 2225 meters (7298 ft)

Chapter 5

Messier 75 to Messier 110

Messier 75

Messier 75 (NGC 6864), shown in Figure 5.1, is a globular cluster located in the constellation of Sagittarius. Even with a fairly large telescope (17-inch aperture), the image is not too impressive and the globular appears small. The reason is that it is a long way away – far beyond the distance to our Galactic center. The distance to M75 is 67,500 ly, which may not seem too far in terms of the distance to massive galaxies, but for an object that is only a small component of our Galaxy, that is a long way. Messier observed it on October 18, 1780 and commented that it was a "Nebula without star, between Sagittarius & the head of Capricorn; seen by M. Méchain on August 27 & 28, 1780." On July 28, 1830, it was observed by John Herschel, who did not seem impressed: "Not bright; small; round; pretty suddenly brighter toward the middle; 2' diameter; mottled, but not resolved. An insignificant object."

The negative image in Figure 5.2 shows Messier 75 and identifies some stars in the field. The bright star towards the top of the image (north is up) is HIP 99046. This is a spectral type G2 star of magnitude 7.99 at a distance of 154.14 ly. The star directly to the west (right) of M75 is Tycho 6326:1260. This star is further away at 210.42 ly, with a magnitude of 9.40. Looking at the periphery of M75 itself is the faint star UCAC4 341:195731. This has a magnitude of 15.47. The apparent size of M75 is 6.8' of arc, so with M75 at a distance of 67,500 ly, it will have an actual diameter of roughly 134 ly.

M75 is shown on the chart in Figure 5.3 in its location in the constellation of Sagittarius. If you draw a line directly from the star 59b01 Sagittarii (HIP 98162)

L. Adam, *Imaging the Messier Objects Remotely from Your Laptop*, The Patrick Moore Practical Astronomy Series, https://doi.org/10.1007/978-3-319-65385-3_5

Fig. 5.1 Image of Messier 75, a globular cluster in Sagittarius

to the star 6-Alpha2 Capricorni (Algedi or HIP 100064), it goes directly through Messier 75. HIP 98162 is a K3 star of magnitude 4.54 at a distance of 1,207.99 ly (Hipparcos Catalog). There is an angular separation of 5° 39' between this star and M75. Algedi is a double star of magnitude 3.58 and spectral type G6/G8, which is 108.68 ly distant. There is an angular separation of roughly 9° 46' between Algedi and Messier 75.

The image was taken on October 29 from Siding Spring at 20h 44m 14s. The corresponding Universal Time was 09h 44m 14s. The Right Ascension of M75 is 20h 07m 03s, and the Local Sidereal Time was 22h 12m 51s. Thus, M75 crossed the meridian 02h 5m 48s prior, making the meridian transit time of M75 18h 38m 26s. The telescope used to take the image of M75 was a Planewave Corrected Dall-Kirkham Astrograph with an aperture of 431 mm and a focal length of 2,912 mm. The mount used was a Planewave Ascension 200HR.

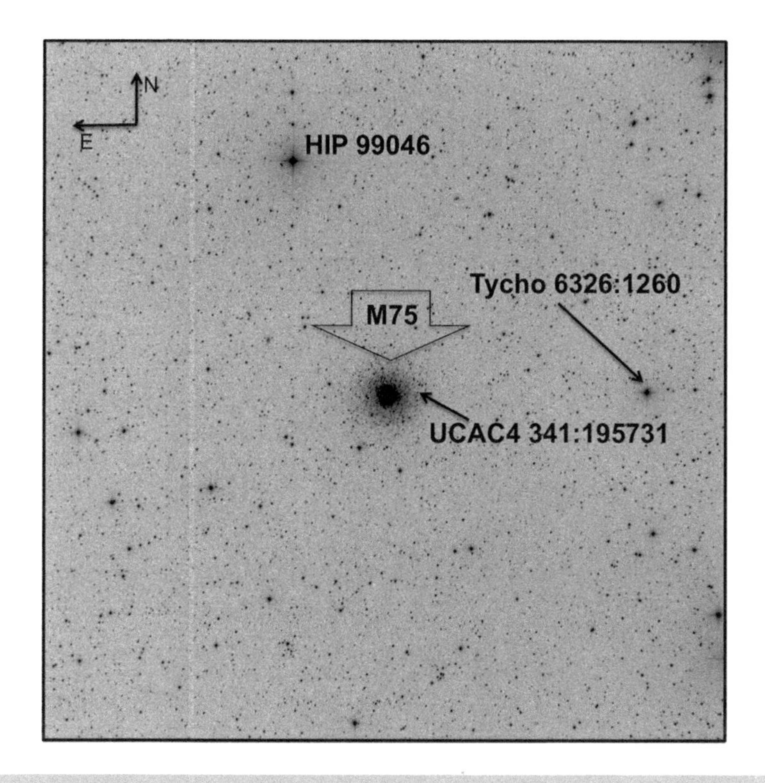

Fig. 5.2 Negative image of Messier 75

Messier 75: Data relating to the image in Figure 5.1

REMOTELY IMAGED FROM NEW SOUTH WALES, AUSTRALIA
NGC 6864
OBJECT TYPE: Globular Cluster

RA: 20h 07m 03s
DEC: -21° 52' 14"
ALTITUDE: +60° 28' 11"
AZIMUTH: 280° 46' 09"
FIELD OF VIEW: 0° 43' 54" x 0° 43' 54"
OBJECT SIZE: 6.8'
POSITION ANGLE: 90° 54' from North
EXPOSURE TIME: 300 s
DATE: 29th October
LOCAL TIME: 20h 44m 14s
UNIVERSAL TIME: 09h 44m 14s
SCALE: 0.64 arcsec/pixel
MOON PHASE: 1.68% (waning)

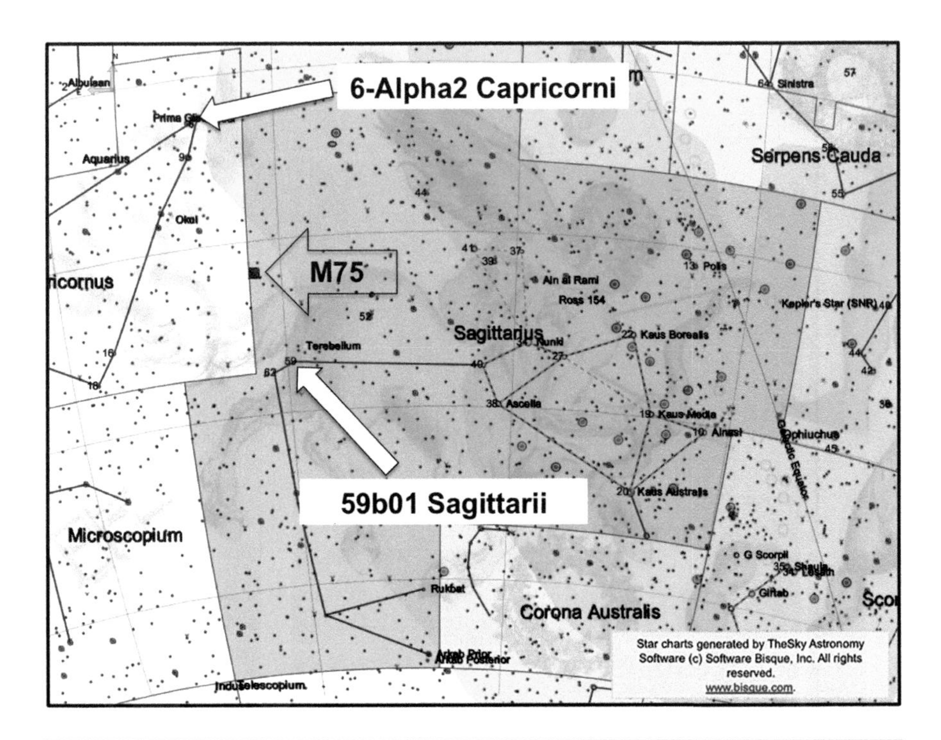

Fig. 5.3 Chart showing the location of M75

Telescope Optics
OTA: Planewave 17" CDK
Optical Design: Corrected Dall-Kirkham Astrograph
Aperture: 431 mm
Focal Length: 2912 mm
F/Ratio: f/6.8
Guiding: Active Guiding Disabled
Mount: Planewave Ascension 200HR

Instrument Package
CCD: FLI Proline 16803
Pixel Size: 9μm square
Sensor: KAF 16803
Cooling: -35° C default
Array: 4096 x 4096 pixels
FOV: 43.2 x 43.2 arcmin

Location
Observatory: Siding Spring
UTC +10.00 (Australia Daylight savings time is observed)

31° 16' 24" South
149 03' 52" East
Elevation: 1122 meters (3681 ft)

Messier 76

The planetary nebula Messier 76 (NGC 650/651), shown in Figure 5.4, is located in the constellation of Perseus. It has two NGC numbers, as it was once regarded as two separate objects. It is usually called the Little Dumbbell Nebula because of its similarity to the Dumbbell Nebula M27. It is also called the Cork Nebula because of its resemblance to a cork pulled from a wine bottle! This is another discovery for Pierre Méchain dated September 5, 1780 and passed on to Charles Messier, who observed it on October 21, 1780. Messier thought that it was "composed of nothing but small stars, containing nebulosity, & that the least light employed to illuminate the micrometer wires causes it disappear: its position was

Fig. 5.4 Image of Messier 76, a planetary nebula in Perseus

determined from the star Phi Andromedae, of fourth magnitude." Observing M76 on November 12, 1787, William Herschel described it as "Two nebulae close together. Both very bright. Distance 2'. One is south preceding and the other north following." This led to Messier 76 being allocated two NGC numbers.

Messier 76 is listed as having a distance of 3,400 ly and apparent dimensions of 2.7' x 1.8' of arc (SEDS). With those figures, the actual dimensions work out (coincidentally) to 2.7 ly x 1.8 ly. However, realistically, the distance to the object is extremely uncertain. The bright orange star labeled in Figure 5.5 is HIP 8063. The orange color indicates that it is a cool star, and it is listed as spectral type K5, corresponding to a temperature in the region of 4,410 K. It has a magnitude of 6.66 and is at a distance of 844.97 ly. On the negative image, two very faint galaxies are indicated. The first is PGC 166409, which has a magnitude of 16.23. The second is PGC 2391257, which has a magnitude of 17.30.

The chart in Figure 5.6 shows the location of Messier 76 in the constellation of Perseus with the plate-solved square image linked to the chart. M76 lies roughly on a line between the star Ruchbah in Cassiopeia and Almach in Andromeda. Ruchbah (HIP 6686) is 37-Delta Cassiopeiae and has a magnitude of 2.66. It is spectral type

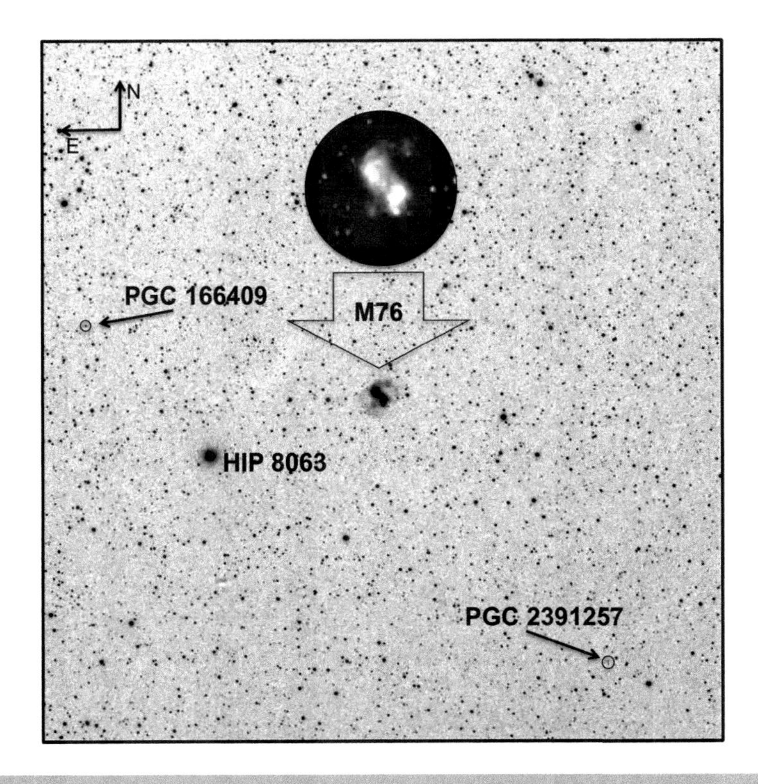

Fig. 5.5 Negative image of Messier 76

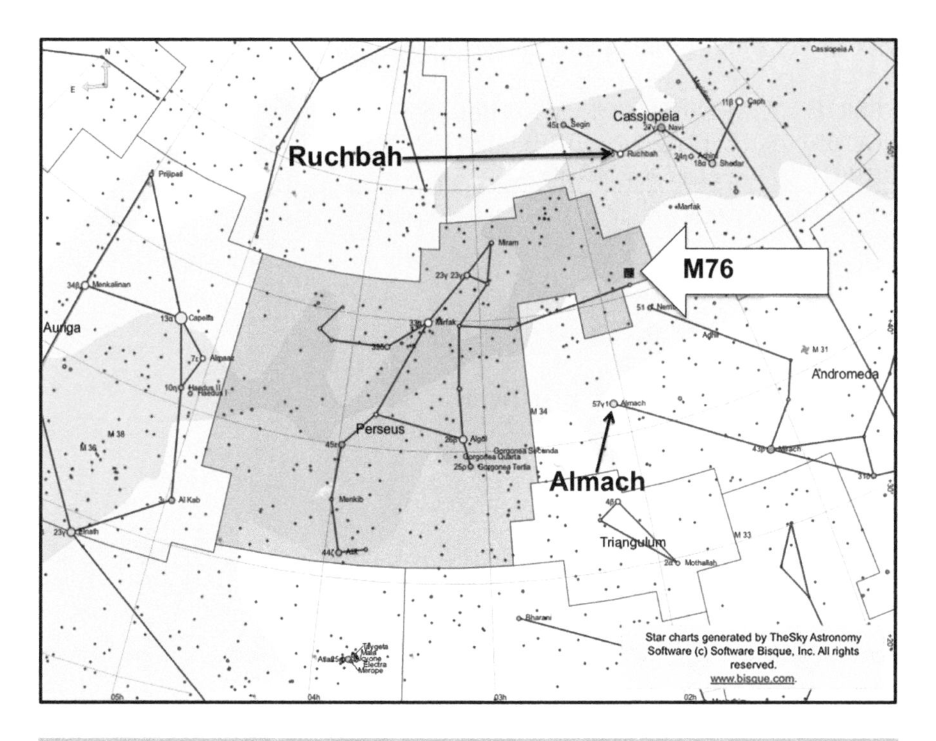

Fig. 5.6 Chart showing the location of M76

A5 and lies at a distance of 99.41 ly. There is a 9° angular separation between M76 and Ruchbah. Almach (HIP 9640) is 57-Gamma1 Andromedae and is magnitude 2.1. It is of spectral type B8 and is at a distance of 354.90 ly. There is an angular separation of 9° 57' between Almach and Messier 76. Notice the meridian, crossing down between Caph and Shedar in Cassiopeia.

The image was taken on September 22 at 01h 21m 55s local time in New Mexico. This corresponded to a Universal Time of 07h 21m 55s. The Right Ascension of M76 is 01h 43m 24s and the Local Sidereal Time was 00h 25m 00s. This means that M76 had not yet crossed the meridian. Subtracting one from the other gives a difference of 1h 18m 24s, indicating the time to go before M76 crossed the meridian. Adding this to the local time when the image was taken gives a meridian transit time of 02h 40m 19s. The telescope used to take the image of M76 was a Takahashi TOA-150 apochromatic refractor with an aperture of 150 mm.

Messier 76: Data relating to the image in Figure 5.4

REMOTELY IMAGED FROM MAYHILL, NEW MEXICO
NGC 650/651 (LITTLE DUMBBELL OR CORK NEBULA)
OBJECT TYPE: Planetary Nebula

RA: 01h 43m 24s
DEC: +51° 39' 12"
ALTITUDE: +66° 21' 13"
AZIMUTH: 31° 15' 14"
FIELD OF VIEW: 47' 04"' X 47' 04"
OBJECT SIZE: 2.7' X 1.8'
POSITION ANGLE: 177° 52' from North
EXPOSURE TIME: 300 s
DATE: 22nd September
LOCAL TIME: 01h 21m 55s
UNIVERSAL TIME: 07h 21m 55s
SCALE: 1.38 arcsec/pixel
MOON PHASE: 62.56% (waning)

Telescope Optics
OTA: Takahashi TOA-150
Optical Design: Apochromatic Refractor
Aperture: 150 mm
Focal Length: 1095 mm
F/Ratio: f/7.3
Guiding: Internal
Mount: Paramount GTS

Instrument Package
SBIG ST-4000M One-Shot Color CCD
A/D Gain: 0.6e-/ADU
Pixel Size: 7.4um square
Resolution: 1.45 arcsec/pixel
Sensor: Frontlit
Cooling: -20°C Winter (-10°C Summer)
Array: 2048 x 2048 (8.3 Megapixels)
FOV: 49.6 x49.6 arcmin

Location
Observatory: New Mexico Skies
UTC Minus 7.00 (Daylight savings time is observed)
Minimum Target Elevation: Approx. 25 – 45 Degrees
(N or S) 32° 54' Decimal: 32.9 North
(W or E) 105° 31' Decimal: 105.5 West
Elevation: 2225 meters (7298 ft)

Messier 77

Messier 77(NGC 1068), shown in Figure 5.7, is a spiral galaxy located in the constellation of Cetus. M77 has the characteristic bright core of a Seyfert galaxy. Seyfert galaxies have extremely active cores surrounded by hot ionized gas. Pierre Méchain discovered the galaxy on October 29, 1780. On December 17, 1780, Messier reported M77 to be a: "cluster of small stars, which contains some nebulosity, in Cetus & on the parallel of the star Delta." W. H. Smyth noted, "This object is wonderfully distant and insulated, with presumptive evidence of intrinsic density in its aggregation; and bearing indication of the existence of a central force, residing either in a central body or in the centre of gravity of the whole system." Smyth was certainly correct in that Seyfert galaxies are believed to have supermassive black holes at their center. The core of the galaxy is also a strong radio source, listed as number 71 in the third Cambridge Catalog of Radio Sources and otherwise known as 3C 71.

Messier 77 has an apparent size of 7' x 6' and is at a distance of 60 million ly. This corresponds with an actual size of approximately 122,000 x 105,000 ly. It is identified on the negative image in Figure 5.8 as is the galaxy NGC 1055, which is a spiral galaxy with apparent dimensions of 7.6' x 2.4' of arc, also lying at 60 mil-

Fig. 5.7 Image of Messier 77, a spiral galaxy in Cetus

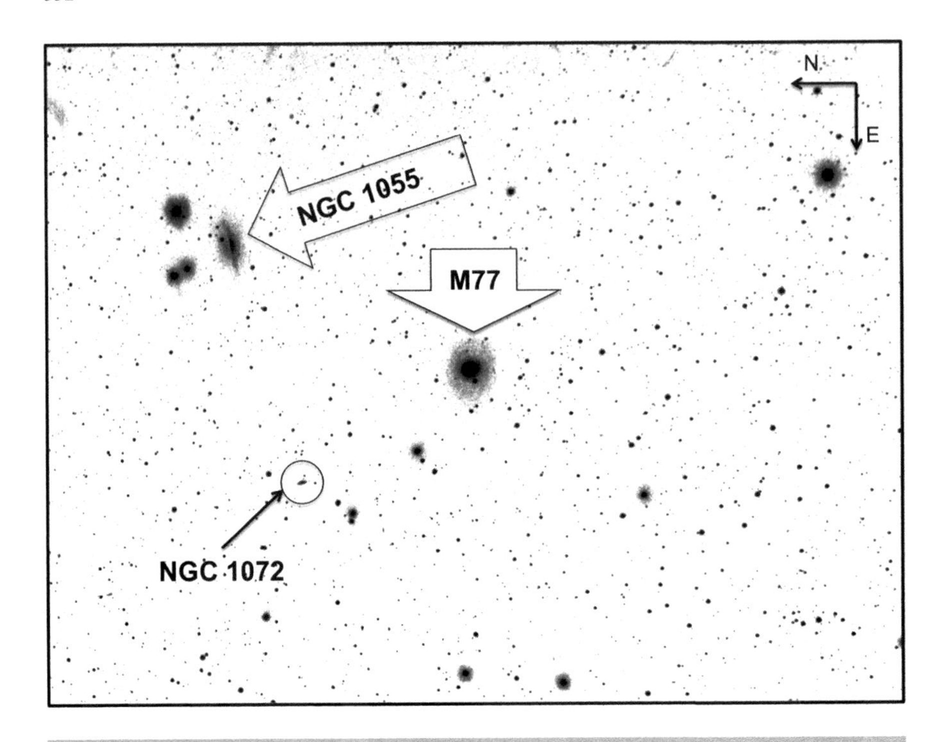

Fig. 5.8 Negative image of Messier 77

lion ly from us. (Data from SEDS). William Herschel discovered M77 in 1783. M77 and NGC 1055 are separated by a mere full Moon's width of about 31'. There is another galaxy just visible in the original field, identified on the negative image. This is NGC 1072, a spiral galaxy with a major axis of 1.5' and a minor axis of 0.5'.

The chart in Figure 5.9 shows the location of Messier 77 in the constellation of Cetus with the plate-solved image embedded on the chart. Note that the chart has north at the top. To the west of the M77 image is the star 82-Delta Ceti, or HIP 12387. This star is magnitude 4.08 and is spectral type B2, lying at a distance of 647.14 ly. The angular separation between M77 and Delta Ceti is roughly 53 arcmin. To the south east of M77 is the star 3-Eta Eridani, HIP 13701. This star has a magnitude of 3.89 and is spectral type K1 at some 133.18 ly distant. There is an angular separation of roughly 9° 30' between M77 and Eta Eridani. The meridian line can be seen to the west of M77.

The image was taken on October 7 from Siding Spring at 01h 25m 38s local time. The corresponding Universal Time was 14h 25m 38s (October 6). The Right Ascension of M77 is 02h 43m 33s, and the Local Sidereal Time was 01h 24m 21s. Thus, M77 would cross the meridian in 1h 19m 12s. Meridian transit time would therefore be 02h 44 50s. The single-shot color camera was used with a 90-mm refractor to obtain the image.

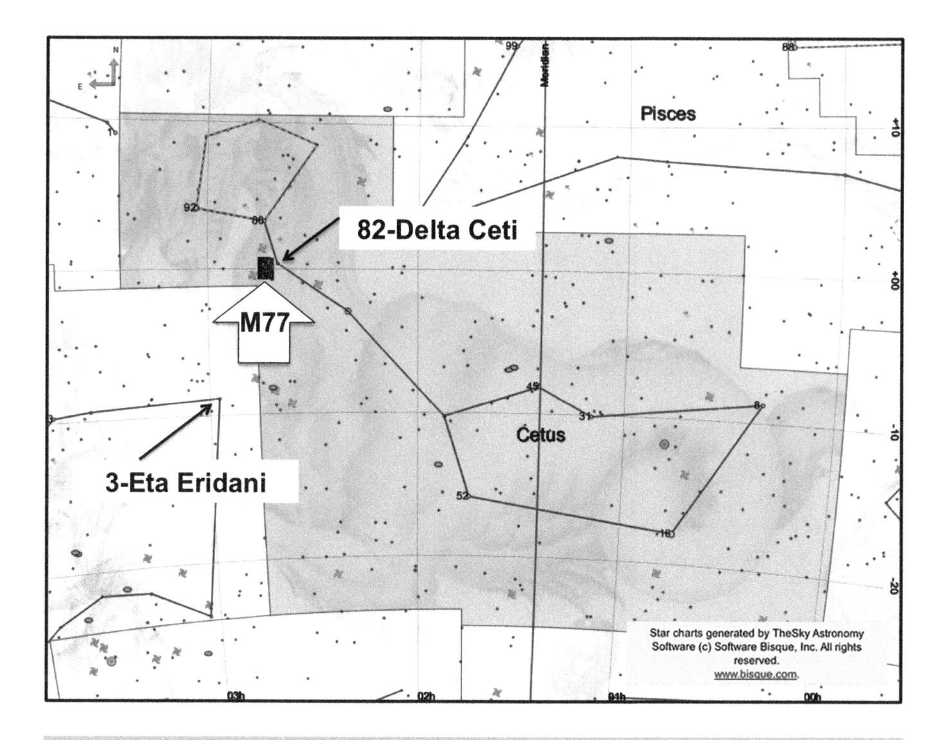

Fig. 5.9 Chart showing the location of M77

Messier 77: Data relating to the image in Figure 5.7

REMOTELY IMAGED FROM NEW SOUTH WALES, AUSTRALIA
NGC 1068
OBJECT TYPE: Spiral Galaxy

RA: 02h 43m 33s
DEC: +00° 03' 29"
ALTITUDE: +53° 30' 24"
AZIMUTH: 34° 43' 18"
FIELD OF VIEW: 1° 37' 44" x 1° 13' 18"
OBJECT SIZE: 7' x 6'
POSITION ANGLE: 270° 10' from North
EXPOSURE TIME: 600 s
DATE: 7th October
LOCAL TIME: 01h 25m 38s
UNIVERSAL TIME: 01h 25m 38s (6th October)
SCALE: 3.67 arcsec/pixel
MOON PHASE: 26.02% (waxing)

Telescope Optics
OTA: Takahashi Sky 90
Optical Design: Apochromatic Refractor
Aperture: 90 mm
Focal Length: 417 mm
F/Ratio: f/5.6
Guiding: None
Mount: Paramount PME

Instrument Package
CCD: SBIG ST2000 XMC
Colour CMOS
Pixel Size: 7.4μm square
Sensor: Frontlit
Cooling: CNA
Array: 1600 x 1200 pixels
FOV: 60.5 x 80.7 arcmin

Location
Observatory: Siding Spring
UTC +10.00 (Australia Daylight savings time is observed)
31° 16' 24" South
149 03' 52" East
Elevation: 1122 meters (3681 ft)

Messier 78

Messier 78 (NGC 2068), shown in Figure 5.10, is a reflection nebula in the constellation of Orion. It is quite a complex region with separate areas defined by their own NGC numbers and even a nebula that has only been visible since 2003. Pierre Méchain originally discovered M78 in 1780. Messier reported on December 17, 1780 a "Cluster of stars, with much nebulosity in Orion & on the same parallel as the star Delta in the belt, which has served to determine its position; the cluster follows the star on the hour wire at 3d 41', & the cluster is above the star by 27'7"." William Herschel noted, "Two bright stars, well defined, within a nebulous glare of light resembling that in Orion's sword. There are also three very faint stars just visible in the nebulous part, which seem to be component particles thereof. I think there is a faint ray near 1/2 degree long towards the east and another towards the south east less extended, but I am not quite so well assured of the reality of these latter phenomena as I could wish, and would rather ascribe them to some deception."

Fig. 5.10 Image of Messier 78, a reflection nebula in Orion

The negative image in Figure 5.11 shows Messier 78 with other identified objects. The two hot B stars that are making the nebula visible by reflecting light from interstellar dust cannot be seen, hidden in the nebulosity. The stars are magnitude 10 HD 38563A and HD 38563B. M78 is 8' x 6' of arc in size and lies at a distance of 1,600 ly. This gives M78 an actual size of 3.7 ly by 2.8 ly. NGC 2067 is a companion nebula, also illuminated by means of a hot spectral type B star. The two other elements of the cloud are NGC 2071 and NGC 2064. Jay W. McNeil, a contemporary amateur astronomer, discovered the object marked as McNeil's Nebula. The International Astronomical Union Circular 8284, dated February 9, 2004, reported the "appearance of a new nebula in a dense region of the Lynds 1630 cloud in Orion.... not present on seven Digitized Sky Survey images from 1951 to 1991... the University of Hawaii confirms that a faint optical counterpart to IRAS 05436-0007 has gone into outburst and has produced a large reflection nebulosity." McNeil used a 3-inch refractor to make the discovery.

The chart in Figure 5.12 shows the location of Messier 78 in the constellation of Orion. It is close to the well-known star Alnitak. This star is 50-Zeta Orionis (HIP 26727) of magnitude 1.74 and spectral type O9.5, lying at a distance of 817.44 ly. (Hipparcos Data). Using the chart, one measures an angular separation of 2° 30' between Alnitak and M78. Alnitak is the eastern star in the Belt of Orion, very close to the Flame Nebula and the Horsehead Nebula. Alnilam is the middle star in Orion's Belt and is 46-Epsilon Orionis (HIP 26311) of magnitude 1.69 and spectral

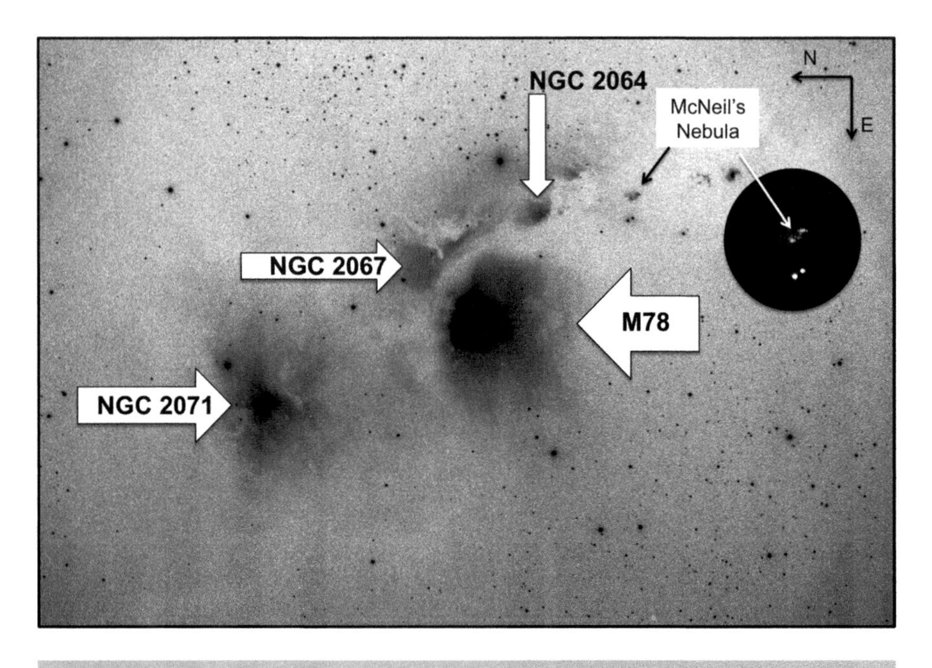

Fig. 5.11 Negative image of Messier 78

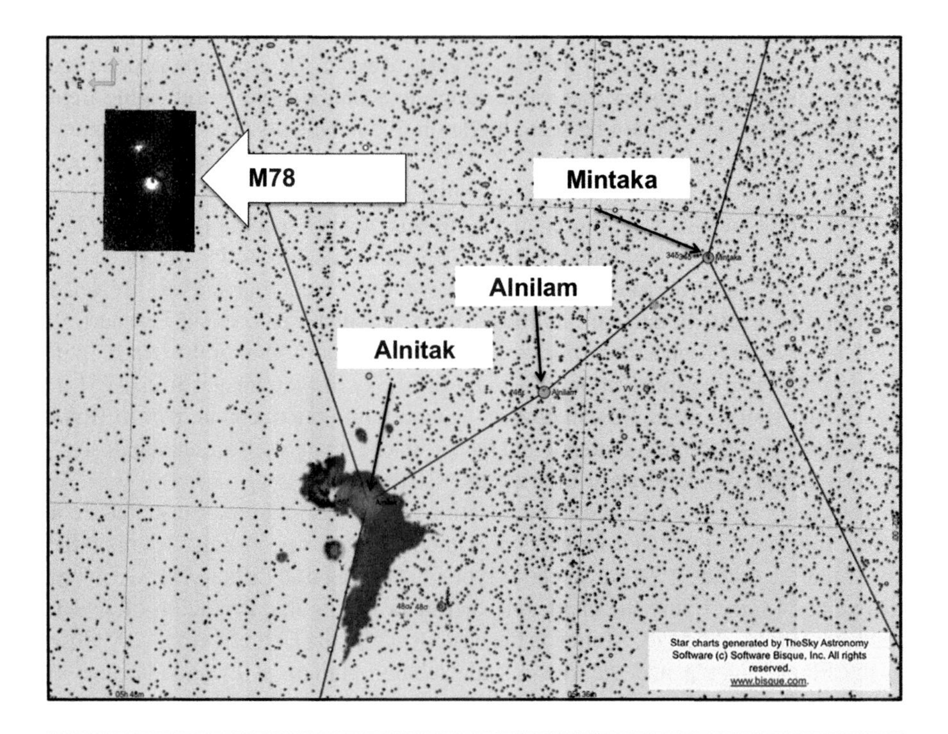

Fig. 5.12 Chart showing the location of M78

type B0, lying at a distance of 1,342.21 ly. The angular separation between M78 and Alnilam is 2° 55'.

The local time in New Mexico when the image of M78 was taken was 05h 37m 56s. The corresponding Universal Time was 11h 37m 56s. The Right Ascension of M78 is 05h 47m 38s, and the Local Sidereal Time was 06h 59m 43s. The difference between these indicates the time since M78 crossed the meridian to be 1h 12m 5s. Thus, M78 transited the meridian at exactly 04h 25m 51s. The telescope used to take the image of M53 was a Corrected Dall-Kirkham Astrograph with an aperture of 510 mm mounted on a Plane Wave Ascension 200HR.

Messier 78: Data relating to the image in Figure 5.10

REMOTELY IMAGED FROM MAYHILL, NEW MEXICO
NGC 2068
OBJECT TYPE: Star-forming Nebula

RA: 05h 47m 38s
DEC: +00° 04' 27"
ALTITUDE: +53° 08' 53"
AZIMUTH: 211° 03' 00"
FIELD OF VIEW: 0° 54' 47" x 0° 36' 31"
OBJECT SIZE: 8' x 6'
POSITION ANGLE: 271° 03' from North
EXPOSURE TIME: 300 s
DATE: 27th October
LOCAL TIME: 05h 37m 56s
UNIVERSAL TIME: 11h 37m 56s
MOON PHASE: 9.41% (waning)

Telescope Optics
OTA: Planewave 20" CDK
Optical Design: Corrected Dall-Kirkham Astrograph
Aperture: 510 mm
Focal Length: 2280 mm (0.66 Focal Reducer Fitted)
F/Ratio: f/4.5
Guiding: Active Guiding Disabled
Mount: Planewave Ascension 200HR

Instrument Package
FLI Proline PL11002M CCD
Camera Pixel Size: 9μm square
Resolution: 0.81 arcsec/pixel
Cooling: -30°C default
Array: 4008 x 2672 (10.7 Megapixels)

Location
Observatory: New Mexico Skies
UTC Minus 7.00 (Daylight savings time is observed)
Minimum Target Elevation: Approx. 25–45 Degrees
(N or S) 32° 54' Decimal: 32.9 North
(W or E) 105° 31' Decimal: 105.5 West
Elevation: 2225 meters (7298 ft)

Messier 79

Messier 79 (NGC 1904), shown in Figure 5.13, is a globular cluster located in the constellation of Lepus. It is suspected that this cluster might not actually belong to our Galaxy, instead lying in a direction away from our Galactic Center and possibly being associated with the Canis Major Dwarf Galaxy that is breaking up and orbiting the Galaxy very closely. The globular is quite small on the image and gives a rough idea of how it would look though smaller telescopes in the 18th Century. Pierre Méchain discovered M79 on October 26, 1780. Messier observed and

Fig. 5.13 Image of Messier 79, a globular cluster in Lepus

recorded it on December 17, 1780: "Nebula without star, situated below Lepus, & on the same parallel as a star of sixth magnitude - this nebula is beautiful; the center brilliant, the nebulosity a little diffuse; its position was determined from the star Epsilon Leporis, of fourth magnitude." W. H. Smyth commented, "the following edge [eastern edge of M79] of whose disc just precedes a line formed by two stars lying across the vertical, and it is followed nearly on the parallel by a 9th-magnitude star. It is a fine object, blazing towards the centre."

Smyth referred to two stars lying across the vertical to the east of M79. This could refer to stars HIP 25458 and Tycho 6480:234, as shown on the negative image. HIP 25458 is magnitude 8.44, and Tycho 6480:234 is magnitude 7.94. Messier referred to a 6th magnitude star on the same parallel. There is the star HIP 25045 shown on the negative image in Figure 5.14, but this is magnitude 5.05 and differs in declination from M79 by 15'. This star has an angular separation from M79 of roughly 36' of arc. This star is also a double star WDS HJ 3752, as shown on the negative image. There are two components of magnitude 5.44 and magnitude 9.19. The magnified image of this double star cropped from the original image has been included. There is an angular separation of 62" between the double star components on the embedded image on the chart. Messier 79 has an apparent size of

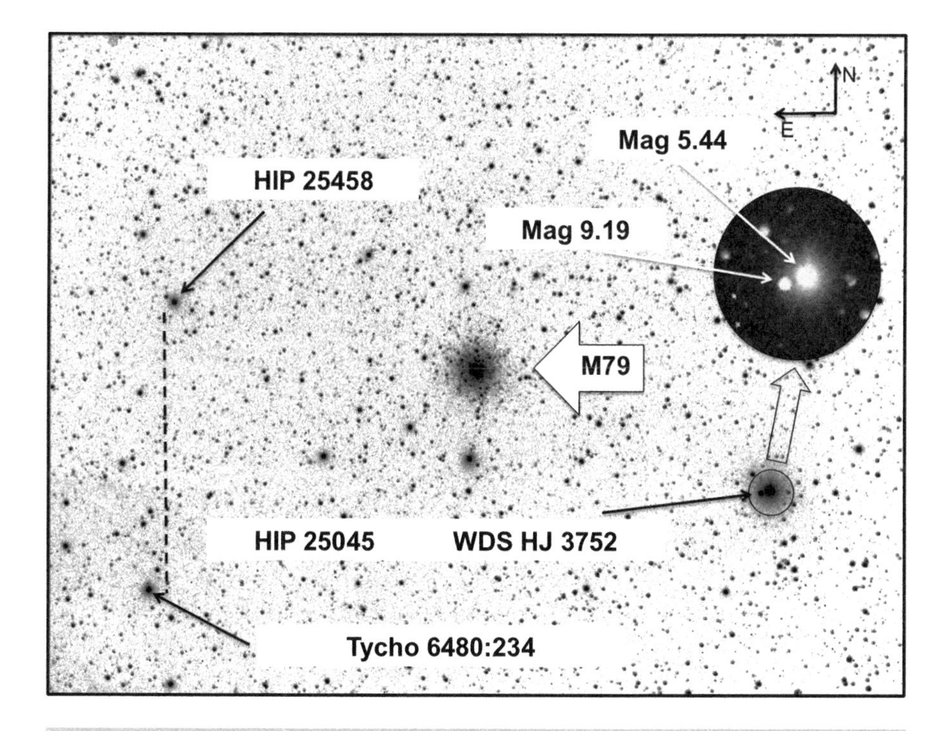

Fig. 5.14 Negative image of Messier 79

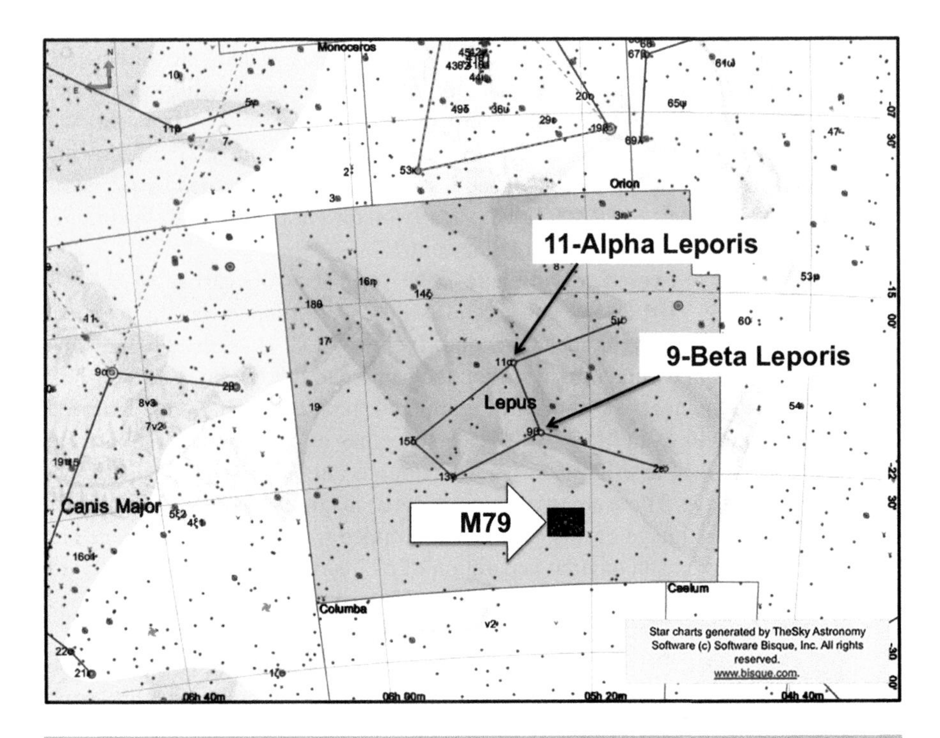

Fig. 5.15 Chart showing the location of M79

9.6' and lies at a distance of 42,100 ly from us. This gives the object an actual diameter of 118 ly.

The chart in Figure 5.15 shows the location of Messier 79 in the constellation of Lepus. To find M79, the two stars 11-Alpha Leporis and 9-Beta Leporis can be used as pointers. From Alpha, project down through Beta and onto M79. The angular separation between Alpha and Beta is roughly 3° 10', and between Beta and M79, 3° 58'. Alpha Leporis is called Arneb and is HIP 25985 with magnitude 2.58. It is a spectral type F0 star at a distance of 1,284.08 ly. Beta Leporis is called Nihal and is HIP 25606 with a magnitude of 2.81. It is a spectral type G5 lying at a distance of 159.18 ly.

Messier 79: Data relating to the image in Figure 5.13

REMOTELY IMAGED FROM NEW SOUTH WALES, AUSTRALIA
NGC 1904
OBJECT TYPE: Globular Cluster
RA: 05h 24m 53s
DEC: -24° 30' 37"
ALTITUDE: +51° 29' 20"
AZIMUTH: 90° 58' 23"

FIELD OF VIEW: 1° 37' 39" x 1° 13' 14"
OBJECT SIZE: 9.6'
POSITION ANGLE: 357° 17' from North
EXPOSURE TIME: 600 s
DATE: 26th November
LOCAL TIME: 23h 12m 46s
UNIVERSAL TIME: 12h 12m 46s
SCALE: 3.66 arcsec/pixel
MOON PHASE: 8.03% (waning)

Telescope Optics
OTA: Takahashi Sky 90
Optical Design: Apochromatic Refractor
Aperture: 90 mm
Focal Length: 417 mm
F/Ratio: f/5.6
Guiding: None
Mount: Paramount PME

Instrument Package
CCD: SBIG ST2000 XMC
Colour CMOS
Pixel Size: 7.4μm square
Sensor: Frontlit
Cooling: CNA
Array: 1600 x 1200 pixels
FOV: 60.5 x 80.7 arcmin

Location
Observatory: Siding Spring
Australia UTC +10.00 (Australia Daylight savings time is observed)
31° 16' 24" South
149 03' 52" East
Elevation: 1122 meters (3681 ft)

Messier 80

Messier 80 (NGC 6093), shown in Figure 5.16, is a globular cluster in the constellation of Scorpius. It appears similar to M79 with an apparent size of 10 arcmin compared to M79's 9.6'. M79 is well away from the Milky Way, whereas M80 is within it – not centrally placed, but only 20° from the Galactic Equator. As a result, the image from the 106-mm telescope in Australia is packed with stars and has

Fig. 5.16 Messier 80, a globular cluster in Scorpius

nebulosity from nearby objects intruding from the east. Messier discovered M80 on January 4, 1781, commenting about a "Nebula without star, in Scorpius, between the stars Rho Ophiuchi and Delta Scorpio, compared to Rho to determine its position: this nebula is round, the center brilliant, & it resembles the nucleus of a small Comet, surrounded with nebulosity." Sir William Herschel noted "A globular cluster of extremely minute and very compressed stars of about 3 or 4 minutes in diameter; very gradually much brighter in the middle; towards the circumference the stars are distinctly seen, and are the smallest imaginable."

M80 lies at a distance of 32,600 ly. With an apparent diameter of 10', that gives an actual diameter of 95 ly, somewhat smaller than M79. The negative image in Figure 5.17 shows M80 at the center but includes a magnified view of the globular cluster taken from the original image. To the east and south (left and down), two bright nebulae are identified. IC 4604 is part of the Rho Ophiuchi Nebula complex and is a reflection nebula lit up by the star Rho Ophiuchus. Some of it intrudes onto the frame from the constellation of Ophiuchus, neighbor to Scorpius. HIP 80462 similarly illuminates IC 4603. The illuminating stars can be seen on the original positive image. Two spiral galaxies are just visible in the frame, and these are

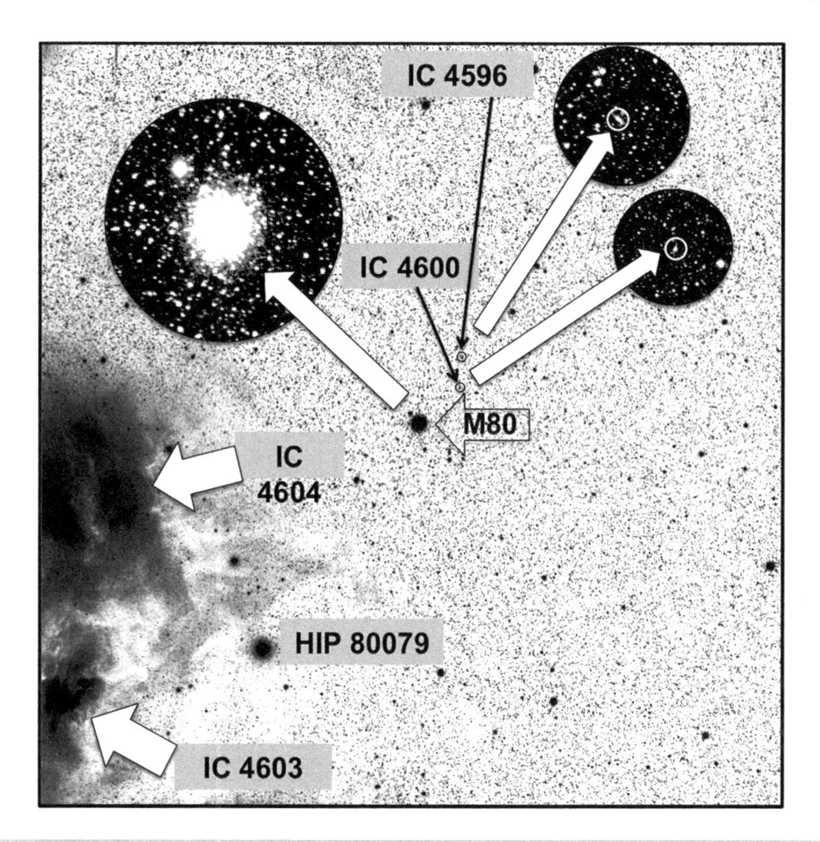

Fig. 5.17 Negative image of Messier 80

shown as enlargements on the negative image. IC 4596 is magnitude 14.70, and IC 4600 is magnitude 15.5.

The chart in Figure 5.18 shows the location of Messier 80 in the constellation of Scorpius. Messier referred to the "nebula without a star" located between Rho Ophiuchus and Delta Scorpio. These stars are highlighted in Figure 5.18. 5-Rho Ophiuchi (HIP 80473) is a magnitude 4.57 star of spectral class B2, which lies at a distance of 394.39 ly. 7-Delta Scorpii (Dschubba or HIP 78401) is a magnitude 2.29 star of spectral class B0, which lies at a distance of 401.67 ly. The angular separation between Rho Ophiuchus and M80 is 2° 1'. The separation between M80 and Dschubba is 3° 52'.

The image was taken on March 28 from Siding Spring at 04h 38m 09s. The corresponding Universal Time was 17h 38m 09s (March 27). The Right Ascension of M80 is 16h 18m 05s, and the Local Sidereal Time was 15h 55m 31s. So, M80 would cross the meridian in 22m 34s. Meridian transit time of M80 would therefore be at 05h 0m 43s. The Takahashi FSQ Astrograph with an aperture of 106 mm was used to obtain the image.

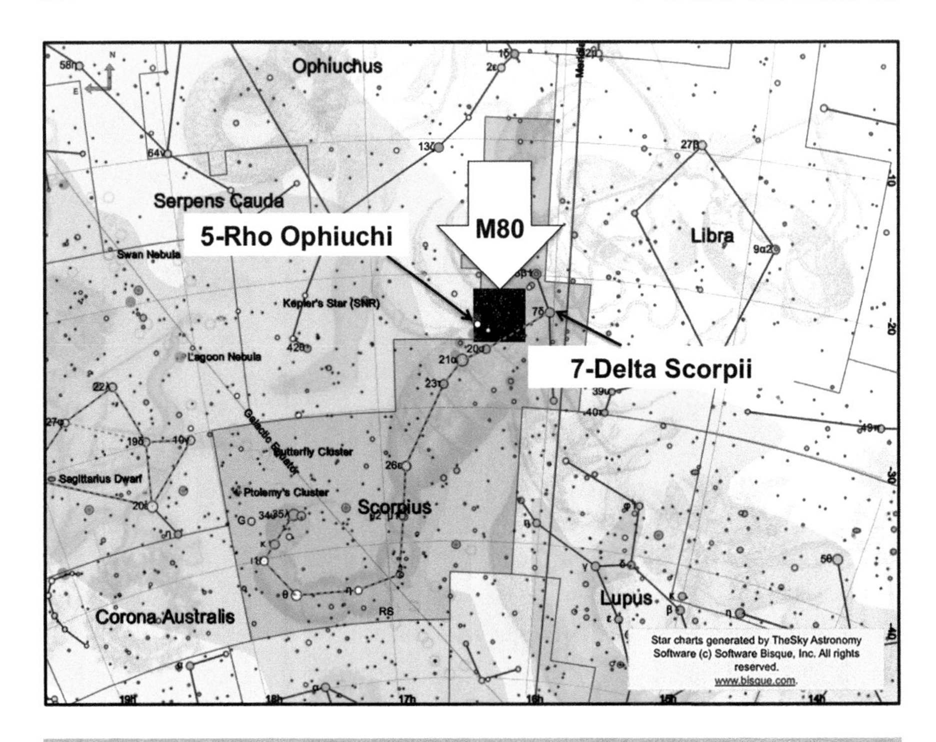

Fig. 5.18 Chart showing the location of M80

Messier 80: Data relating to the image in Figure 5.16

REMOTELY IMAGED FROM NEW SOUTH WALES, AUSTRALIA
NGC 6093
OBJECT TYPE: Globular Cluster

RA: 16h 18m 05s
DEC: -23° 00' 50"
ALTITUDE: +80° 22' 03"
AZIMUTH: 32° 43' 30"
FIELD OF VIEW: 3° 58' 48" x 3° 58' 48"
OBJECT SIZE: 10'
POSITION ANGLE: 0° 42' from North
EXPOSURE TIME: 300 s
DATE: 28th March
LOCAL TIME: 04h 38m 09s
UNIVERSAL TIME: 17h 38m 09s (27th March)
SCALE: 3.5 arcsec/pixel
MOON PHASE: 0.26% (waning)

Telescope Optics
OTA: Takahashi FSQ ED
Optical Design: Petzval Apochromatic Astrograph
Aperture: 106 mm
Focal Length: 530 mm
F/Ratio: f/5.0
Guiding: External
Mount: Paramount PME

Instrument Package
CCD: FLI Microline 16803
Pixel Size: 9µm square
Sensor: KAF – 16803 Frontlit
Cooling: -25°C Summer (-30°C Winter)
Array: 4096 x 4096 pixels
FOV: 238.8 x 238.8 arcmin

Location
Observatory: Siding Spring
Australia UTC +10.00 (Australia Daylight savings time is observed)
31° 16' 24" South
149 03' 52" East
Elevation: 1122 meters (3681 ft)

Messier 81

Messier 81 (NGC 3031) is a spectacular spiral galaxy located in the constellation of Ursa Major, shown in Figure 5.19. Commonly known as Bode's Galaxy, it is part of the M81 group of galaxies and is very close to Messier 82, which is just outside the frame of the image. German astronomer Johann Bode first discovered the galaxy on December 31, 1774. Messier made an observation on February 9 1781, noting "A nebula near the ear of the great Bear (Ursa Major), on the parallel of the star d (24d Ursae Majoris), of fourth or fifth magnitude: its position was determined from that star. This nebula is a little oval, the center clear, & one can see it well in an ordinary telescope of 3.5 feet." In 1810, William Herschel reported, "I viewed the nebula with the large 10 feet. It has a bright, resolvable nucleus, certainly consisting either of 3 or 4 stars or something resembling them. It is about 15 or 16' long."

The negative image in Figure 5.20 shows Messier 81 and the measurements made from the plate-solved image superimposed on the chart. The major axis of M81 was measured at 21' 25" of arc and the minor axis 10' 3". The distance to M81 is 12 million ly (SEDS Data), from which the actual size of the major axis is calculated to be about 75,000 ly and the minor axis 35,000 ly. In reality, the galaxy

Fig. 5.19 Image of Messier 81, a spiral galaxy in Ursa Major

will be larger, as its full extent was not captured with the 300-second exposure. The field of view (frame size) is 47' 08" x 47' 08". There are a number of faint galaxies in the field of view. Most of them are too faint to be detected with this telescope and only a 300-second exposure time, but two of these, spotted as faint smudges on the image, have been identified. Because the image was plate-solved and linked to the chart, it was only necessary to click on the smudges to identify them. The first of these is PGC 28848. This galaxy has a major axis of 0.6' and a minor axis of 0.3'. It has a magnitude of 16.20 in the Principal Galaxy Catalog. The second galaxy is PGC 28505 with axes of 0.8' x 0.5' and a magnitude of 15.71.

The chart in Figure 5.21 shows Messier 81 in the constellation of Ursa Major. Note that north is at the top and the well-known star Dubhe in the Big Dipper asterism is identified to help your orientation. M81 sits just below Messier 82. Messier referred to the star "d," which is 24d Ursae Majoris. This star is identified on the chart. It is not quite on the parallel of M81, as there is difference of 47' between their declinations. There is an angular separation between M81 and 24d of 2° and between M81 and Dubhe of 10° 7'. The star 24d Ursae Majoris (HIP 46977) has a magnitude of 4.54 and is spectral type G4, lying at a distance of 105.59 ly. Dubhe

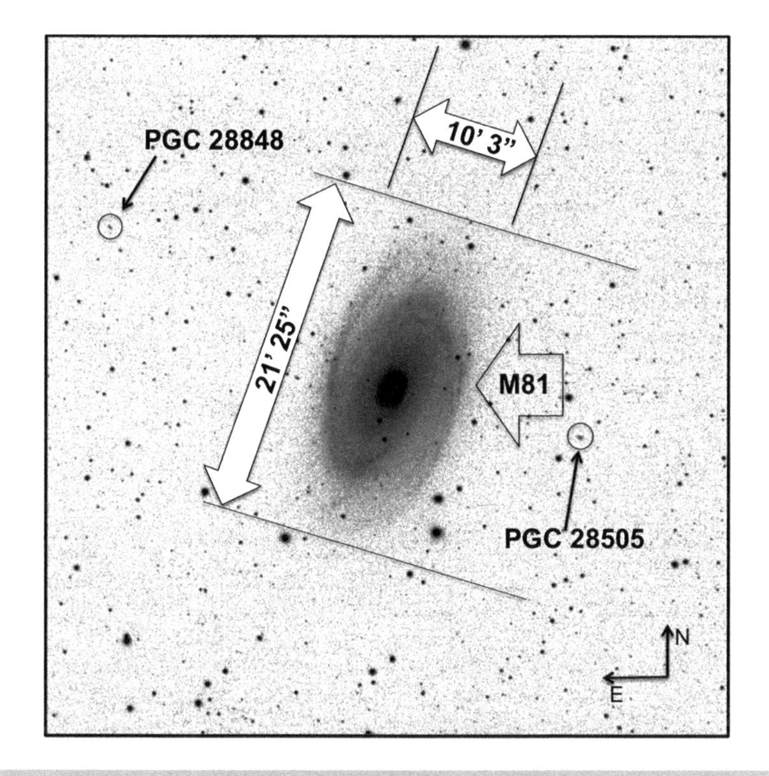

Fig. 5.20 Negative image of Messier 81

(HIP 54061) has a magnitude of 1.81 and is spectral type F7, lying at a distance of 123.64 ly.

The image was taken on January 18 at 23h 05m 34s local time in New Mexico. This corresponded to a Universal Time of 06h 05m 34s (January 19). The Right Ascension of M81 is 09h 56m 58s, and the Local Sidereal Time was 06h 57m 37s. This means that M81 had yet to cross the meridian. Subtracting one from the other gives a difference of 2h 59m 21s, indicating the time needed before M81 crossed the meridian. Adding this to the local time when the image was taken gives a meridian transit time of 02h 04m 55s. The telescope used to take the image of M81 was a Takahashi TOA-150 apochromatic refractor with an aperture of 150 mm.

Messier 81: Data relating to the image in Figure 5.19

REMOTELY IMAGED FROM MAYHILL, NEW MEXICO
NGC 3031 (BODE'S GALAXY)
OBJECT TYPE: Spiral Galaxy

RA: 09h 56m 58s
DEC: +68° 58' 55"

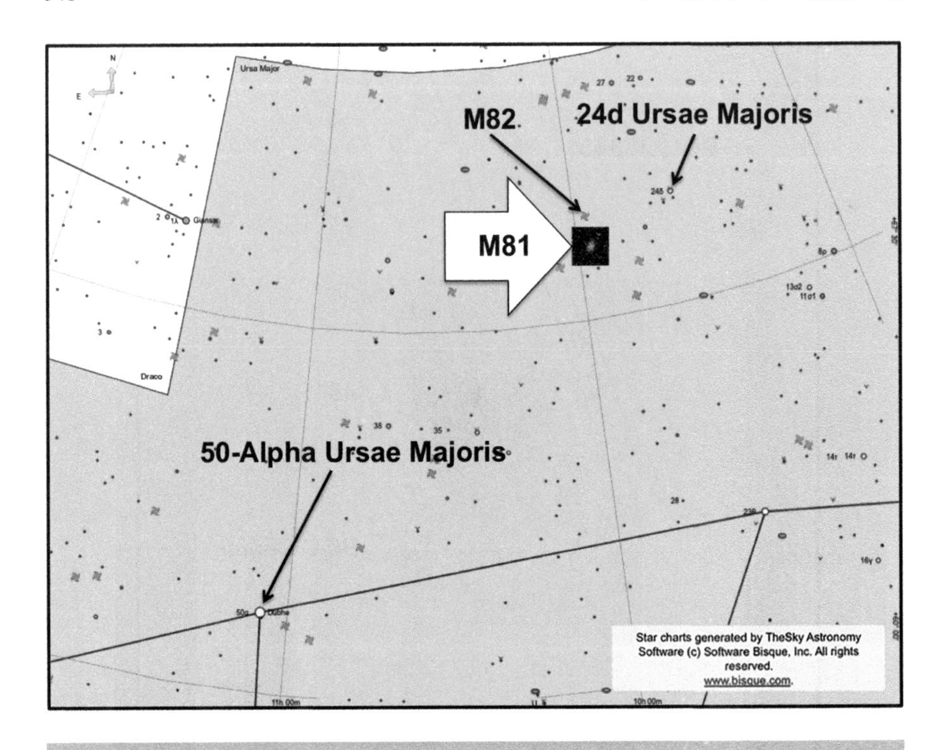

Fig. 5.21 Chart showing the location of M81

ALTITUDE: +45° 59' 36"
AZIMUTH: 21° 20' 48"
FIELD OF VIEW: 47’ 08”’ X 47’ 08”
OBJECT SIZE: 21’ X 10’
POSITION ANGLE: 177° 45’ from North
EXPOSURE TIME: 300 s
DATE: 18th January
LOCAL TIME: 23h 05m 34s
UNIVERSAL TIME: 06h 05m 34s (19th January)
SCALE: 1.38 arcsec/pixel
MOON PHASE: 56.57% (waning)

Telescope Optics
OTA: Takahashi TOA-150
Optical Design: Apochromatic Refractor
Aperture: 150 mm
Focal Length: 1095 mm
F/Ratio: f/7.3
Guiding: Internal
Mount: Paramount GTS

Instrument Package
SBIG ST-4000M One-Shot Color CCD
A/D Gain: 0.6e-/ADU
Pixel Size: 7.4um square
Resolution: 1.45 arcsec/pixel
Sensor: Frontlit
Cooling: -20°C Winter (-10°C Summer)
Array: 2048 x 2048 (8.3 Megapixels)
FOV: 49.6 x49.6 arcmin

Location
Observatory: New Mexico Skies
UTC Minus 7.00 (Daylight savings time is observed)
Minimum Target Elevation: Approx 25–45 Degrees
(N or S) 32° 54' Decimal: 32.9 North
(W or E) 105° 31' Decimal: 105.5 West
Elevation: 2225 meters (7298 ft)

Messier 82

Messier 82 (NGC 3034), shown in Figure 5.22, is a close companion of M81 in the constellation of Ursa Major. It is shaped rather like a spindle and is known as an irregular galaxy. It is also called the Cigar Galaxy. It was Johann Bode first discovered M82 on December 31, 1774, the same night that he discovered Messier 81. It is believed that the distorted shape of M82 is linked to interaction with its M81 companion. Messier observed M82 on February 9, 1781 and noted a "Nebula without star, near the preceding [M81], both are appearing in the same field of the telescope, this one is less distinct than the preceding; its light faint & is elongated: at its extremity is a telescopic star." Messier 82 is a strong infrared source and is one of the brightest infrared galaxies to be observed.

The negative image in Figure 5.23 shows M82 in the center of the 47' square field. I measured the dimensions of M82 as shown as 9' 21" x 3' 32". The distance to M82 is 12 million ly, which gives the galaxy, based on my image, actual dimensions of roughly 32,400 ly x 12,200 ly.There are a number of galaxies within the field of view when I check on the chart of the area but these are too faint too have been recorded, having magnitudes of 17 to 18 or fainter. The brightest star in the field, the orange star HIP 48486, is magnitude 7.92 and is spectral class K0, which would correspond to a surface temperature in the region of 4720K. It lies at a distance of 886 ly. In contrast the star HIP 48573 is magnitude 9.35 and is spectral class G5, which would have a surface temperature of 5010K and is only 266 ly from us.

The chart in Figure 5.24 shows the location of M82 in Ursa Major, just above the galaxy M81. M82 is one of the galaxies in the M81 group, which contains per-

Fig. 5.22 Image of Messier 82, an irregular galaxy in Ursa Major

haps over 20 galaxies in total. Two nearby members shown on the chart are NGC 3077 and NGC 2976. The angular separation of M81 and M82 is 36' 58", between M82 and NGC 3077 is 1° 9' 20" and between M82 and NGC 2976 is 1° 55' 45". Ursa Major also contains the Messier objects M40, M97, M101, M108 and M109.

The image was taken on January 18 at 23h 25m 51s local time in New Mexico. This corresponded to a Universal Time of 06h 25m 51s (January 19). The Right Ascension of M82 is 09h 57m 19s, and the Local Sidereal Time was 07h 17m 57s. Thus, M82 had yet to cross the meridian. Subtracting one from the other gives a difference of 2h 39m 22s and provides a meridian transit time of 02h 05m 13s. The telescope used to take the image of M82 was a Takahashi TOA-150 apochromatic refractor with an aperture of 150 mm.

Messier 82: Data relating to the image in Figure 5.22

REMOTELY IMAGED FROM MAYHILL, NEW MEXICO
NGC 3034 (CIGAR GALAXY)
OBJECT TYPE: Irregular Galaxy

RA: 09h 57m 19s
DEC: +69° 35' 52"

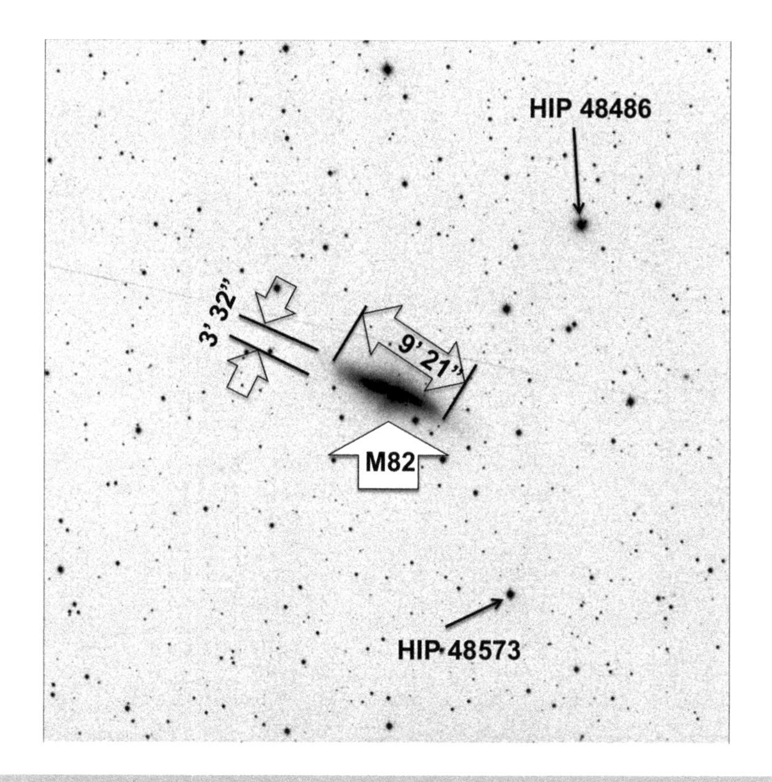

Fig. 5.23 Negative image of Messier 82

ALTITUDE: +47° 05' 56"
AZIMUTH: 19° 09' 12"
FIELD OF VIEW: 47' 08"' X 47' 08"
OBJECT SIZE: 9' X 4'
POSITION ANGLE: 177° 43' from North
EXPOSURE TIME: 300 s
DATE: 18th January
LOCAL TIME: 23h 25m 51s
UNIVERSAL TIME: 06h 25m 51s (19th January)
SCALE: 1.38 arcsec/pixel
MOON PHASE: 56.44% (waning)

Telescope Optics
OTA: Takahashi TOA-150
Optical Design: Apochromatic Refractor
Aperture: 150 mm
Focal Length: 1095 mm
F/Ratio: f/7.3

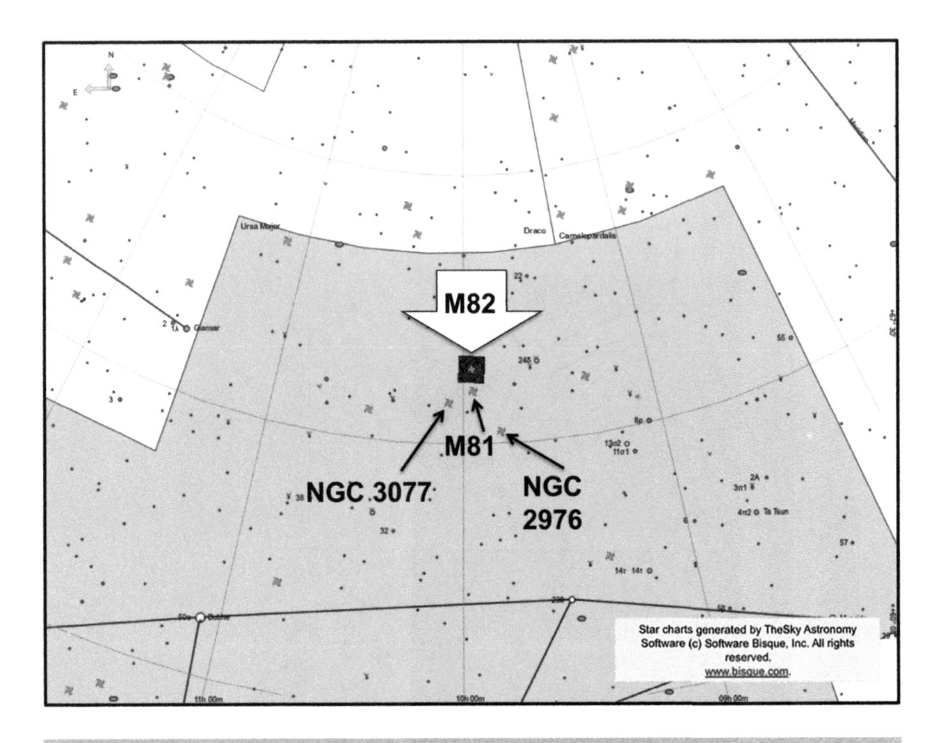

Fig. 5.24 Chart showing the location of M82

Guiding: Internal
Mount: Paramount GTS

Instrument Package
SBIG ST-4000M One-Shot Color CCD
A/D Gain: 0.6e-/ADU
Pixel Size: 7.4um square
Resolution: 1.45 arcsec/pixel
Sensor: Frontlit
Cooling: -20°C Winter (-10°C Summer)
Array: 2048 x 2048 (8.3 Megapixels)
FOV: 49.6 x49.6 arcmin

Location
Observatory: New Mexico Skies
UTC Minus 7.00 (Daylight savings time is observed)
Minimum Target Elevation: Approx 25–45 Degrees
(N or S) 32° 54' Decimal: 32.9 North
(W or E) 105° 31' Decimal: 105.5 West
Elevation: 2225 meters (7298 ft)

Messier 83

The spectacular spiral galaxy Messier 83 (NGC 5326), shown in Figure 5.25 in the constellation of Hydra is known as the Southern Pinwheel Galaxy. (The "northern" Pinwheel Galaxy is M101). French astronomer Nicholas-Louis de La Caille discovered M83 on February 23, 1752 from his observatory at the Cape of Good Hope in South Africa. The southern declination of M83 (almost -30°) makes it difficult for northern observers. He was the first astronomer to make a systematic survey of the southern sky. Charles Messier, based in Paris, struggled to observe M83 on February 17, 1781: "One is only able with the greatest concentration to see it at all. Nebula without star, near the head of Centaurus: it appears as a faint & even glow, but it is difficult to see in the telescope, as the least light to illuminate the micrometer wires makes it disappear. One is only able with the greatest concentration to see it at all: it forms a triangle with two stars estimated of sixth & seventh magnitude: determined from the stars i, k and h in the head of Centaurus." Checking back, it appears that M83 was only slightly more than 12° in altitude at its best that night from Paris.

Fig. 5.25 Image of Messier 83, a spiral galaxy in Hydra

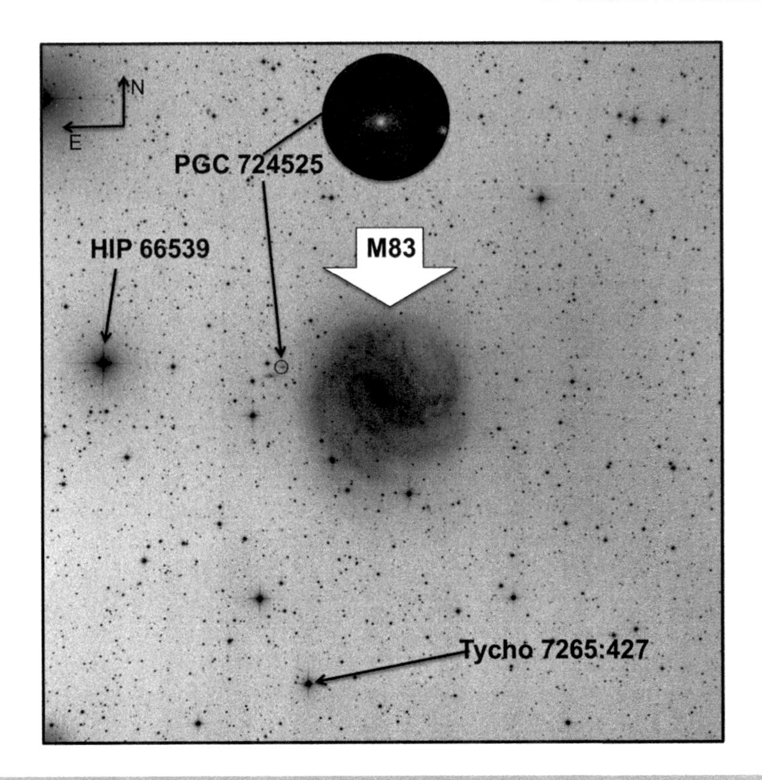

Fig. 5.26 Negative image of Messier 83

The galaxy is shown on the negative image in Figure 5.26 with its 11' x 10' of arc apparent size taking up about a quarter of the 44' field of view. Messier 83 lies at a distance of 15 million ly, so its actual dimensions from our viewpoint are 48,000 x 44,000 ly. One apparently tiny galaxy in the frame has been pointed out on the negative image. PGC 724525 is magnitude 16.66 with apparent dimensions 0.7'x 0.3'. The bright star in the frame, HIP 66539, is magnitude 7.15 and is spectral type F2, lying at a distance of 415.49 ly. The star Tycho 7265:427 is magnitude 9.24 at 194.14 ly and is just across the border in the constellation of Centaurus.

The chart in Figure 5.27 shows the location of M83 in the southern part of eastern Hydra, with the plate-solved image running down into Centaurus. The star 49-Pi Hydrae (HIP 68895) is a magnitude 3.25 star of spectral type K2 lying at a distance of 101.39 ly. There is an angular separation of 7° 12' between M83 and 49-Pi Hydrae. The star 5-Theta Centauri (Menkent or HIP 68933) is a magnitude 2.06 star of spectral type K0, which has a distance of 60.94 ly. The angular separation between Menkent and Messier 83 is 9° 1'. Messier referred to stars i, k and h in the head of Centaurus. These are identified on the chart.

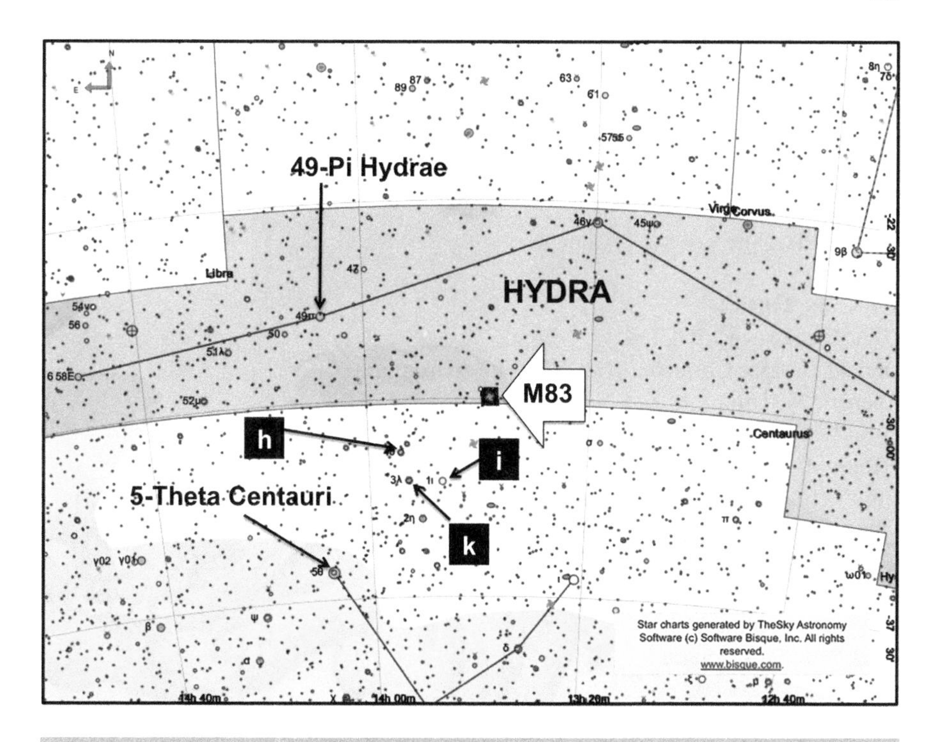

Fig. 5.27 Chart showing the location of M83

The image was taken at local time 02h 52m 03s on February 21. The corresponding Universal Time was 15h 52m 03s (February 20). Messier 83 has a Right Ascension of 13h 37m 59s, and the Local Sidereal Time was 11h 51m 08s, This means that M83 would the meridian from east to west in just less than 2 hours. The difference is exactly 1h 46m 51s, providing a meridian transit time of 04h 38m 54s. The telescope used was a 431-mm Planewave instrument.

Messier 83: Data relating to the image in Figure 5.25

REMOTELY IMAGED FROM NEW SOUTH WALES, AUSTRALIA
NGC 5236 (SOUTHERN PINWHEEL GALAXY)
OBJECT TYPE: Spiral Galaxy

RA: 13h 37m 59s
DEC: -29° 57' 08"
ALTITUDE: +67° 01' 36"
AZIMUTH: 93° 44' 35"
FIELD OF VIEW: 0° 43' 54" x 0° 43' 54"
OBJECT SIZE: 11’ X 10’

POSITION ANGLE: 90° 48' from North
EXPOSURE TIME: 300 s
DATE: 21st February
LOCAL TIME: 02h 52m 03s
UNIVERSAL TIME: 15h 52m 03s (20th February)
SCALE: 0.64 arcsec/pixel
MOON PHASE: 32.98% (waning)

Telescope Optics
OTA: Planewave 17" CDK
Optical Design: Corrected Dall-Kirkham Astrograph
Aperture: 431 mm
Focal Length: 2912 mm
F/Ratio: f/6.8
Guiding: Active Guiding Disabled
Mount: Planewave Ascension 200HR

Instrument Package
CCD: FLI Proline 16803
Pixel Size: 9μm square
Sensor: KAF 16803
Cooling: -35°C default
Array: 4096 x 4096 pixels
FOV: 43.2 x 43.2 arcmin

Location
Observatory: Siding Spring
UTC +10.00 (Australia Daylight savings time is observed)
31° 16' 24" South
149 03' 52" East
Elevation: 1122 meters (3681 ft)

Messier 84

Messier 84 (NGC 4374), shown in Figure 5.28, is an elliptical galaxy in the constellation of Virgo. Although here it is described as elliptical, it could potentially be a face-on lenticular galaxy. It is part of the Virgo cluster, located in an area crowded with galaxies. As well as the visible galaxies, there are dozens of galaxies in the field of view that are too faint to be captured. Messier 84 was discovered by Charles Messier on March 18, 1781. He commented on a "Nebula without star, in Virgo; the center it is a bit brilliant, surrounded with a slight nebulosity: its brightness & its appearance resemble that of those in this Catalog [M59 and M60]." The field of

Fig. 5.28 Image of Messier 84, an elliptical galaxy in Virgo

view of the image also includes the Messier object M86, another member of the Virgo Cluster.

Some galaxies captured in the 600-second image are identified on the negative image in Figure 5.29. M84 has an apparent diameter of 5' of arc and is believed to be at a distance of 60 millionly. This gives it an actual diameter of over 87,000 ly. The spiral galaxy NGC 4388 has apparent dimensions of 5.6' x 1.5', whereas NGC 4387 is a mere 1.7' x 1.1'. NGC 4413 has dimensions of 2.3' x 1.4', and NGC 4402 is 3.9' x 1.1'. IC 3303 is a relatively tiny elliptical galaxy with dimensions of 1.0' x 0.6'. Note that the image is tilted with respect to north. The plate solution gave a position angle of 11° 21' from north, so the camera would need to be rotated through that angle clockwise to get north at the top.

The chart in Figure 5.30 shows M84 in Virgo and the camera tilt is clearly seen from the plate-solved image. The star 42-Alpha Comae Berenices (Diadem or HIP 64241) lies to the northeast of M84. The angular separation between M84 and Diadem is 11° 43'. The star 9-Omicron Virginis lies to the southwest of Messier 84. The angular separation between them is 6° 25'. To the western side of the chart (right), you can see the meridian line dropping down from Ursa Major through Leo and then into the constellation of Crater.

The image was taken on March 5 at 00h 28m 18s local time in New Mexico. This corresponded to a Universal Time of 07h 28m 18s. The Right Ascension of M84 is 12h 25m 56s, and the Local Sidereal Time was 11h 17m 59s. The LST is

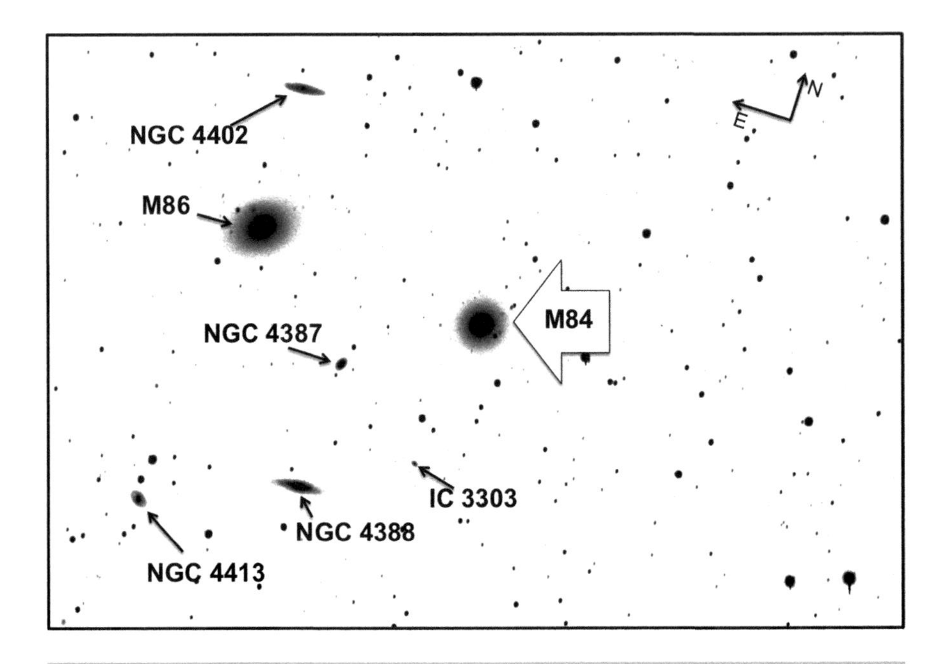

Fig. 5.29 Negative image of Messier 84

lower than the RA, so, as you can see on the chart, M84 had yet to cross the meridian from east to west. Subtracting one from the other gives a difference of 1h 7m 57s. Adding this to the local time when the image was taken gives a meridian transit time of 01h 36m 15s. The telescope used to take the image of M84 was a Takahashi Epsilon Astrograph with an aperture of 250 mm.

Messier 84: Data relating to the image in Figure 5.28

REMOTELY IMAGED FROM MAYHILL, NEW MEXICO
NGC 4374
OBJECT TYPE: Elliptical Galaxy

RA: 12h 25m 56s
DEC: +12° 47' 25"
ALTITUDE: +64° 41' 01"
AZIMUTH: 138° 13' 02"
FIELD OF VIEW: 1° 00’ 34”’ X 0° 40’ 49”
OBJECT SIZE: 5’
POSITION ANGLE: 11° 21’ from North
EXPOSURE TIME: 600 s
DATE: 5th March

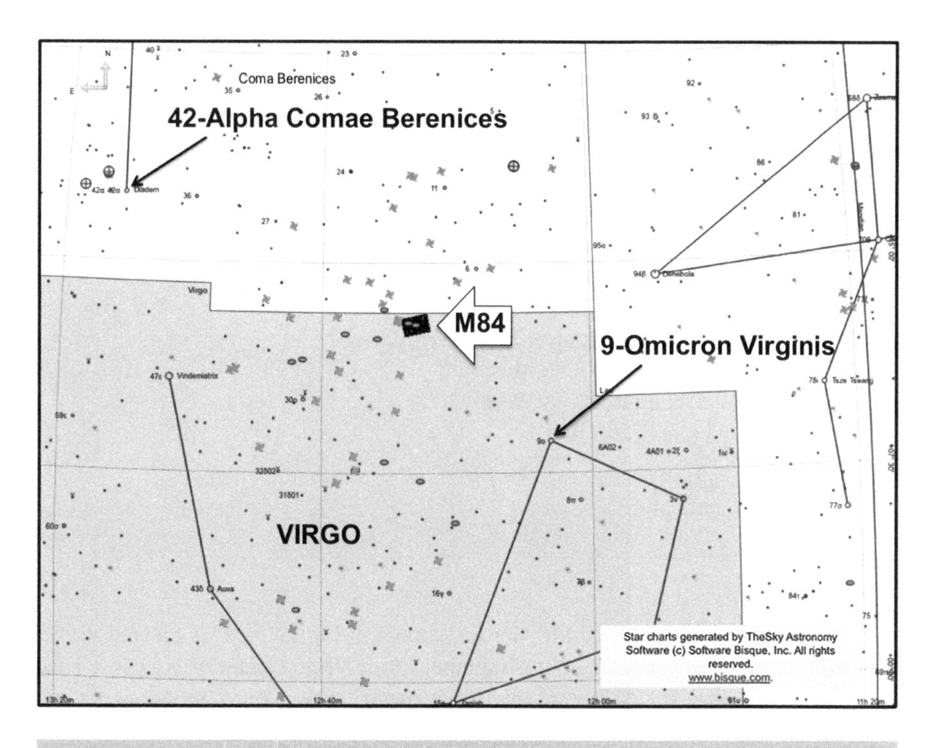

Fig. 5.30 Chart showing the location of M84

LOCAL TIME: 00h 28m 18s
UNIVERSAL TIME: 07h 28m 18s
SCALE: 1.66 arcsec/pixel
MOON PHASE: 48.17% (waxing)

Telescope Optics
OTA: Takahashi Epsilon 250
Optical Design: Hyperbolic Flat-Field Astrograph
Aperture: 250 mm
Focal Length: 850 mm
F/Ratio: f/3.4
Guiding: External
Mount: Paramount ME

Instrument Package
SBIG ST-10XME
A/D Gain: 1.3e-/ADU
Pixel Size: 6.8um Square

Resolution: 1.65 arcsec/pixel
Sensor: Frontlit
Cooling: Set to -15°C default
Array: 2184 x 1472 (3.2 Megapixels)
FOV: 40.4 x 60 arcmin

Location
Observatory: New Mexico Skies
UTC Minus 7.00 (Daylight savings time is observed)
Minimum Target Elevation: Approx 25–45 Degrees
(N or S) 32° 54' Decimal: 32.9 North
(W or E) 105° 31' Decimal: 105.5 West
Elevation: 2225 meters (7298 ft)

Messier 85

Although the lenticular galaxy Messier 85, shown in Figure 5.31, is located in the constellation of Coma Berenices, it is member of the Virgo cluster of galaxies. This is another discovery by Pierre Méchain that made its way onto Messier's list of objects. Méchain made the discovery on March 4, 1781, and Messier observed it on March 18, 1781 as a "Nebula without star, above & near to the ear of the Virgin, between the two stars in Coma Berenices, No's 11 & 14 of the Catalog of Flamsteed: this nebula is very faint." In 1833, John Herschel wrote that it was "Very bright; round; brighter toward the middle; 2' diameter; has a star at position angle 80° north preceding at distance 30" from edge."

On the negative image in Figure 5.32, some other galaxies are labeled. The galaxy close to M85 is NGC 4394 and is a barred spiral. It is probably physically associated with M85, as they have the same recessional velocity. M85 has apparent dimensions of 7.1' x 5.2' and lies at a distance of 60 million ly. This distance gives actual dimensions for M85 of 124,000 ly x 91,000 ly. NGC 4394 has dimensions 3.4' x 3.2', so at the same distance would be 59,000 ly x 56,000 ly. Galaxies labeled in Figure 5.32 go down to about magnitude 16.

The chart in Figure 5.33 shows the location of M85 in the constellation of Coma Berenices. Two reference stars have been labeled. The first, to the east of M85, is the optical double star 24 Comae Berenices (HIP 61418/ HIP 61415). HIP 61418 has a magnitude of 5.03 and is of spectral type K2 at a distance of 614.23 ly. HIP 61415 is magnitude 6.05 and is spectral type A7 at a distance of 2,630.30 ly. The angular separation between HIP 61415 and M85 is 2° 17'. The second reference star is 94-Beta Leonis (Denebola or HIP 57632) positioned to the southwest of M85, a magnitude 2.14 star of spectral type A3 at a distance of 36.18 ly, separated from M85 by 9° 43'.

Fig. 5.31 Image of Messier 85, a lenticular galaxy in Coma Berenices

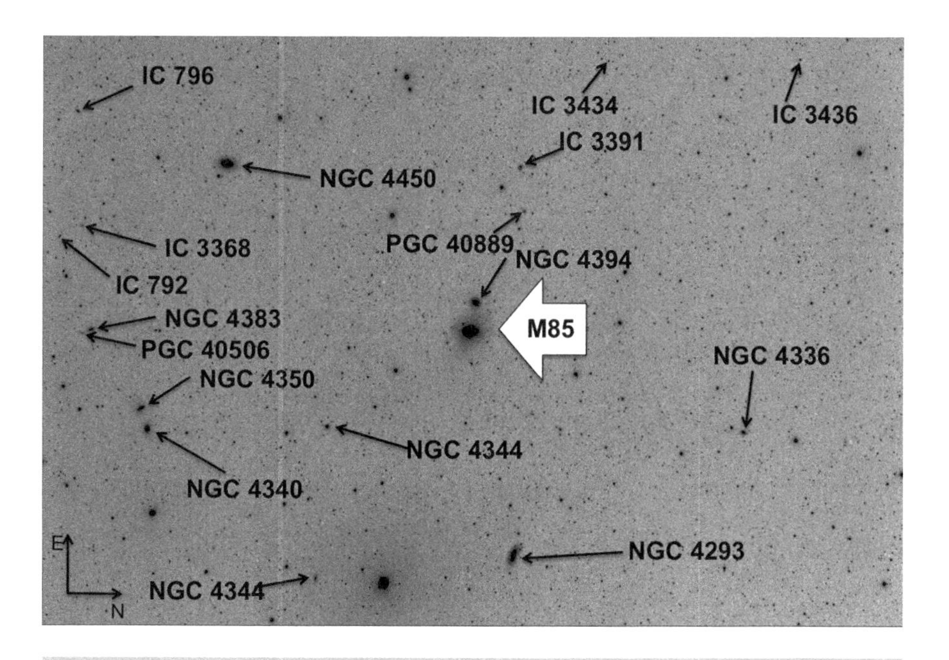

Fig. 5.32 Negative image of Messier 85

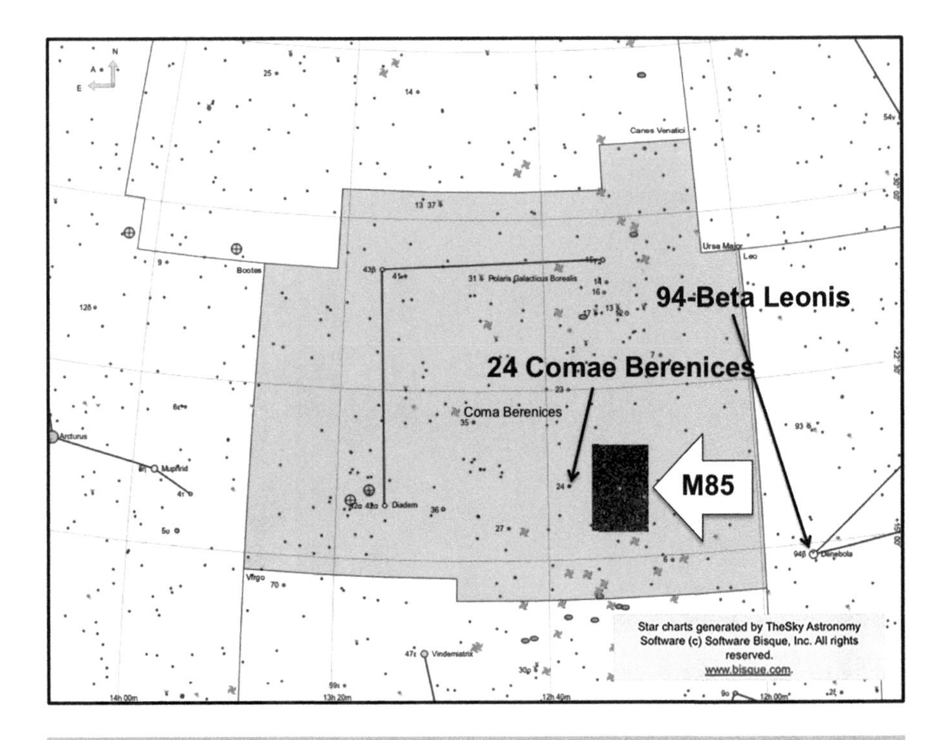

Fig. 5.33 Chart showing the location of M85

The image was taken on the morning of February 2 at 01h 19m 53s local time in New Mexico. The Universal Time was 08h 19m 53s. The Right Ascension of M85 is 12h 26m 16s, and the Local Sidereal Time was 10h 07m 30s. The spiral galaxy had not yet crossed the meridian. It would take 2h 18m 46s for that to happen, giving M85 a meridian transit time of 03h 38m 39s. The telescope was a Takahashi FSQ-ED with an aperture of 106 mm and a focal length of 530 mm. The telescope was mounted on a Paramount ME.

Messier 85: Data relating to the image in Figure 5.31

REMOTELY IMAGED FROM MAYHILL, NEW MEXICO
NGC 4382
OBJECT TYPE: Lenticular Galaxy
RA: 12h 26m 16s
DEC: +18° 05' 40"
ALTITUDE: +55° 36' 47"
AZIMUTH: 106° 40' 34"
FIELD OF VIEW: 3° 54' 12" x 2° 36' 08"

OBJECT SIZE: 7.1' X 5.2'
POSITION ANGLE: 89° 56' from North
EXPOSURE TIME: 300 s
DATE: 2nd February
LOCAL TIME: 01h 19m 53s
UNIVERSAL TIME: 08h 19m 53s
SCALE: 3.51 arcsec/pixel
MOON PHASE: 29.92% (waxing)

Telescope Optics
OTA: Takahashi FSQ - ED
Optical Design: Petzval Apochromatic Astrograph
Aperture: 106 mm
Focal Length: 530 mm
F/Ratio: f/5.0
Guiding: External
Mount: Paramount PME

Instrument Package
CCD: SBIG STL-11000M
Pixel Size: 9μm square
Sensor: Frontlit
Cooling: -15°C
Array: 4008 x 2672 pixels
FOV: 188.8 x 233.7 arcmin
Position Angle: 092.7°

Location
Observatory: New Mexico Skies
UTC Minus 7.00 (Daylight savings time is observed)
Minimum Target Elevation: Approx. 25 – 45 Degrees
(N or S) 32° 54' Decimal: 32.9 North
(W or E) 105° 31' Decimal: 105.5 West
Elevation: 2225 meters (7298 ft)

Messier 86

Messier 86 (NGC 4406) in the constellation of Virgo, shown in Figure 5.34, may be classified as either an elliptical or a lenticular galaxy. M86 has been found to contain thousands of globular clusters. Additionally, it appears to be approaching us rather than receding from us - the latter is the case with most galaxies. Charles Messier discovered M86 on the night of March 18, 1781 -- a busy night for him, as

Fig. 5.34 M86, a galaxy in Virgo

he made a number of discoveries in this galaxy filled area of the sky. He commented: "Nebula without star, in Virgo, on the parallel & very near to the nebula above, [M84], their appearances are the same, & both appear together in the same field of the telescope." John Herschel reported that Messier 86 was: "Very bright; large; round; gradually brighter toward the middle where there is a nucleus; mottled."

The negative image in Figure 5.35 shows M86 at the center with M84 in the field of view towards the southwest. M86 has apparent dimensions of 7.5' x 5.5' of arc. With a distance of 60 million ly, this gives actual dimensions of 131,000 ly x 96,000 ly. The field is filled with galaxies, some of which have been identified. Spiral galaxy NGC 4425 has a major axis of 2.8' and a minor axis of 1.0'. Elliptical galaxy NGC 4387 has a major axis of 1.7' and a minor axis of 1.1'. IC 3355 is an irregular galaxy with a major axis of 1.1' and a minor axis of 0.5'.

Messier 86 is located at the northern edge of the constellation of Virgo, as can be seen on the chart in Figure 5.36. Two reference stars are labeled on the chart. The first is 20 Virginis, which is a magnitude 6.29 star of spectral type G8 lying at

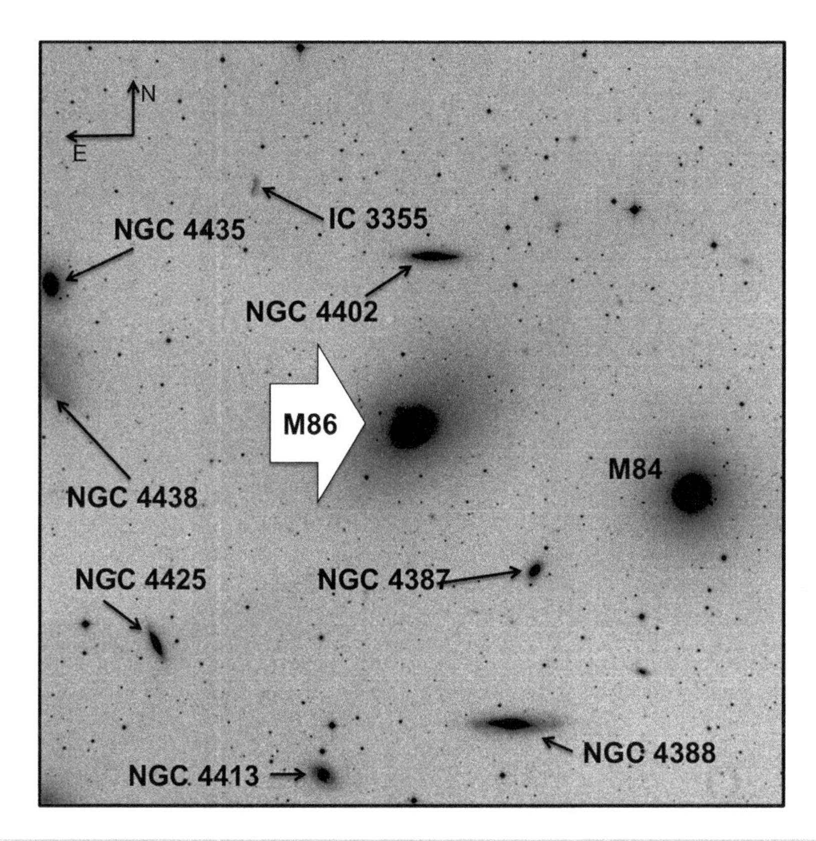

Fig. 5.35 Negative image of Messier 86

a distance of 471.33 ly. There is an angular separation of 3° 7' between 20 Virginis and Messier 86. The second reference is 12 Virginis, which is a magnitude 5.85 star of spectral type A2 lying at a distance of 161.62 ly. The meridian was to the west of M86 when the image was taken.

The image was taken on February 21 from Siding Spring at 03h 18m 18s. The corresponding Universal Time was 16h 18m 18s (February 20). The Right Ascension of M86 is 12h 27m 04s, and the Local Sidereal Time was 12h 17m 27s. Thus, there were 9m 37s to go before M86 crossed the meridian, making the meridian transit time 03h 27m 55s. The telescope used to take the image of M86 was a Planewave Corrected Dall-Kirkham Astrograph with an aperture of 431 mm and a focal length of 2,912 mm. The mount used was a Planewave Ascension 200HR.

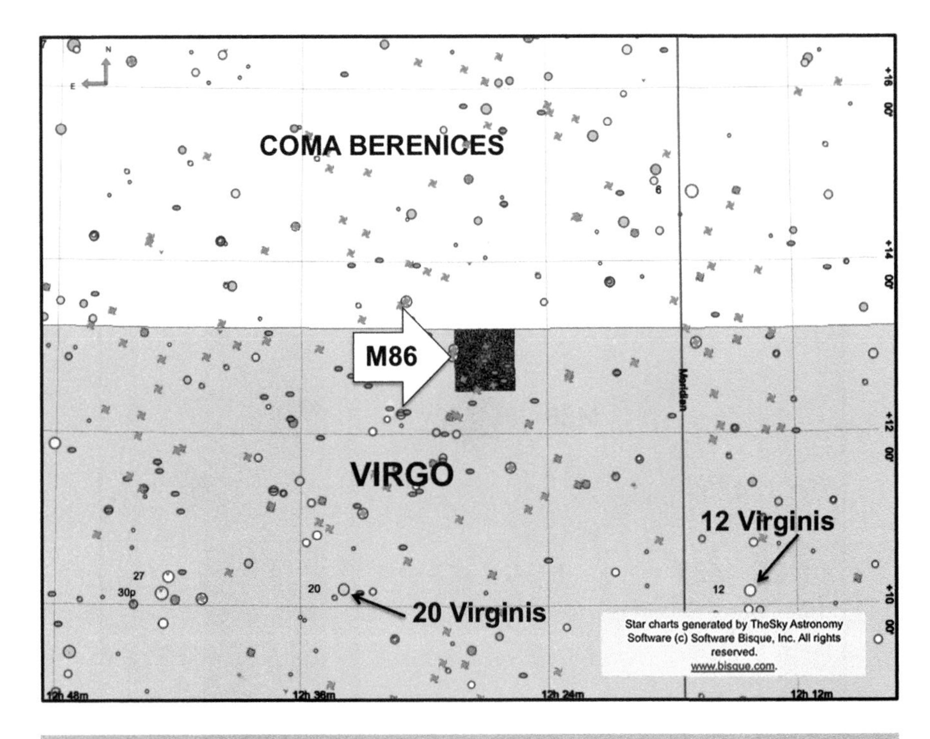

Fig. 5.36 Chart showing the location of M86

Messier 86: Data relating to the image in Figure 5.34

REMOTELY IMAGED FROM NEW SOUTH WALES, AUSTRALIA
NGC 4406
OBJECT TYPE: Elliptical / Lenticular galaxy

RA: 12h 27m 04s
DEC: +12° 51' 00"
ALTITUDE: +45° 50' 59"
AZIMUTH: 3° 21' 51"
FIELD OF VIEW: 0° 43' 54" x 0° 43' 54"
OBJECT SIZE: 7.5’ x 5.5’
POSITION ANGLE: 90° 49' from North
EXPOSURE TIME: 300 s
DATE: 21st February
LOCAL TIME: 03h 18m 18s
UNIVERSAL TIME: 16h 18m 18s (20th February)
SCALE: 0.64 arcsec/pixel
MOON PHASE: 32.82% (waning)

Telescope Optics
OTA: Planewave 17" CDK
Optical Design: Corrected Dall-Kirkham Astrograph
Aperture: 431 mm
Focal Length: 2912 mm
F/Ratio: f/6.8
Guiding: Active Guiding Disabled
Mount: Planewave Ascension 200HR

Instrument Package
CCD: FLI Proline 16803
Pixel Size: 9μm square
Sensor: KAF 16803
Cooling: -35°C default
Array: 4096 x 4096 pixels
FOV: 43.2 x 43.2 arcmin

Location
Observatory: Siding Spring
UTC +10.00 (Australia Daylight savings time is observed)
31° 16' 24" South
149 03' 52" East
Elevation: 1122 meters (3681 ft)

Messier 87

Messier 87 (NGC 4486), shown in Figure 5.37, is a massive elliptical galaxy in the Virgo galaxy cluster. M87 was yet another discovery made by Charles Messier on March 18, 1781 as he investigated the Virgo cluster. M87 has a very large system of globular clusters and an impressive jet thousands of light-years long that is lost in the glare in this image. This was Messier's comment: "Nebula without star, in Virgo, below & very near a star of eighth magnitude, the star having the same Right Ascension as the nebula, & its Declination was 13d 42' 21" north. This nebula appears at the same luminosity as the two nebulae M84 and M86." H. D. Curtis from Lick Observatory discovered the jet in 1918. He noted: "Exceedingly bright; the sharp nucleus shows well in a 5 minute exposure. The brighter central portion is about 0.5' in diameter, and the total diameter is about 2', nearly round. No spiral structure is discernible. A curious straight ray lies in a gap in the nebulosity in p.a. 20deg, apparently connected with the nucleus by a thin line of matter. The ray is brightest at its inner end."

Messier 87 has an apparent diameter of 7' of arc and is believed to lie at a distance of 60 million ly. This equates to an actual diameter of 122,000 ly, however, it

Fig. 5.37 M87, an elliptical galaxy in Virgo

is now thought that the extent of M87 is much larger, and that the galaxy could be as much as half a million ly in extent. Messier referred to a magnitude 8 star with the same Right Ascension as M87 immediately above it. This has to be HIP 61051, which is magnitude 8.65 and has a Right Ascension of 12h 31m 41.5s. Messier 87 has a Right Ascension of 12h 31m 42s! The negative of the original image, shown in Figure 5.38, labels some of the galaxies in the same field of view.

Messier 87 lies roughly on a line between 94-Beta Leonis (Denebola or HIP 57632), and 47-Epsilon Virginis (Vindemiatrix or HIP 63608) as shown in Figure 5.39. Denebola is magnitude 2.14 and has a spectral classification of A3, lying at a distance of 36.18 ly. There is an angular separation of 10° 24' between M87 and Denebola. Vindemiatrix is magnitude 2.85 and has a spectral classification of G8, lying at a distance of 102.24 ly. There is an angular separation of 7° 49' between Vindemiatrix and Messier 87.

The image was taken on February 21 from Siding Spring at 03h 36m 51s. The corresponding Universal Time was 16h 36m 51s (February 20). The Right Ascension of M87 is 12h 31m 42s, and the Local Sidereal Time was 12h 36m 04s. So, M87 crossed the meridian 4m 22s prior, making the meridian transit time of

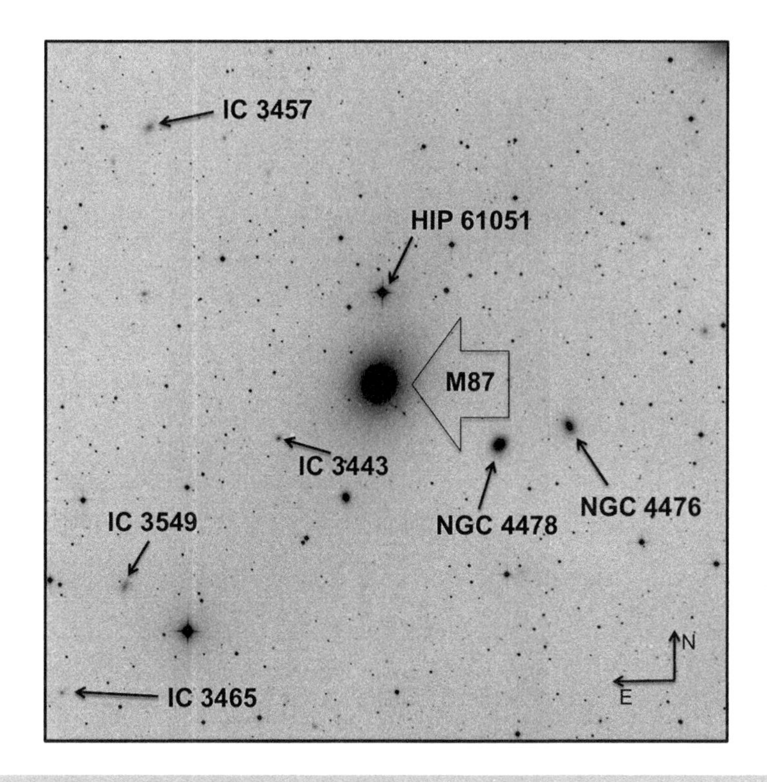

Fig. 5.38 Negative image of Messier 87

M87 03h 32m 29s. The telescope used to take the image of M87 was a Planewave Corrected Dall-Kirkham Astrograph with an aperture of 431 mm and a focal length of 2912 mm.

Messier 87: Data relating to the image in Figure 5.37

REMOTELY IMAGED FROM NEW SOUTH WALES, AUSTRALIA
NGC 4486
OBJECT TYPE: Elliptical Galaxy

RA: 12h 31m 42s
DEC: +11° 57' 42"
ALTITUDE: +46° 46' 24"
AZIMUTH: 357° 49' 03"
FIELD OF VIEW: 0° 43' 54" x 0° 43' 54"
OBJECT SIZE: 7'
POSITION ANGLE: 90° 46' from North
EXPOSURE TIME: 300 s
DATE: 21st February

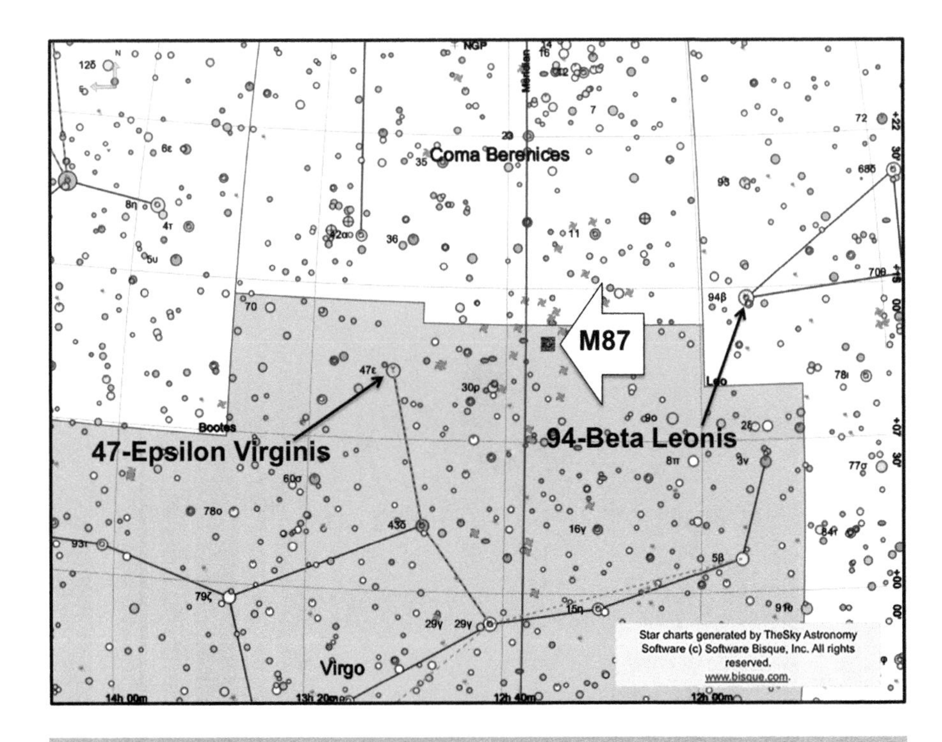

Fig. 5.39 Chart showing the location of M87

LOCAL TIME: 03h 36m 51s
UNIVERSAL TIME: 16h 36m 51s (21st February)
SCALE: 0.64 arcsec/pixel
MOON PHASE: 32.70% (waning)

Telescope Optics
OTA: Planewave 17" CDK
Optical Design: Corrected Dall-Kirkham Astrograph
Aperture: 431 mm
Focal Length: 2912 mm
F/Ratio: f/6.8
Guiding: Active Guiding Disabled
Mount: Planewave Ascension 200HR

Instrument Package
CCD: FLI Proline 16803
Pixel Size: 9μm square
Sensor: KAF 16803
Cooling: -35° C default

Array: 4096 x 4096 pixels
FOV: 43.2 x 43.2 arcmin

Location
Observatory: Siding Spring
UTC +10.00 (Australia Daylight savings time is observed)
31° 16' 24" South
149 03' 52" East
Elevation: 1122 meters (3681 ft)

Messier 88

Messier 88 (NGC 4501), shown in Figure 5.40, is a spiral galaxy in the constellation of Coma Berenices. It is very close to another messier galaxy, M91, and is part of the Virgo-Coma Cluster of galaxies. This area was observed by W. H. Smyth in May 1836, who noted, "This is a wonderful nebulous region, and the diffused matter occupies an extensive space, in which several of the finest objects of Messier and the Herschels will readily be picked up by the keen observer in extraordinary

Fig. 5.40 Image of Messier 88, a spiral galaxy in Coma Berenices

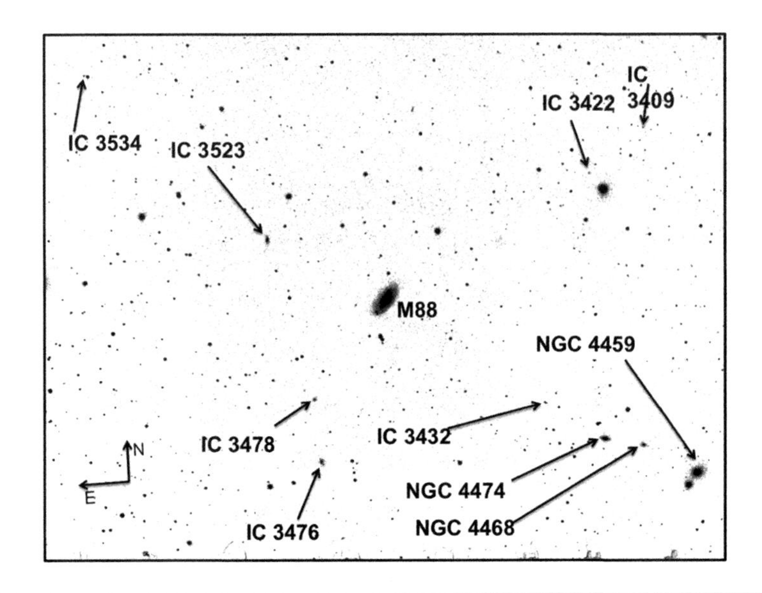

Fig. 5.41 Negative image of Messier 88

proximity." M88 was one of 8 galaxies discovered by Charles Messier on the night of March 18, 1781. Remember that when referring to "nebulae" before galaxies were understood, Messier described it as a nebula not containing any stars. After observing M88, he commented, "Nebula without star, in Virgo, between two small stars & one star of the sixth magnitude, which appear at the same time as the nebula in the field of the telescope. Its luminosity is one of the faintest, & resembles the one reported in Virgo, Messier 58."

There are many galaxies visible on the 300-second image, most of which have been labeled in the negative image in Figure 5.41. Messier 88 itself has two little stars at the southeastern end of the galaxy that always help to identify the object. Messier 88 has dimensions of 7' x 4' of arc and lies at a distance of 60 million ly, giving it an actual size of about 122,000 x 70,000 ly. One of the larger galaxies visible on the image is NGC 4459 at the lower right (southwest from M88). This galaxy has a major axis of 3.5' and a minor axis of 2.7'. The elliptical galaxy IC 3422 (top right or northwest from M88), on the other hand, only has a diameter of 0.5'.

The chart in Figure 5.42 shows the location of M88 in Coma Berenices, and you can see just how close it is to M91. There is an angular separation of just less than 51' between M88 and M91. The North Galactic Pole (NGP) is to be found in Coma Berenices, and its location has been labeled. The angular difference between M88 and the NGP is 13° 27'. A nearby reference star 25 Comae Berenices is also labeled. This star is HIP 61571, a magnitude 5.70 star of spectral type K5 lying at a distance of 518.53 ly. Using the measuring tool provided with the charting

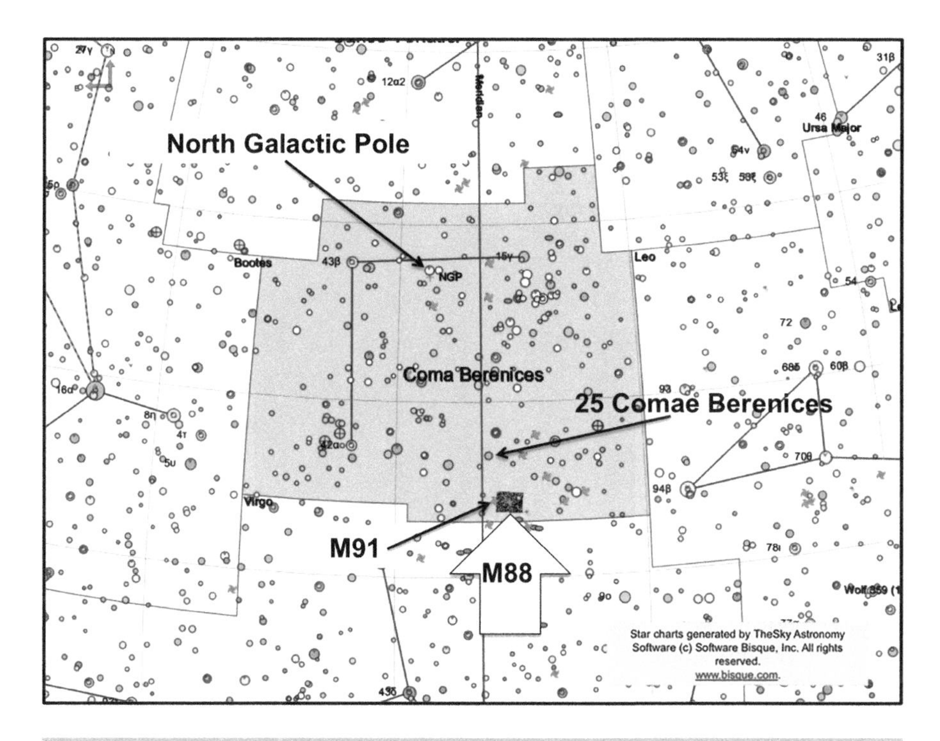

Fig. 5.42 Chart showing the location of M88

software, I measure an angular separation of 2° 55' between HIP 61571 and Messier 88.

The image was taken on June 1 from Siding Spring at 20h 02m 59s. The corresponding Universal Time was 10h 02m 59s. The Right Ascension of M88 is 12h 32m 52s, and the Local Sidereal Time was 12h 39m 19s. Thus, M88 crossed the meridian 6m 27s prior. The meridian transit time of M88 was at 19h 56m 32s. The single-shot color camera was used with a 90-mm refractor to obtain the image.

Messier 88: Data relating to the image in Figure 5.40

REMOTELY IMAGED FROM NEW SOUTH WALES, AUSTRALIA
NGC 4501
OBJECT TYPE: Spiral Galaxy

RA: 12h 32m 52s
DEC: +14° 19' 32.34"
ALTITUDE: +44° 24' 29"
AZIMUTH: 357° 48' 43"
FIELD OF VIEW: 1° 37' 44" x 1° 13' 18"
OBJECT SIZE: 7' x 4'

POSITION ANGLE: 177° 04' from North
EXPOSURE TIME: 300 s
DATE: 1st June
LOCAL TIME: 20h 02m 59s
UNIVERSAL TIME: 10h 02m 59s
MOON PHASE: 48.95% (waxing)

Telescope Optics
OTA: Takahashi Sky 90
Optical Design: Apochromatic Refractor
Aperture: 90 mm
Focal Length: 417 mm
F/Ratio: f/5.6
Guiding: None
Mount: Paramount PME

Instrument Package
CCD: SBIG ST2000 XMC
Colour CMOS
Pixel Size: 7.4μm square
Sensor: Frontlit
Cooling: CNA
Array: 1600 x 1200 pixels
FOV: 60.5 x 80.7 arcmin

Location
Observatory: Siding Spring
UTC +10.00 (Australia Daylight savings time is observed)
31° 16’ 24” South
149 03’ 52” East
Elevation: 1122 meters (3681 ft)

Messier 89

Messier 89 (NGC 4552), shown in Figure 5.43, is an elliptical galaxy that appears perfectly circular from our viewpoint. It is a member of the Virgo cluster of galaxies and was discovered by Charles Messier on the same night (March 18, 1781) that he discovered a number of other cluster members. Messier described his observation of M89: “Nebula without star, in Virgo, a little of distance from & on the same parallel as the nebula reported above, [Messier 87]. Its light was extremely faint & pale, & and it is not without difficulty that one can distinguish it." A telescope at Siding Spring took the image, adjacent to the Anglo Australian Telescope and the

Fig. 5.43 Image of Messier 89, an elliptical galaxy in Virgo

UK Schmidt Telescope (UKST). In 1979, the astronomer David Malin used the 1.2-meter aperture UKST to take deep images, which showed the presence of an extended envelope around the elliptical galaxy and also an optical jet, which extended 10 arcmin from M89.

Messier 89 is identified at the center of the negative image in Figure 5.44. M89 has a diameter of 4 arcmin and lies at a distance of 6 million ly. This means that M89 has an actual diameter of 70,000 ly. The field of view of the image is roughly 44' x 44'. Three galaxies in the frame are labeled. At the bottom of the image (south), there are two galaxies – one a lenticular galaxy and one elliptical. The lenticular galaxy is NGC 4550, which has a major axis of 3.3' of arc and a minor axis of 0.9'. This galaxy is of great interest following a 1992 discovery by the American astronomer Vera Rubin who published a paper describing the way in which NGC 4550 contained stars rotating in two directions at the same time, rather like two contra-rotating disks. The elliptical galaxy is NGC 4551, which has a major axis of 1.8 arcmin and a minor axis of 1.4'. The third galaxy is IC 3586, an elliptical galaxy with dimensions of 1.1' x 1.0'.

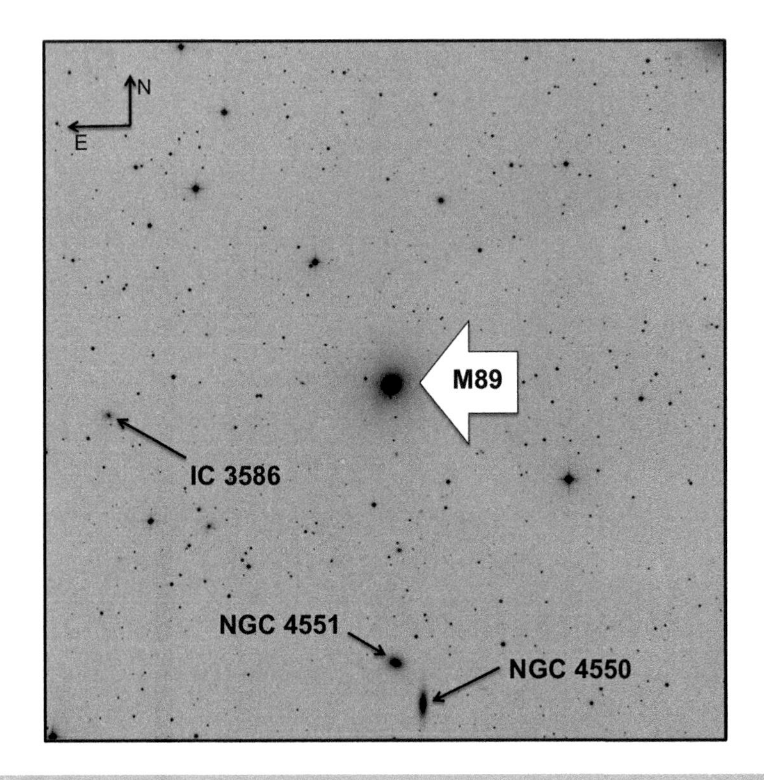

Fig. 5.44 Negative image of Messier 89

The chart in Figure 5.45 shows the position of Messier 89 in relation to some of the other Messier galaxies in the constellation of Virgo. The angular separation between Messier 89 and Messier 87 is roughly 1° 12'. Between M89 and M86 is about 2° 21', and between M89 and M84 is 2° 36'. The star 20 Virginis is labeled on the negative image. This star (HIP 61246) is magnitude 6.29 and spectral classification G8. It lies at a distance of 471.33 ly. The angular separation between Messier 89 and 20 Virginis is 2° 21'.

The image was taken on February 21 from Siding Spring at 03h 51m 47s. The corresponding Universal Time was 16h 51m 47s (February 20). The Right Ascension of M89 is 12h 36m 32s, and the Local Sidereal Time was 12h 51m 02s. So, M89 crossed the meridian 14m 30s before, making the meridian transit of M89 03h 37m 17s. The telescope used to take the image of Messier 89 was a Planewave Corrected Dall-Kirkham Astrograph with an aperture of 431 mm and a focal length of 2,912 mm. The mount used was a Planewave Ascension 200HR.

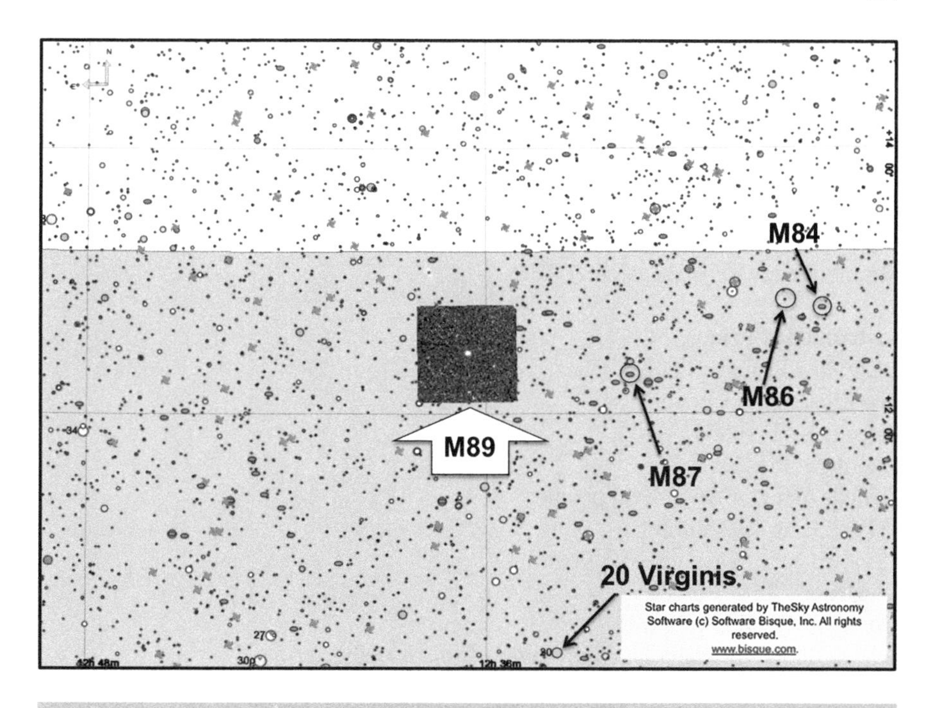

Fig. 5.45 Chart showing the location of M89

Messier 89: Data relating to the image in Figure 5.43

REMOTELY IMAGED FROM NEW SOUTH WALES, AUSTRALIA
NGC 4552
OBJECT TYPE: Elliptical Galaxy

RA: 12h 36m 32s
DEC: +12° 27' 37"
ALTITUDE: +46° 09' 41"
AZIMUTH: 354° 53' 18"
FIELD OF VIEW: 0° 43' 54" x 0° 43' 54"
OBJECT SIZE: 4'
POSITION ANGLE: 90° 46' from North
EXPOSURE TIME: 300 s
DATE: 21st February
LOCAL TIME: 03h 51m 47s
UNIVERSAL TIME: 16h 51m 47s (20th February)
SCALE: 0.64 arcsec/pixel
MOON PHASE: 32.61% (waning)

Telescope Optics
OTA: Planewave 17" CDK
Optical Design: Corrected Dall-Kirkham Astrograph
Aperture: 431 mm
Focal Length: 2912 mm
F/Ratio: f/6.8
Guiding: Active Guiding Disabled
Mount: Planewave Ascension 200HR

Instrument Package
CCD: FLI Proline 16803
Pixel Size: 9μm square
Sensor: KAF 16803
Cooling: -35° C
Default Array: 4096 x 4096 pixels
FOV: 43.2 x 43.2 arcmin

Location
Observatory: Siding Spring
UTC +10.00 (Australia Daylight savings time is observed)
31° 16' 24" South
149 03' 52" East
Elevation: 1122 meters (3681 ft)

Messier 90

Messier 90 (NGC 4569), shown in Figure 5.46, is a spiral galaxy in the constellation of Virgo. This is another member of the Virgo Group of galaxies, as well as another of the discoveries made on the night of March 18, 1781. He found nine new objects in total. Of M90, Messier noted, "Nebula without star, in Virgo: its light is as faint as the preceding [M89]." A paper published in 2016 by contemporary astronomer Boselli et. al. referred to a narrow band Hydrogen-Alpha image that had been obtained of M90: "The image reveals the presence of long tails of diffuse ionized gas without any associated stellar component extending from the disc of the galaxy." The tails extend for over 300,000 ly, longer than the size of Messier 90 itself.

Messier 90 has an apparent size of 9.5' x 4.5' of arc and lies at a distance of 60 million ly. This gives it an actual size of 166,000 x 79,000 ly. On the negative of the image in Figure 5.47, the relatively bright star HIP 61600 to the southeast of M90 is labeled. This is a magnitude 8.22 star of spectral type G5, which lies at a distance of 2,090.75 ly. To the northwest of M90, the irregular galaxy IC 3583 has been identified. Just at the eastern edge of the frame is NGC 4584, a spiral galaxy.

Fig. 5.46 Image of Messier 90, a spiral galaxy in Virgo

The chart in Figure 5.48 shows the location of Messier 90 in the constellation of Virgo. The image taken actually straddles Virgo and Coma Berenices as you can see. Three nearby galaxies are labeled. The first galaxy is NGC 4531, which is a spiral measuring 3.1' x 2'. NGC 4531 is about 38' to the west of M90. IC 3631 lies roughly to the southeast of M90 and is another spiral with axes of 0.8' and 0.5'. It has an angular separation of 45' from M90. IC 3611 is to the northeast of M90 and is also a spiral with axes of 1.4' x 0.8'. You can see that M90 is approaching the meridian to the west (right) of the chart.

The image was taken on February 21 from Siding Spring at 03h 25m 58s. The corresponding Universal Time was 16h 25m 58s (February 20). The Right Ascension of M90 is 12h 37m 42s, and the Local Sidereal Time was 12h 25m 09s. So, M90 would cross the meridian in 12m 33s, making the meridian transit of M90 03h 38m 31s. The telescope used to take the image of Messier 90 was a Planewave Corrected Dall-Kirkham Astrograph with an aperture of 431 mm and a focal length of 2,912 mm. The mount used was a Planewave Ascension 200HR.

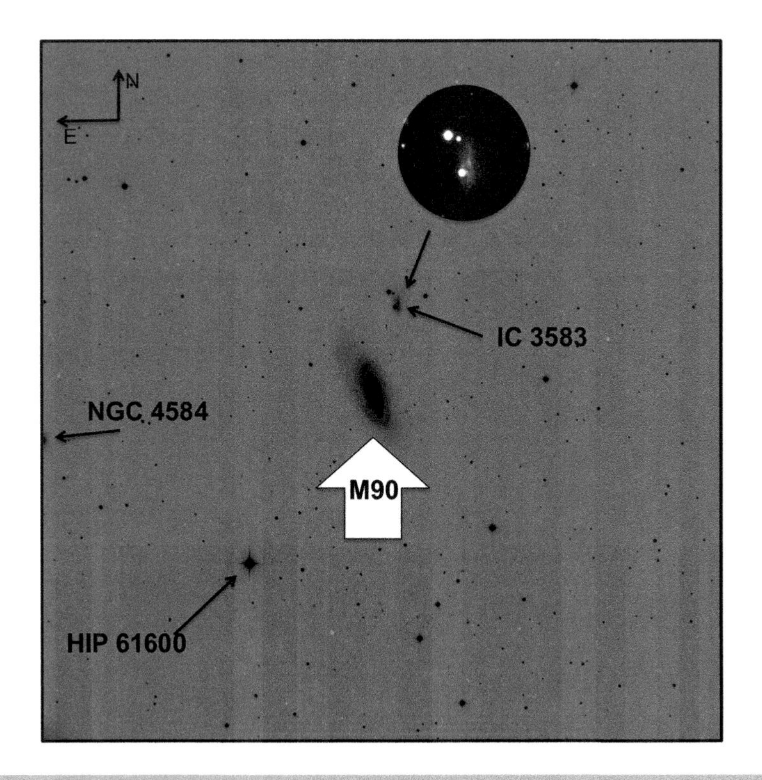

Fig. 5.47 Negative image of Messier 90

Messier 90: Data relating to the image in Figure 5.46

REMOTELY IMAGED FROM NEW SOUTH WALES, AUSTRALIA
NGC 4569
OBJECT TYPE: Spiral Galaxy

RA: 12h 37m 42s
DEC: +13° 04' 05"
ALTITUDE: +45° 35' 22"
AZIMUTH: 4° 22' 23"
FIELD OF VIEW: 0° 43' 54" x 0° 43' 54"
OBJECT SIZE: 9.5’ x 4.5’
POSITION ANGLE: 90° 49' from North
EXPOSURE TIME: 300 s
DATE: 21st February
LOCAL TIME: 03h 25m 58s
UNIVERSAL TIME: 16h 25m 58s
SCALE: 0.64 arcsec/pixel
MOON PHASE: 32.77% (waning)

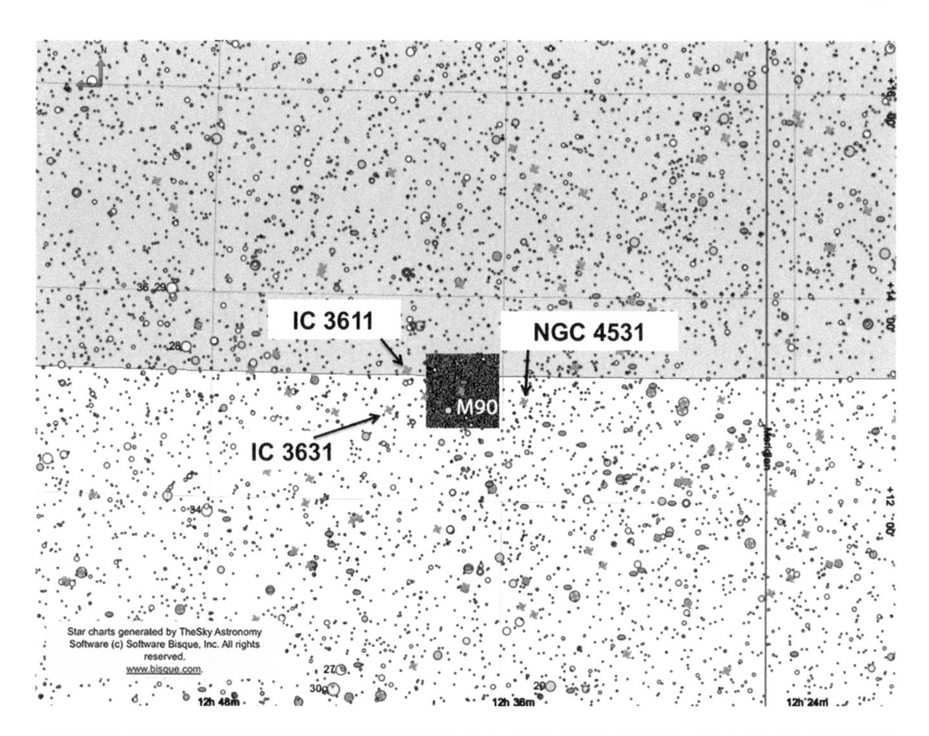

Fig. 5.48 Chart showing the location of M90

Telescope Optics

OTA: Planewave 17" CDK
Optical Design: Corrected Dall-Kirkham Astrograph
Aperture: 431 mm
Focal Length: 2912 mm
F/Ratio: f/6.8
Guiding: Active Guiding Disabled
Mount: Planewave Ascension 200HR

Instrument Package

CCD: FLI Proline 16803
Pixel Size: 9μm square
Sensor: KAF 16803
Cooling: -35° C default
Array: 4096 x 4096 pixels
FOV: 43.2 x 43.2 arcmin

Location

Observatory: Siding Spring
UTC +10.00 (Australia Daylight savings time is observed)

31° 16' 24" South
149 03' 52" East
Elevation: 1122 meters (3681 ft)

Messier 91

Messier 91 (NGC 4548), shown in Figure 5.49, is a barred spiral galaxy in the constellation of Coma Berenices. M91 is a member of the Coma-Virgo Cluster of galaxies. Messier made a number of discoveries on the night of March 18, 1781, but when he recorded the position of what is now believed to be NGC 4548, he made an error, putting down a location that did not correspond to an appropriate object. Messier commented that M91 was a "Nebula without star, in Virgo, above the preceding [M90] its light is still fainter than that of the above." It was not until 1969 in a letter to Sky and Telescope from William C. Williams that a solution was proposed. Williams wrote: "It can be simply demonstrated that the lost Messier object M91 is very probably the galaxy NGC 4548.... the Skalnate Pleso Atlas Catalogue gives the visual magnitudes of NGC 4548 and M90 as 10.8 and 10.0, respectively. This checks with Messier's statement that M91 was the fainter of the two."

The dimensions of M91 are given by SEDS as 5.4' x 4.4' of arc. It is believed to lie at a distance of 60 million ly, giving the galaxy actual dimensions of 94,000 ly x 77,000 ly. There are a number of small galaxies visible on the negative image

Fig. 5.49 Image of Messier 91, a barred spiral galaxy in Coma Berenices

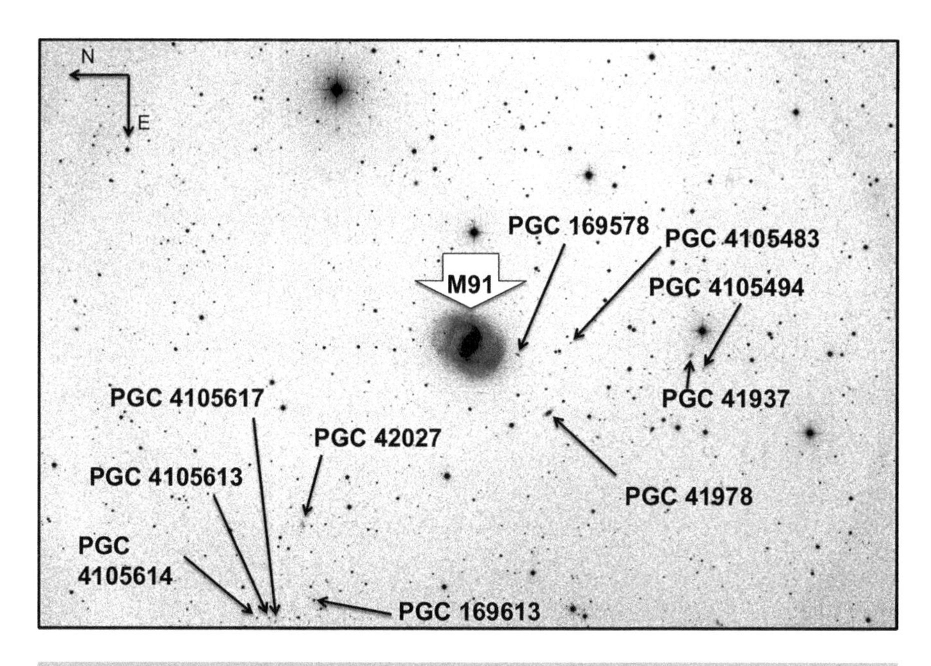

Fig. 5.50 Negative image of Messier 91

in Figure 5.50. Close to M91 is PGC 169578 (from the Principal Galaxy Catalog). The catalog gives it a major axis of 0.6' and a minor axis of 0.3'. The catalog lists it at magnitude 17.86. The brighter galaxy PGC 41978 is cataloged at magnitude 15.24 with dimensions 0.7' x 0.3'. The faint galaxy PGC 169613 has a magnitude of 18.16 and dimensions of 0.2' x 0.2'. Notice that in the image, north is to the left and east is down.

The chart in Figure 5.51 shows the position of Messier 91 at the lower end of Coma Berenices. Notice that north is up and east is to the left on the chart. Two interesting bright stars can be located on the chart if you look closely. To the east (left) is the bright star Arcturus. Arcturus (HIP 69673) is 16-Alpha Bootis with a magnitude of -0.05 and is of spectral type K2 lying at a distance of 36.71 ly. Measuring the angular separation between Arcturus and Messier 91 on the chart gives a value of 24° 28'. To the west of the chart, the star Zosma can be found. Zosma (HIP 54872) is 68-Delta Leonis with a magnitude of 2.56 and is of spectral type A4 lying at a distance of 57.71 ly. There is an angular separation between Messier 91 and Zosma of 20° 21'.

The image was taken on May 2 from New Mexico at 00h 21m 21s. The corresponding Universal Time was 06h 21m 21s. The Right Ascension of M91 is 12h 36m 19s, and the Local Sidereal Time was 13h 59m 31s. M91 therefore crossed the meridian 1h 23m 12s before, making the meridian transit time 22h 58m 09s. The telescope used to take the image of M91 was a Planewave Corrected Dall-Kirkham

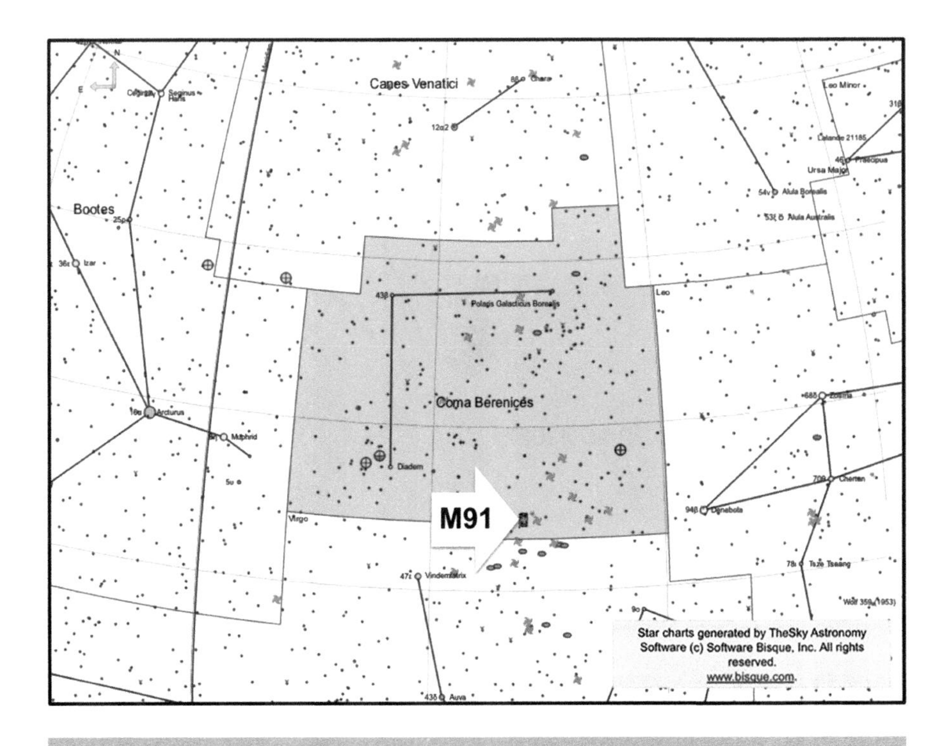

Fig. 5.51 Chart showing the location of M91

Astrograph with an aperture of 510 mm and a focal length of 2,280 mm. The mount used was a Planewave Ascension 200HR.

Messier 91: Data relating to the image in Figure 5.49

REMOTELY IMAGED FROM MAYHILL, NEW MEXICO
NGC 4548
OBJECT TYPE: Barred Spiral Galaxy
RA: 12h 36m 19s
DEC: +14° 24' 06"
ALTITUDE: +63° 37' 28"
AZIMUTH: 230° 44' 12"
FIELD OF VIEW: 0° 54' 47" x 0° 36' 31"
OBJECT SIZE: 5.4’ x 4.4’
POSITION ANGLE: 271° 08' from North
EXPOSURE TIME: 300 s
DATE: 2nd May 2nd
LOCAL TIME: 00h 21m 21s
UNIVERSAL TIME: 06h 21m 21s

MOON PHASE: 40.74% (waxing)

Telescope Optics
OTA: Planewave 20" CDK
Optical Design: Corrected Dall-Kirkham Astrograph
Aperture: 510 mm
Focal Length: 2280 mm (0.66 Focal Reducer Fitted)
F/Ratio: f/4.5
Guiding: Active Guiding Disabled
Mount: Planewave Ascension 200HR

Instrument Package
FLI Proline PL11002M CCD Camera
Pixel Size: 9-μm square
Resolution: 0.81 arcsec/pixel
Cooling: -30°C default
Array: 4008 x 2672 (10.7 Megapixels)

Location
Observatory: New Mexico Skies
UTC Minus 7.00 (Daylight savings time is observed)
Minimum Target Elevation: Approx. 25–45 Degrees
(N or S) 32° 54' Decimal: 32.9 North
(W or E) 105° 31' Decimal: 105.5 West
Elevation: 2225 meters (7298 ft)

Messier 92

Messier 92 (NGC 6341), shown in Figure 5.52, is a globular cluster in the constellation of Hercules. This cluster is so bright it can be seen with the naked eye under the right conditions. It is overshadowed by the spectacular globular Messier 13, which is not too far away in the same constellation and is even brighter. Johann Bode discovered M92, writing, "On this occasion, I also want to announce that on December 27, 1777 I have discovered a new nebula in Hercules, not known to me, southwest below the star *s* in his foot, which shows up in a mostly round figure with a pale glimmer of light." Charles Messier observed it on his extremely busy night of March 18, 1781, noting: "Nebula, fine, distinct, & very bright, between the knee & the left leg of Hercules, it can be seen very well in a telescope of one foot. It contains no star; the center is clear & brilliant, surrounded by nebulosity & resembles the nucleus of a large Comet" (SEDS).

The brightest star in the image is HIP 84658, which has a magnitude of 8.65 and is of spectral type K0. It lies at a distance of 813.36 ly. M92 itself lies at a distance

Fig. 5.52 Image of Messier 92, a globular cluster in Hercules

of 26,700 ly, which is roughly 33 times the distance to that star. M92 has an angular diameter of 14' of arc, which gives it an actual diameter of 109 ly. The field of view of the image is 0° 54' 47" x 0° 36' 28". The negative image in Figure 5.53 labels some of the galaxies in the frame of the image. The nearest identified galaxy to M92 is PGC 59984 with a major axis of 1' and a minor axis of 0.5'. The galaxy PGC 2204313 is even smaller with dimensions of 30" x 18", and the galaxy PGC 2204180 is 24" x 18" (Principal Galaxy Catalog Data).

The chart in Figure 5.54 shows the location of M92 and the plate-solved image in the constellation of Hercules. You can see that the globular cluster Messier 13 is not too far away. There is an angular separation of 9° 33' between M92 and M13. The position of another nearby globular cluster, NGC 6229, has been marked. It has a diameter of only 4.5' in comparison with M92 at 14' and M13 at 20'. NGC 6229 is 6° 54' from Messier 92. M92 lies roughly on a line between the stars 85-Iota Herculis and 44-Eta Herculis. 85-Iota Herculis (HIP 86414) has a magnitude of 3.82 and is of spectral type B3, lying at a distance of 495.68 ly. 44-Eta Herculis (HIP 81833) has a magnitude of 3.48 and spectral type G8 with a distance of 112.04 ly. The angular separation between M92 and 85-Iota Herculis is 6° 21' and between M92 and 44-Eta Herculis is 7° 44'.

The image was taken on April 7 from New Mexico at 04h 09m 12s in the morning. The corresponding Universal Time was 10h 09m 12s. The Right Ascension of M92 is 17h 17m 40s, and the Local Sidereal Time was 16h 09m 26s. M92 would

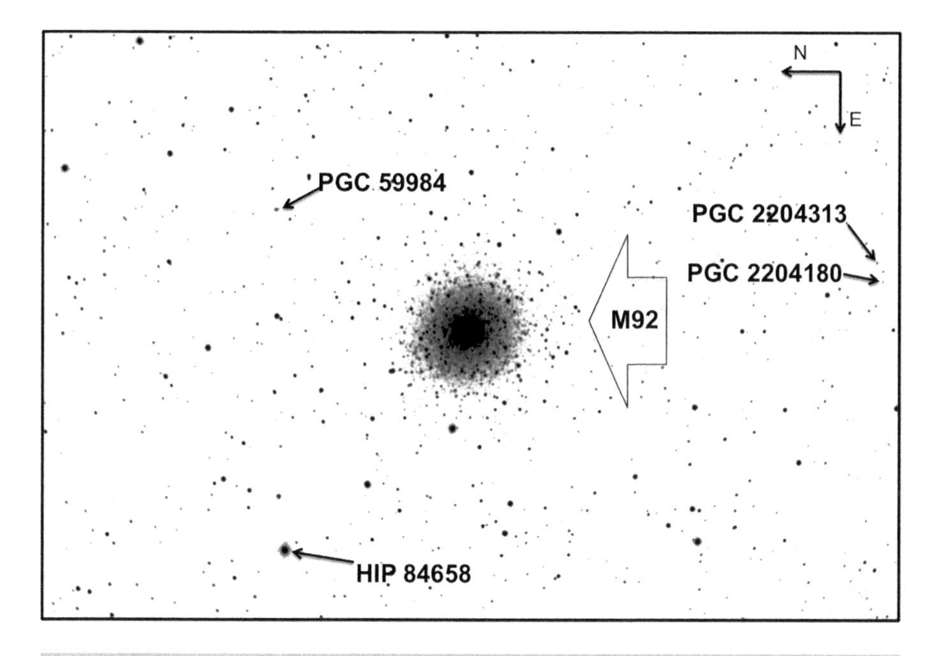

Fig. 5.53 Negative image of Messier 92

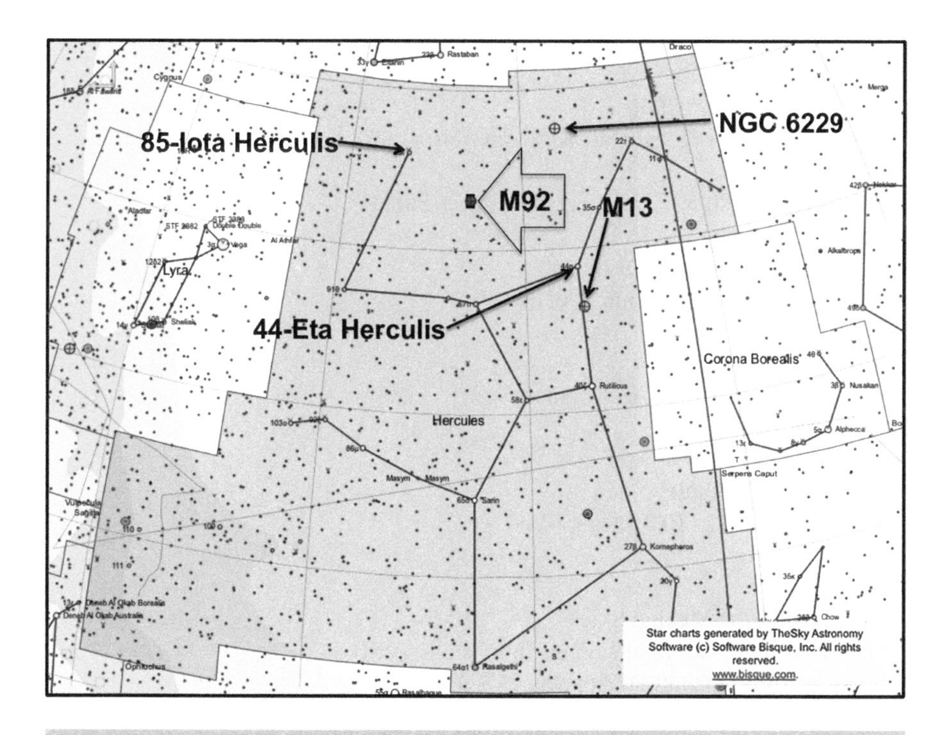

Fig. 5.54 Chart showing the location of M92

therefore cross the meridian in 1h 8m 14s, providing a transit time of 05h 17m 26s The telescope used to take the image of M92 was a Planewave Corrected Dall-Kirkham Astrograph with an aperture of 510 mm and a focal length of 2,280 mm. The mount used was a Planewave Ascension 200HR.

Messier 92: Data relating to the image in Figure 5.52

REMOTELY IMAGED FROM MAYHILL, NEW MEXICO
NGC 6341
OBJECT TYPE: Globular Cluster

RA: 17h 17m 40s
DEC: +43° 07' 003
ALTITUDE: +73° 05' 41"
AZIMUTH: 47° 25' 06"
FIELD OF VIEW: 0° 54' 47" x 0° 36' 28"
OBJECT SIZE: 14'
POSITION ANGLE: 271° 18' from North
EXPOSURE TIME: 300 s
DATE: 7th April
LOCAL TIME: 04h 09m 12s
UNIVERSAL TIME: 10h 09m 12s
MOON PHASE: 85.64% (waxing)

Telescope Optics
OTA: Planewave 20" CDK
Optical Design: Corrected Dall-Kirkham Astrograph
Aperture: 510 mm
Focal Length: 2280 mm (0.66 Focal Reducer Fitted)
F/Ratio: f/4.5
Guiding: Active Guiding Disabled
Mount: Planewave Ascension 200HR

Instrument Package
FLI Proline PL11002M CCD Camera
Pixel Size: 9µm square
Resolution: 0.81 arcsec/pixel
Cooling: -30°C default
Array: 4008 x 2672 (10.7 Megapixels)

Location
Observatory: New Mexico Skies
UTC Minus 7.00 (Daylight savings time is observed)
Minimum Target Elevation: Approx. 25 – 45 Degrees
(N or S) 32° 54' Decimal: 32.9 North
(W or E) 105° 31' Decimal: 105.5 West
Elevation: 2225 meters (7298 ft)

Messier 93

Messier 93 (NGC 2447), shown in Figure 5.55, is an open cluster located in the constellation of Puppis. It was discovered on March 20, 1781 by Charles Messier, who commented: "Cluster of small stars, without nebulosity, between the Greater Dog and the prow of the ship." In 1783, it was observed by William Herschel's sister Caroline Herschel, who thought it was a new discovery as the report of Messier's discovery was not published until the 1784 "Connoissance de Temps" (pg. 265). M93 is not a particularly exciting cluster, but Messier listed it to make sure it was not confused with a potential comet. John Herschel described M93 as a "Cluster; large; pretty rich; little compressed; stars from 8th to 13th magnitude."

Messier 93 extends to 22 arcmin and lies at a distance of 3,600 ly containing in the region of 80 stars. The size and distance of M93 give it an actual size of 23 ly. The field of view of the image is 1° 37' 41" x 1° 13' 16". The orange star HIP 37729, identified in Figure 5.56, is a magnitude 8.18 star of spectral type K2 and lies at a distance of 724.79 ly, so it is not a physical part of the cluster. (Hipparcos Data). This is the brightest star in the cluster, and its magnitude corresponds to John Herschel's estimate of the magnitude range from 8 to 13. The star HIP 37880,

Fig. 5.55 Image of Messier 93, an open cluster in Puppis

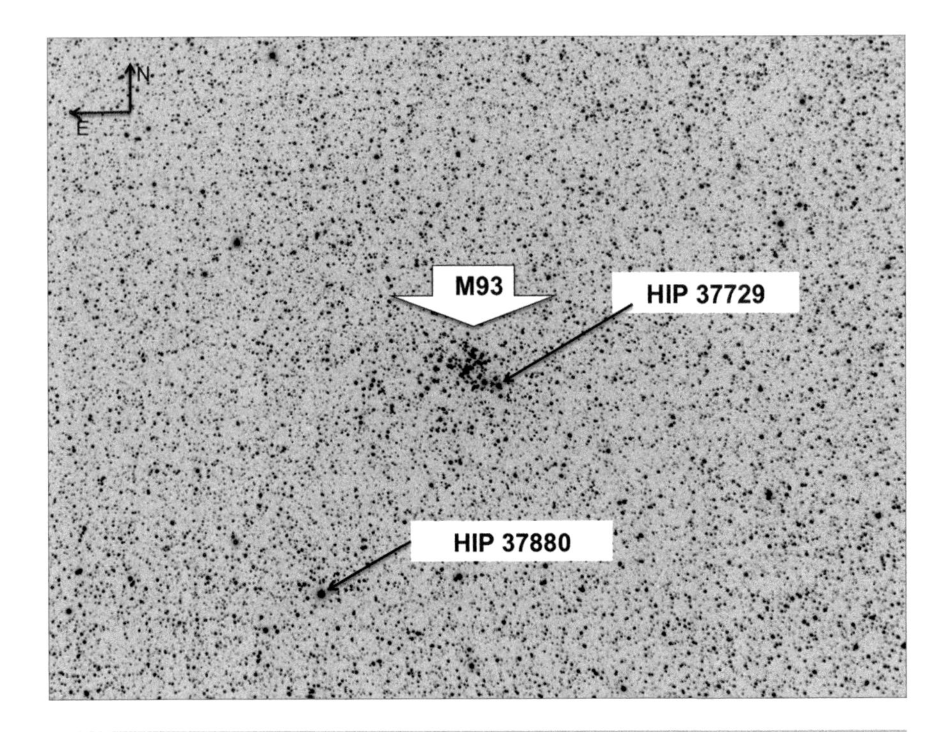

Fig. 5.56 Negative image of Messier 93

separated from the cluster, is magnitude 6.73 and spectral type B3 lying at a distance of 2,836.15 ly. The cluster is not very impressive and almost looks like a random concentration of stars in a very crowded star field. The WEBDA page for M93 gives the cluster a log age of 8.588, which corresponds to a calculated age of 387 million years, but most sources quote the distance as 100 million years. That would correspond to a log age of 8.0. Either way, when compared to the age of the Sun at 5 billion years -- this is indeed a young cluster!

The chart in Figure 5.57 shows the location of Messier 93 in the constellation of Puppis. Constellations were formally reorganized around 1930, so they have changed since Messier's day. M93 lies roughly on a line between 24-Omicron2 Canis Majoris in the constellation of 'the Greater Dog' and 12 Puppis. 24-Omicron2 Canis Majoris (HIP 33977) has a magnitude of 3.02 and is of spectral type B3, at a distance of 2,568.17 ly. It has an angular separation of 9° 30' from Messier 93. 12 Puppis (HIP 39023) has a magnitude of 5.09 and is spectral type G8, at a distance of 785.92 ly. (Hipparcos Catalog Data). The angular separation between Messier 93 and 24-Omicron2 Canis Majoris is 9° 30'. Between M93 and 12 Puppis, the angular separation is 3° 24'.

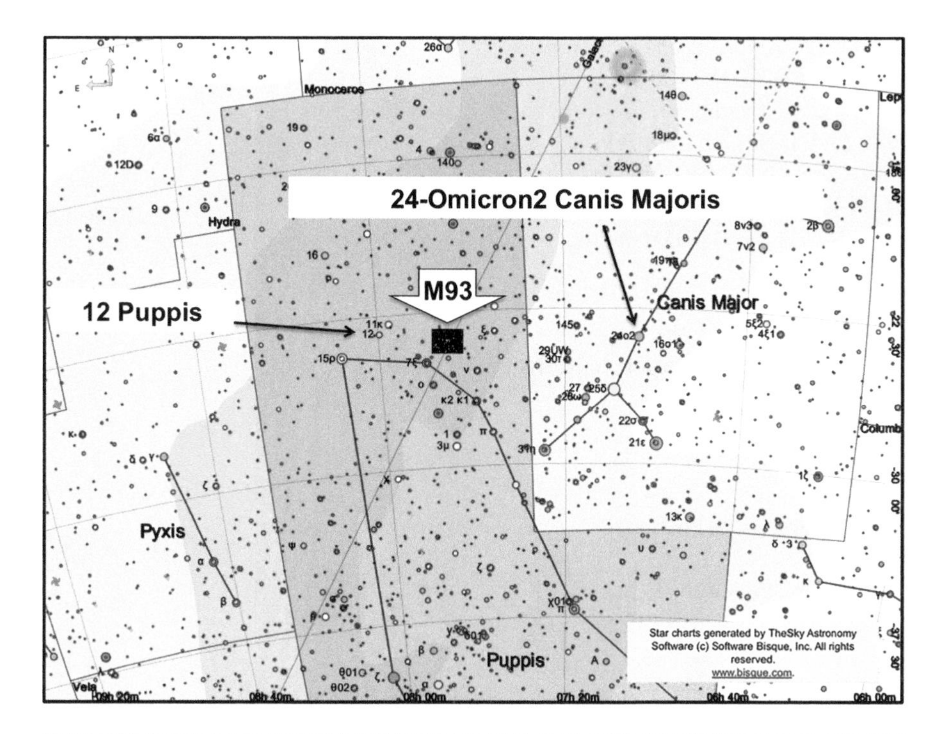

Fig. 5.57 Chart showing the location of M93

The image was taken on November 28 from Siding Spring at 23h 48m 38s. The corresponding Universal Time was 12h 48m 38s. The Right Ascension of M93 is 07h 45m 13s, and the Local Sidereal Time was 03h 16m 02s. M93 would cross the meridian in 4h 29m 11s, making the meridian transit time of M93 04h 17m 49s. The 90-mm telescope, a Takahashi Sky90, took the image.

Messier 93: Data relating to the image in Figure 5.55

REMOTELY IMAGED FROM NEW SOUTH WALES, AUSTRALIA
NGC 2447
OBJECT TYPE: Open Cluster

RA: 07h 45m 13s
DEC: -23° 53' 38"
ALTITUDE: +30° 47' 04"
AZIMUTH: 100° 57' 24"
FIELD OF VIEW: 1° 37' 41" x 1° 13' 16"
OBJECT SIZE: 22'

POSITION ANGLE: 357° 06' from North
EXPOSURE TIME: 300 s
DATE: 28th November
LOCAL TIME: 23h 48m 38s
UNIVERSAL TIME: 12h 48m 38s
SCALE: 3.66 arcsec/pixel
MOON PHASE: 1.05% (waning)

Telescope Optics
OTA: Takahashi Sky 90
Optical Design: Apochromatic Refractor
Aperture: 90 mm
Focal Length: 417 mm
F/Ratio: f/5.6
Guiding: None
Mount: Paramount PME

Instrument Package
CCD: SBIG ST2000 XMC
Colour CMOS
Pixel Size: 7.4μm square
Sensor: Frontlit
Cooling: CNA
Array: 1600 x 1200 pixels
FOV: 60.5 x 80.7 arcmin

Location
Observatory: Siding Spring
UTC +10.00 (Australia Daylight savings time is observed)
31° 16’ 24” South
149 03’ 52” East
Elevation: 1122 meters (3681 ft)

Messier 94

Messier 94 (NGC 4736), shown in Figure 5.58, is a spiral galaxy in the constellation of Canes Venatici. I initially took an image with a 106-mm refractor from Siding Spring, but it only recorded the central bright core, so I moved to the 610-mm (24-inch) telescope based in Auberry, California. This gave a much better result. Pierre Méchain discovered M94 on March 22, 1781 and passed the location to Charles Messier, who observed it two nights later. He recorded: “Nebula without star, above the Heart of Charles [Alpha Canum Venaticorum], on the parallel of the

Fig. 5.58 Image of Messier 94, a spiral galaxy in Canes Venatici

star no. 8, of sixth magnitude of the Hunting Dogs [Canes Venatici], according to Flamsteed: In the center it is brilliant & the nebulosity a bit diffuse. It resembles the nebula, which is below Lepus [Messier 79], but this on is more beautiful & brighter." John Herschel noted it was, "Very bright; large; irregularly round; very suddenly very much brighter toward the middle where there is a bright nucleus; mottled."

The negative image in Figure 5.59 shows Messier 94 and identifies the location of some of the smaller galaxies spotted in the field. North is up in this image. The galaxy at top left (northeast of M94) is PGC 2180031, which has dimensions of 18" of arc x 12". It has a magnitude of 17.14 and a radial velocity of 21407 km/s. The galaxy PGC 2175000 also has dimensions of 18" x 12" and a magnitude of 18.32. Other galaxies are arrowed, but their identifications but are omitted for anyone interested in tracking them down. M94 has dimensions of 7' x 3', not large in comparison with Messier 81, for example, which has dimensions 21' x 10'. M94 is at a distance of 14.5 million ly (SEDS provides this number, but there are differing views on the distance), so the actual dimensions are 30,000 ly x 13,000 ly.

The chart in Figure 5.60 shows M94 in the constellation of Canes Venatici. The positions of other Messier objects in the constellation are labeled. These are M51,

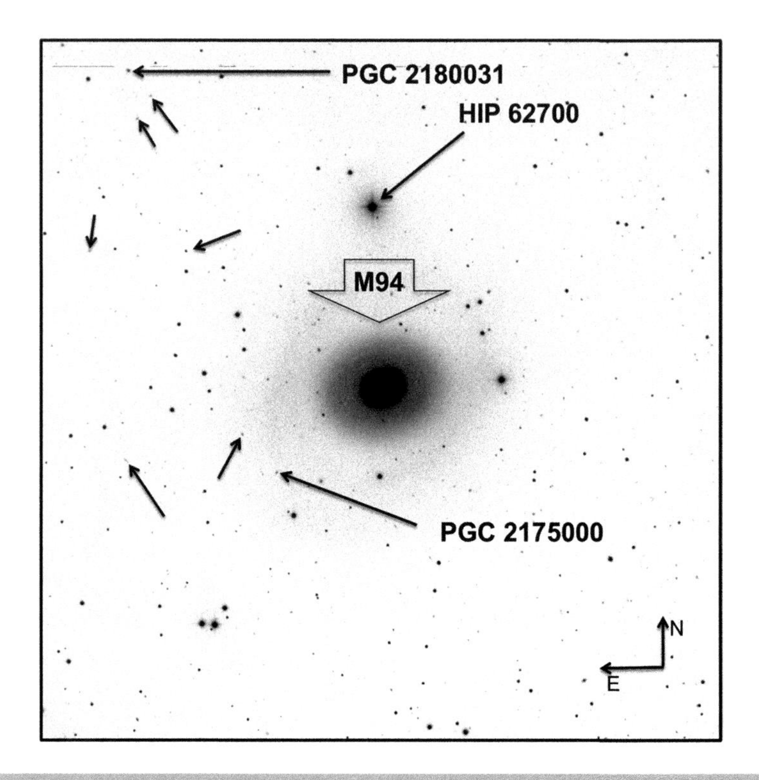

Fig. 5.59 Negative image of Messier 94

M63, M106, which are spiral galaxies, and M3, which is a globular cluster. Messier referred to "The Heart of Charles," which is the star Cor Coroli. This is the star that is marked on the chart as 12-Alpha1 Canum Venaticorum (HIP 63121). Cor Coroli has a magnitude of 5.61 and is spectral type F0, lying at a distance of 81.64 ly. He also referred to star number 8 in Canes Venatici. This is the star Chara (HIP 61317), or 8-Beta Canum Venaticorum. Chara has a magnitude of 4.24 and is spectral type G0, lying at a distance of 27.30 ly. The angular separation of M84 and Cor Coroli is 3°, and between M94 and Chara is 3° 16'.

The image was taken on March 2 from Auberry, California at 23h17m 06s. The corresponding Universal Time was 07h 17m 06s (March 3). The Right Ascension of M94 is 12h 51m 42s, and the Local Sidereal Time was 10h 04m 32s. Thus, M94 would cross the meridian in 2h 47m 10s at a meridian transit time of 02h 4m 17s. The large aperture telescope located at 4,610 ft. in the foothills of the Sierra Nevada mountain range in Auberry, California was used to take the image.

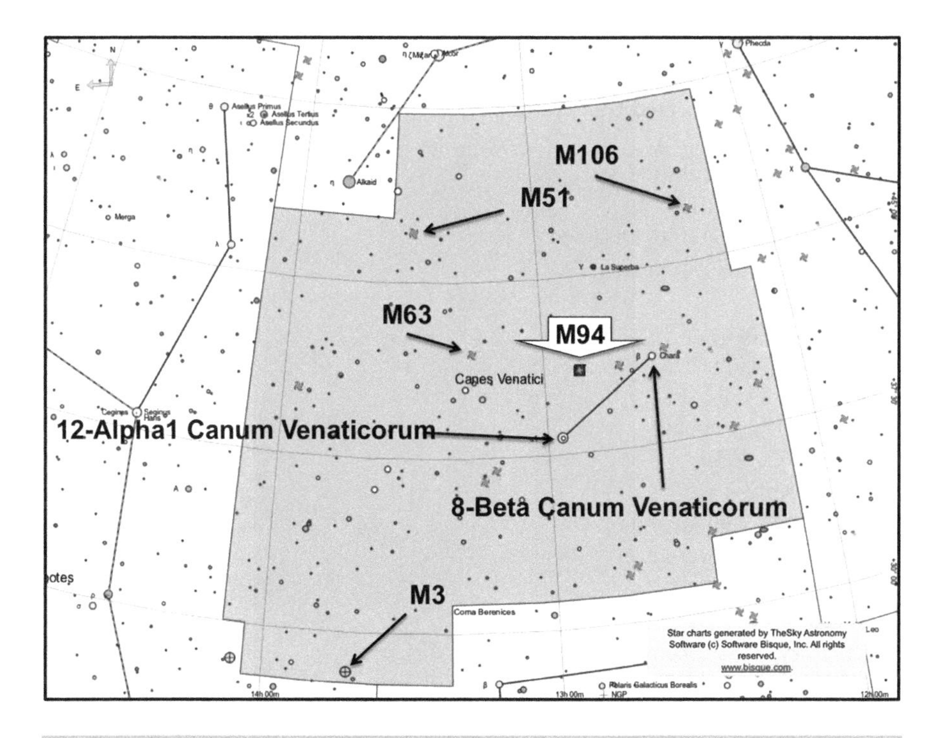

Fig. 5.60 Chart showing the location of M94

Messier 94: Data relating to the image in Figure 5.58

REMOTELY IMAGED FROM AUBERRY CALIFORNIA USA
NGC 4736
OBJECT TYPE: Spiral Galaxy

RA: 12h 51m 42s
DEC: +41° 01' 33"
ALTITUDE: +57° 36' 57"
AZIMUTH: 69° 49' 45"
FIELD OF VIEW: 0° 32' 08" x 0° 32' 08"
OBJECT SIZE: 7’ X 3’
POSITION ANGLE: 88° 19’ from North
EXPOSURE TIME: 300 s
DATE: 2nd March
LOCAL TIME: 23h 17m 06s
UNIVERSAL TIME: 07h 17m 06s (3rd March)
SCALE: 0.63 arcsec/pixel
MOON PHASE: 26.04% (waxing)

Telescope Optics
OTA: Planewave 24" (0.61m)
Optical Design: Corrected Dall-Kirkham Astrograph
Aperture: 610 mm
Focal Length: 3962 mm
F/Ratio: f/6.5
Guiding: Active Guiding Disabled
Mount: Planewave Ascension 200HR

Instrument Package
FLI-PL09000 CCD
A/D Gain: e-/ADU
Pixel Size: 12μm square
Array: 3056 x 3956
Sensor: Frontlit
Cooling: -35°C default
FOV: 31.8 x 31.8 arcmin

Location
Observatory: Sierra Remote Observatory – MPC U69
UTC Minus -8.00 (Daylight savings time is observed)
Minimum Target Elevation: Approx 25 Degrees
37.07°N, 119.4W
Elevation: 1405 meters (4610 ft)

Messier 95

Messier 95 (NGC 3351), shown in Figure 5.61, is a barred spiral galaxy in the constellation of Leo. Along with M95, the image has captured two other Messier objects M96 and M105. These galaxies together with some fainter galaxies make up the M96 or Leo 1 group of galaxies. Pierre Méchain discovered M95 on March 20, 1781. On March 24, Messier observed M95 and commented: "Nebula without star, in the Lion, above star l [53 Leonis]: its light is very faint." William Herschel commented on March 18, 1784: "A fine, bright nebula, much brighter in the middle than at the extremes, of a pretty considerable extent, perhaps 3 or 4' or more. The middle seems to be of the magnitude of 3 or 4 stars joined together, but not exactly round; from the brightest part of it there is a sudden transition to the nebulous part, so that I should call it cometic."

Messier 95 and nearby galaxies are labeled on the negative image in Figure 5.62. M95 has dimensions 4.3' x 3.3' - relatively small compared to some of the Messier galaxies. It lies at a distance of 38 million ly, which would give an actual size of 48,000 ly x 36,000 ly. Messier referred to M95's proximity to the star 53 l Leonis.

Fig. 5.61 Image of Messier 95, a barred spiral galaxy in Leo

This star was captured at the edge of the frame and is labeled on the negative image. M95 is 1° 43' 45" from 53 Leonis. The field of view of this telescope is large at 3° 54' 12" x 2° 36' 08", allowing the three Messier objects to be captured on a single frame. The galaxy NGC 3389 is receding from us at a greater rate then the other members of the Leo 1 group, so it is believed to be at a greater distance as a background object. The galaxy PGC 32251, which has a major axis of 2.8' and a minor axis of 0.9', is also a background galaxy, as indicated by its high radial velocity.

The chart in Figure 5.63 shows the location of M95 in the constellation of Leo. The large field of view is evident from the plate-solved image in relation to the chart. The star 53 Leonis is labeled on the border of the frame. M95 lies roughly on a line drawn between 78-Iota Leonis (Tsze Tseang or HIP 55642) and 32-Alpha Leonis (Regulus or HIP 49669). 78-Iota Leonis is a magnitude 4.0 star of spectral type F2 that lies at a distance of 79.05 ly. The angular separation of Messier 95 and Tsze Tseang is 9° 52'. Regulus is a magnitude 1.36 star of spectral type B7 and lies at a distance of 77.49 ly. The angular separation between Regulus and Messier 95 is 8° 45'.

The image was taken on February 2 from New Mexico at 02h 18m 44s. The corresponding Universal Time was 09h 18m 44s. The Right Ascension of M94 is 10h 44m 53s, and the Local Sidereal Time was 11h 06m 30s. M95 therefore crossed the 21m 37 ago at a meridian transit time of 01h 57m 07s. The telescope used to capture Messier 63 was a wide-angle instrument with an aperture of 106 mm on a Paramount ME mount.

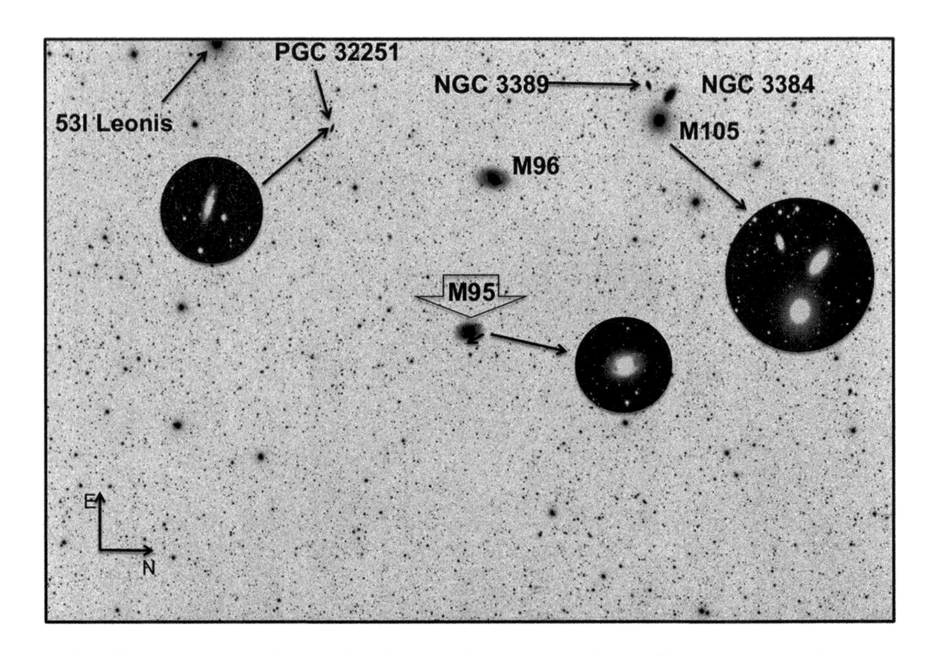

Fig. 5.62 Negative image of Messier 95

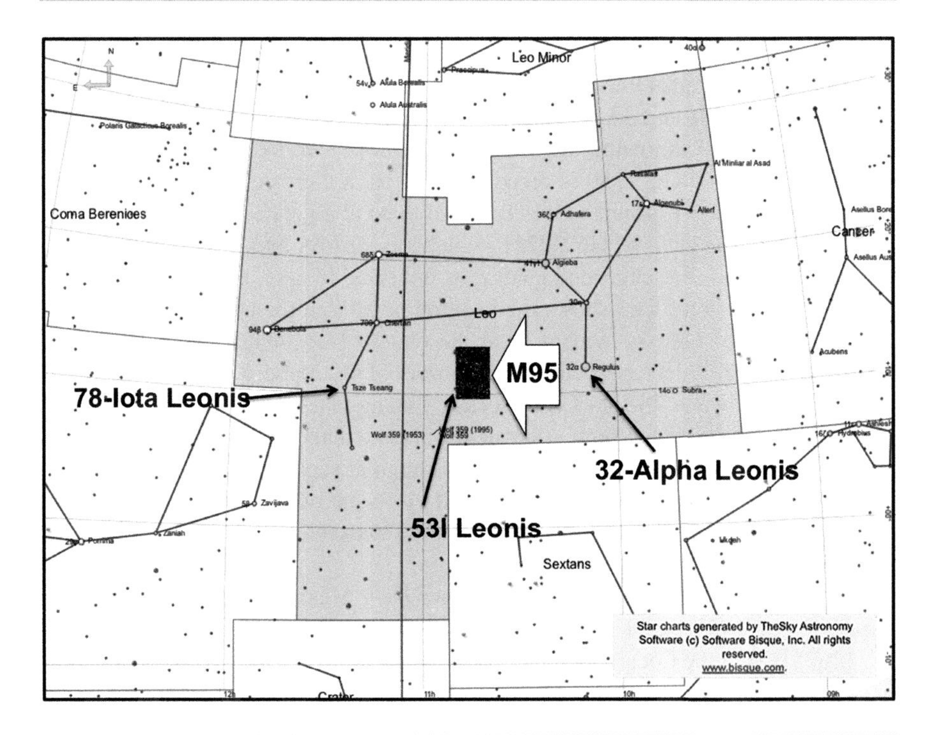

Fig. 5.63 Chart showing the location of M95

Messier 95: Data relating to the image in Figure 5.61

REMOTELY IMAGED FROM MAYHILL, NEW MEXICO
NGC 3351
OBJECT TYPE: Barred Spiral Galaxy

RA: 10h 44m 53s
DEC +11° 36' 40"
ALTITUDE: +68° 15' 14"
AZIMUTH: 194° 25' 36"
FIELD OF VIEW: 3° 54' 12" x 2° 36' 08"
OBJECT SIZE: 4.4' X 3.3'
POSITION ANGLE: 89° 41' from North
EXPOSURE TIME: 300 s
DATE: 2nd February
LOCAL TIME: 02h 18m 44s
UNIVERSAL TIME: 09h 18m 44s
SCALE: 3.51 arcsec/pixel
MOON PHASE: 30.34% (waxing)

Telescope Optics
OTA: Takahashi FSQ - ED
Optical Design: Petzval Apochromatic Astrograph
Aperture: 106 mm
Focal Length: 530 mm
F/Ratio: f/5.0
Guiding: External
Mount: Paramount PME

Instrument Package
CCD: SBIG STL-11000M
Pixel Size: 9μm square
Sensor: Frontlit
Cooling: -15°C
Array: 4008 x 2672 pixels
FOV: 188.8 x 233.7 arcmin
Position Angle: 092.7°

Location
Observatory: New Mexico Skies
UTC Minus 7.00 (Daylight savings time is observed)
Minimum Target Elevation: Approx. 25 – 45 Degrees
(N or S) 32° 54' Decimal: 32.9 North
(W or E) 105° 31' Decimal: 105.5 West
Elevation: 2225 meters (7298 ft)

Messier 96

Messier 96, shown in Figure 5.64, is a spiral galaxy in the constellation of Leo, heading the M96 Group of galaxies in that area. M96 was captured in the image for Messier 95, as a wide-angle telescope was being used, but this image is taken with a larger telescope with a smaller field of view. Further, the M95/M96 wide-angle image was taken from New Mexico, whereas this image of M96 was taken from New South Wales. Not surprisingly, because of their proximity, Messier 96 was discovered on the same night as M95 by Pierre Méchain on March 20 1781. Méchain passed the details on to Charles Messier, who, on March 24, 1781, noted, "Nebula without star, in the Lion, near the preceding [Messier 95]: this one is less distinct, both are on the same parallel of Regulus: they resemble the two nebulae in the Virgin, M84 and M86."

The negative image in Figure 5.65 identifies Messier 96 and the galaxy NGC 3389, both of which were captured on the image of Messier 95. NGC 3389 is not a member of the M96 group, as it is a background galaxy. M96 has dimensions 6' of arc x 4' and lies at a distance of 38 million ly. These figures give actual dimensions for M96 of 66,000 x 44,000 ly. There are spiral arm fragments in a ring around the galaxy extending its diameter from these figures. The bright star to the south (left) of M96 is HIP 52707, a magnitude 8.77 star of spectral type A0 lying at a distance of 838.45 ly. The star slightly to the north of east of Messier 96 is HIP 52833, a magnitude 9.73 star of spectral type G0 at a distance of 392.02 ly.

Fig. 5.64 Image of Messier 96, a spiral galaxy in Leo

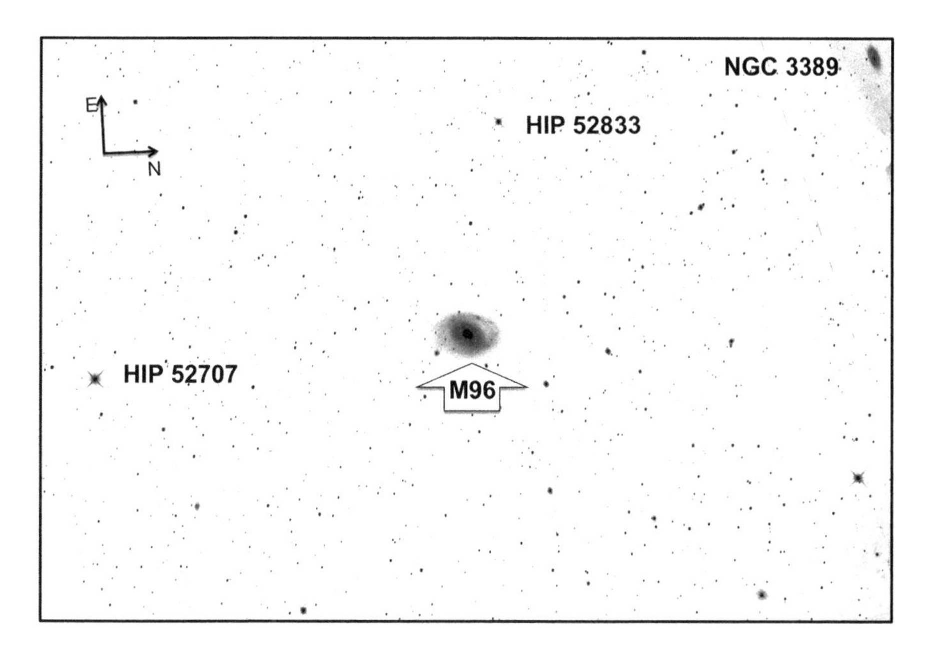

Fig. 5.65 Negative image of Messier 96

The chart in Figure 5.66 shows Messier 96 in the constellation of Leo. The smaller field of view of the frame is evident from the size of the plate solved embedded image in comparison with the similar chart for M95. The M96 image has a field of view of 1° 26' 35" x 0° 57' 32". M96 lies roughly between the stars 70-Theta Leonis and 47-Rho Leonis. 70-Theta Leonis (Chertan or HIP 54879) is a magnitude 3.33 star of spectral type A2 lying at a distance of 177.65 ly. The angular separation between Chertan and M96 is 7° 39’. 47-Rho Leonis (HIP 51624) is a magnitude 3.84 star of spectral type B1 lying at a distance of 5,722.05 ly. This star is a blue supergiant star, which is about 36 times the size of our own star, the Sun. Every February/March, there is the Rho-Leonid meteor shower with a radiant near this star. The angular separation between M96 and 47-Rho Leonis is 4° 12’.

The image was taken on March 25 from Siding Spring at 20h 55m 32s. The corresponding Universal Time was 09h 55m 32s The Right Ascension of M96 is 10h 47m 41s, and the Local Sidereal Time was 08h 03m 56s. M96 would cross the meridian in 2h 43m 56s. The meridian transit time of M96 would thus be at 23h 39m 28s. The 406-mm Fast Newtonian Astrograph with a focal length of 1,425 mm was used to take the image from Siding Spring.

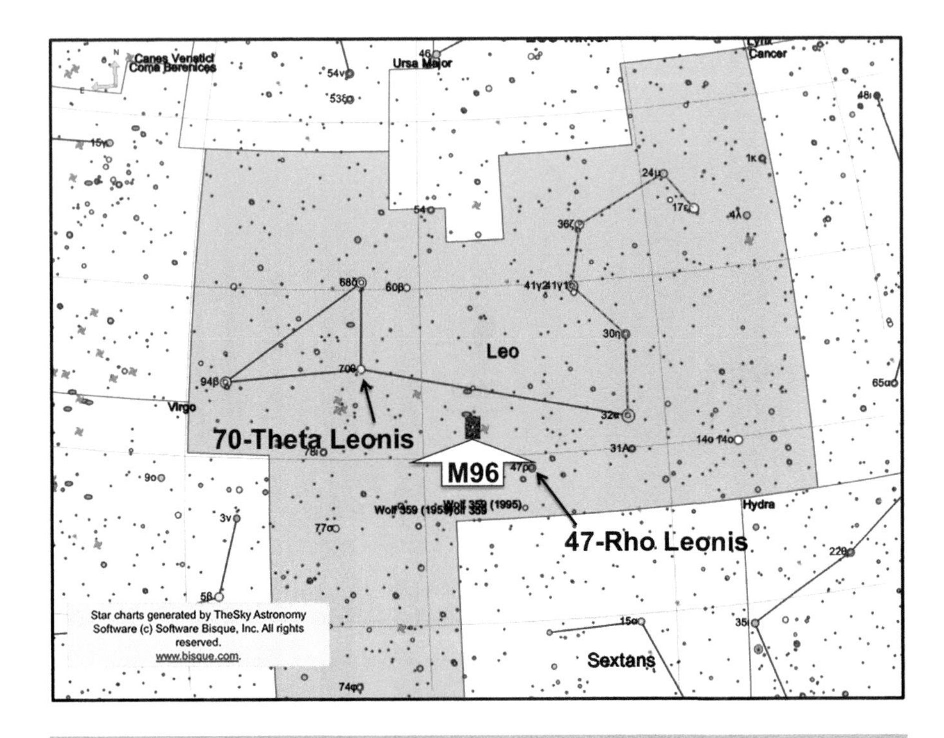

Fig. 5.66 Chart showing the location of M96

Messier 96: Data relating to the image in Figure 5.64

REMOTELY IMAGED FROM NEW SOUTH WALES, AUSTRALIA
NGC 3368
OBJECT TYPE: Spiral Galaxy

RA: 10h 47m 41s
DEC: +11° 43' 38"
ALTITUDE: +31° 46' 23"
AZIMUTH: 49° 03' 21"
FIELD OF VIEW: 1° 26' 35" x 0° 57' 32"
OBJECT SIZE: 6’ x 4’
POSITION ANGLE: 86° 39' from North
EXPOSURE TIME: 300 s
DATE: 25th March
LOCAL TIME: 20h 55m 32s
UNIVERSAL TIME: 09h 55m 32s
SCALE: 1.07 arcsec/pixel
MOON PHASE: 9.11% (waning)

Telescope Optics
OTA: 16" 0.4 m Astro Systeme Austria
Optical Design: Fast Newtonian Astrograph
Aperture: 400 mm
Focal Length: 1425 mm
F/Ratio: f/3.5
Guiding: Off-Axis Guiding
Mount: Paramount PME

Instrument Package
CCD: Apogee Aspen CG16070 Class 1 CCD
Pixel Size: 7.4μm square
Sensor: KAI – 16070 ABG
Cooling: -30°C default
Array: 4864 x 3232 pixels
FOV: 88.3 x 58.7 arcmin

Location
Observatory: Siding Spring
UTC +10.00 (Australia Daylight savings time is observed)
31° 16' 24" South
149 03' 52" East
Elevation: 1122 meters (3681 ft)

Messier 97

Messier 97 (NGC 3587), shown in Figure 5.67, is a planetary nebula, known as the Owl Nebula as a result of its similarity to the face of an owl. Anglo-Irish astronomer William Parsons, the 3rd Earl of Rosse coined the name in 1848. The image, taken from southern Spain, also includes the spiral galaxy M108 in the field of view. M97 was yet another discovery for Messier's colleague Pierre Méchain, dated February 16, 1781. Messier observed M97 on March 24, 1781 and noted: "Nebula in the great Bear, near Beta: It is difficult to see, reports M. Méchain, especially when one illuminates the micrometer wires: its light is faint, without a star. M. Méchain saw it the first time on Feb 16, 1781, & the position is that given by him. Near this nebula he has seen another one, which has not yet been determined [M108] and also a third, which is near Gamma of the Great Bear [M109]."

The negative image in Figure 5.68 shows the Owl Nebula and incorporates a magnified view of it from the original image. Messier 108 is also identified, M97 has dimensions 3.4' of arc x 3.3' and lies at a distance of 2,600 ly, giving it an actual size of 2.57 ly by 2.5 ly. On the enlarged image you can just spot the central star UCAC4 726:47396, which is magnitude 16.02 (UCAC4 Catalog Data). Two

Fig. 5.67 Image of Messier 97, a planetary nebula in Ursa Major

small galaxies are identified in the frame, PGC 34410 and PGC 34403. Both of these galaxies have high radial velocities and are a long way out!

The chart in Figure 5.69 shows the position of M97 in Ursa Major. The image has been rotated 90° to get north at the top. M97 sits roughly between the stars 64-Gamma Ursae Majoris and 48-Beta Ursae Majoris, both of which are part of the Big Dipper asterism. 64-Gamma Ursae Majoris is Phecda or HIP 58001 and is a magnitude 2.41 star of spectral type A0 lying at a distance of 83.65 ly (Hipparcos Data). The angular separation between Messier 97 and Phecda is roughly 5° 51'. 48-Beta Ursae Majoris is Merak or HIP 53910 and is a magnitude 2.34 star of spectral type A1 lying at a distance of 79.41 ly.

The image was taken on February 1 from Nerpio, Spain at 06h 10m 27s. The corresponding Universal Time was 05h 10m 27s. The Right Ascension of M97 is 11h 15m 47s, and the Local Sidereal Time was 13h 47m 36s. Thus, M97 crossed the meridian 2h 31m 49s prior, giving it a meridian transit time of 03h 38m 38s. A 152-mm Apochromatic Refractor with a focal length of 1,095 mm was used to take the image.

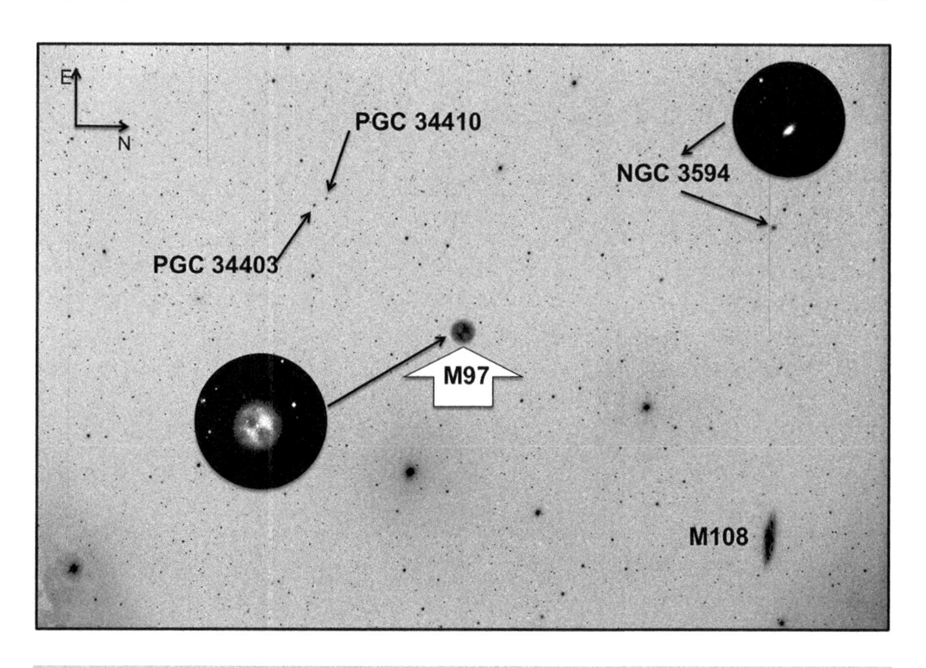

Fig. 5.68 Negative image of Messier 97

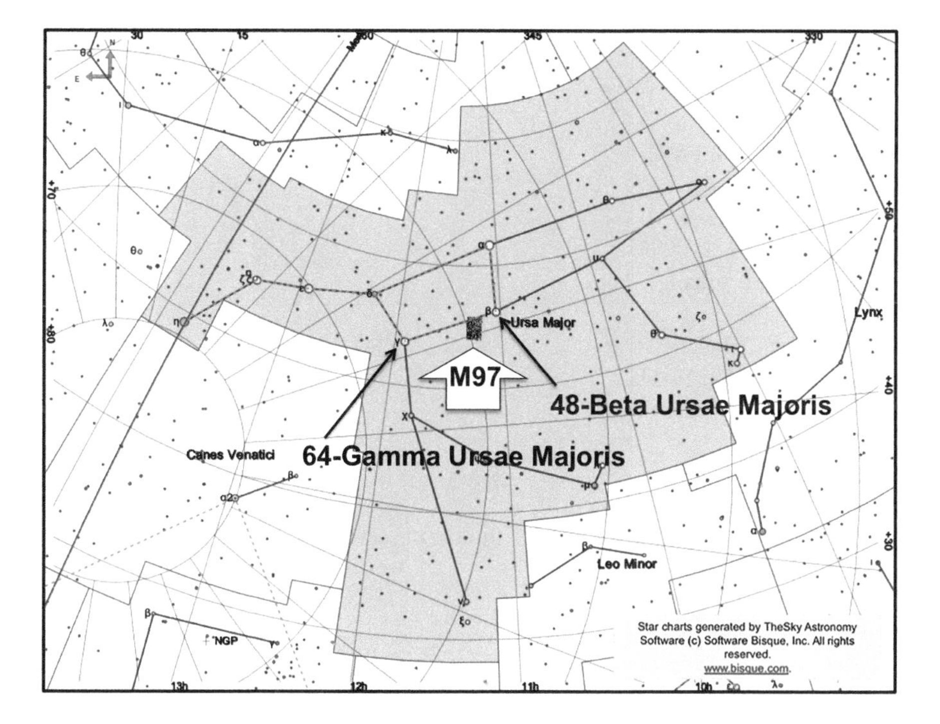

Fig. 5.69 Chart showing the location of M97

Messier 97: Data relating to the image in Figure 5.67

REMOTELY IMAGED FROM NERPIO, SPAIN
NGC 3587 (OWL NEBULA)
OBJECT TYPE: Planetary Nebula

RA: 11h 15m 47s
DEC: +54° 55' 21"
ALTITUDE: +59° 31' 42"
AZIMUTH: 315° 48' 57"
FIELD OF VIEW: 1° 51' 41" x 01° 14' 28"
OBJECT SIZE: 3.4' x 3.3'
POSITION ANGLE: 88° 11' from North
EXPOSURE TIME: 300 s
DATE: 1ST February
LOCAL TIME: 06h 10m 27s
UNIVERSAL TIME: 05h 10m 27s
SCALE: 1.67 arcsec/pixel
MOON PHASE: 19.10% (waxing)

Telescope Optics
OTA: Takahashi TOA-150
Optical Design: Apochromatic Refractor
Aperture: 150 mm
Focal Length: 1095 mm
F/Ratio: f/7.3
Guiding: Internal
Mount: Paramount PME

Instrument Package
CCD: SBIG STL-11000M
Pixel Size: 9μm square
Sensor: Frontlit
Cooling: -20°C default
Array: 4008 x 2672 pixels
FOV: 75.4 x 113.1 arcmin

Location
Observatory: AstroCamp
UTC +1.00 (Madrid Daylight savings time is observed)
38° 09' North
02° 19' West
Elevation: 1650 meters (5413 ft)

Messier 98

The spiral galaxy Messier 98 (NGC 4192), shown in Figure 5.70, is located in Coma Berenices and is a member of the Coma – Virgo cluster of galaxies. This was another discovery for Pierre Méchain. It is not particularly bright, and the image taken with the 106-mm aperture telescope from New Mexico resulted in quite a small size for the galaxy in the full image. However, the resolution of the camera allowed for substantial enlargement, and the wide field of view contained a large number of identifiable galaxies. Messier 99 is the galaxy at the top of the image, above M98, just making it onto the frame. Pierre Méchain discovered M98 on March 15, 1781. Charles Messier observed it on April 13, 1781 and commented: "Nebula without star, of an extremely faint light, above the northern wing of Virgo, on the parallel & near to the star no. 6 of fifth magnitude, of Coma Berenices, according to Flamsteed."

The negative image in Figure 5.71 incorporates an enlargement of M98 from the original image. Messier referred to its proximity to the star 6 magnitude 5 in Coma Berenices, labeled on the negative image (6 Comae Berenices). Note that east is up on the image. This star is HIP 59819 and is given a magnitude 5.09 in the Hipparcos Catalog. It is of spectral type A3 and lies at a distance of 197.79 ly. The image contains dozens of galaxies in the field, a few of which are identified on the nega-

Fig. 5.70 Image of Messier 98, a spiral galaxy in Coma Berenices

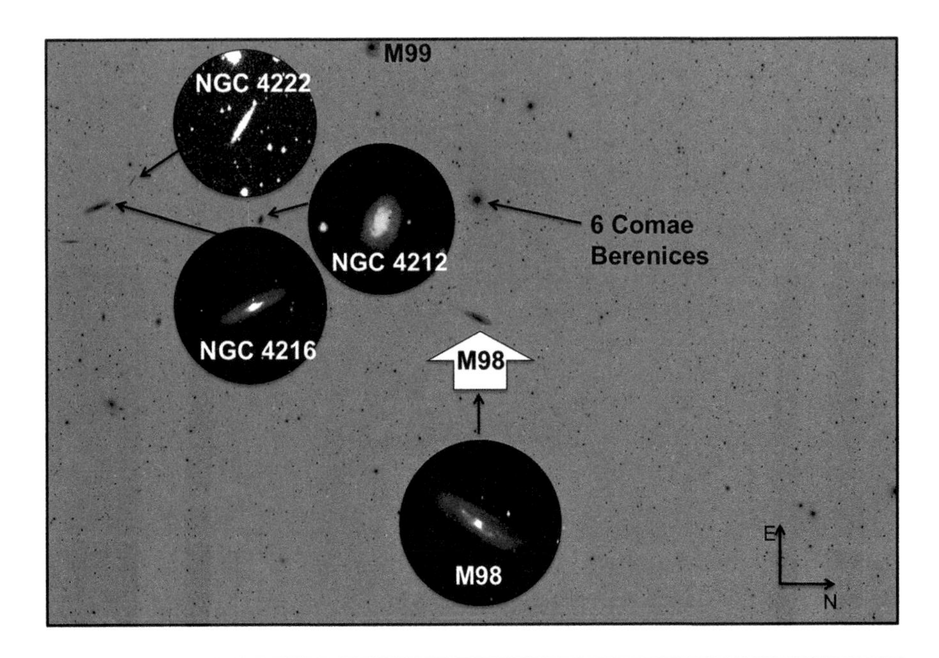

Fig. 5.71 Negative image of Messier 98

tive image. M98 has a major axis of 9.5' of arc and a minor axis of 3.2' (SEDS). It is at a distance of 60 million ly (the distance to the Virgo Cluster), which gives it actual dimensions of 165,800 ly x 56,000 ly, based on the SEDS data.

The chart in Figure 5.72 shows the location of M98 in the constellation of Coma Berenices. The large size of the plate-solved image shows that it is a wide-angle shot with a field of view of 3° 54' 00" x 2° 36' 00". Messier 98 sits roughly on a line between the stars 15-Gamma Comae Berenices and 9-Omicron Virginis. 15-Gamma Comae Berenices (HIP 60742) is a star of magnitude 4.34 of spectral type K2 lying at a distance of 170.05 ly. The angular separation between this star and M98 is 13° 49'. 9-Omicron Virginis (HIP 58948) is a magnitude 4.12 star of spectral type G8 lying at a distance of 170.94 ly. The angular separation between this star and M98 is 6° 28'. On the chart, the meridian can be seen running down the eastern edge of the Coma Berenices border.

The image was taken on February 1 at 04h 52m 15s local time in New Mexico. The Universal Time was 11h 52m 15s. The Right Ascension of M98 is 12h 14m 40s, and the Local Sidereal Time was 13h 36m 30s. So, the spiral galaxy crossed the meridian 1h 21m 50s before with a meridian transit time of 03h 30m 25s. The telescope used to take the image is a Takahashi FSQ-ED with an aperture of 106 mm and a focal length of 530 mm.

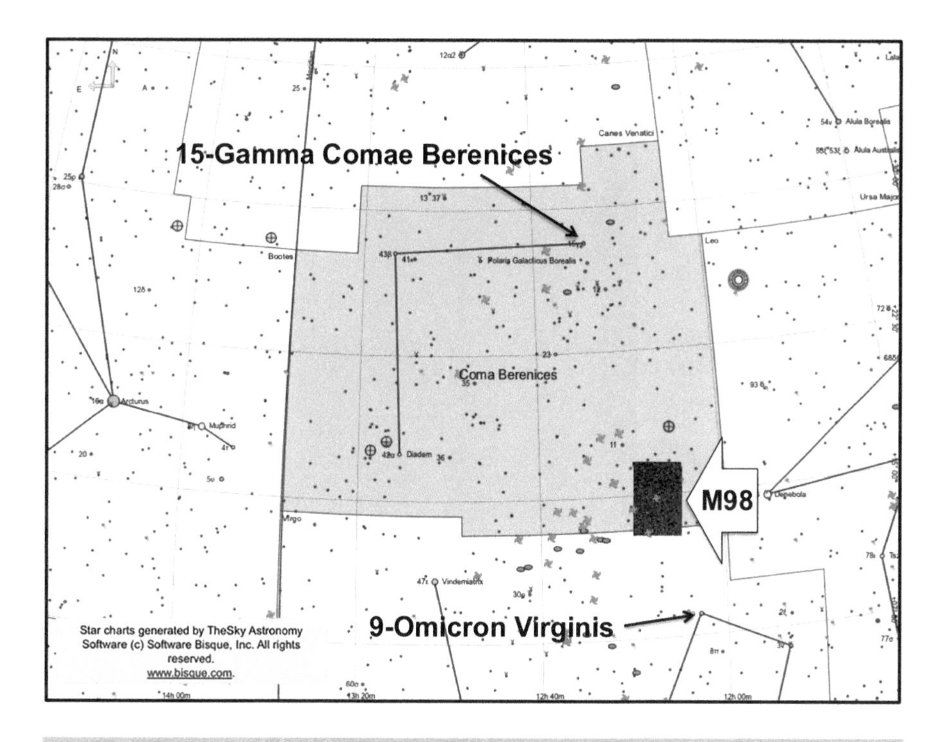

Fig. 5.72 Chart showing the location of M98

Messier 98: Data relating to the image in Figure 5.70

REMOTELY IMAGED FROM MAYHILL, NEW MEXICO
NGC 4192
OBJECT TYPE: Spiral Galaxy

RA: 12h 14m 40s
DEC: +14° 48' 10"
ALTITUDE: +64° 08' 42"
AZIMUTH: 230° 47' 26"
FIELD OF VIEW: 3° 54' 00" x 2° 36' 00"
OBJECT SIZE: 9.5' X 3.2'
POSITION ANGLE: 89° 42' from North
EXPOSURE TIME: 300 s
DATE: 1st February
LOCAL TIME: 04h 52m 15s
UNIVERSAL TIME: 11h 52m 15s
SCALE: 3.5 arcsec/pixel
MOON PHASE: 21.60% (waxing)

Telescope Optics
OTA: Takahashi FSQ - ED
Optical Design: Petzval Apochromatic Astrograph
Aperture: 106 mm
Focal Length: 530 mm
F/Ratio: f/5.0
Guiding: External
Mount: Paramount PME

Instrument Package
CCD: SBIG STL-11000M
Pixel Size: 9μm square
Sensor: Frontlit
Cooling: Set to -15°C
Array: 4008 x 2672 pixels
FOV: 188.8 x 233.7 arcmin
Position Angle: 092.7°

Location
Observatory: New Mexico Skies
UTC Minus 7.00 (Daylight savings time is observed)
Minimum Target Elevation: Approx. 25–45 Degrees
(N or S) 32° 54' Decimal: 32.9 North
(W or E) 105° 31' Decimal: 105.5 West
Elevation: 2225 meters (7298 ft)

Messier 99

Messier 99 (NGC 4254), shown in Figure 5.73, is a spiral galaxy located in the constellation of Coma Berenices and is a member of the Coma-Virgo group of galaxies. M99 had been captured in the image for Messier 98, as they are separated by only 1° 19'. M98 is also in this image frame. M99 is receding from us with a recession velocity of 2324 km/s (SEDS Data), in comparison with M98, which is actually moving towards us. In fact M99 has the highest recession velocity of any Messier object. M99 is a Pierre Méchain discovery dated March 15 1781. As usual, this was followed up by Charles Messier's own observations on April 13, 1781: "Nebula without star, of a very pale light, nevertheless a little clearer than the preceding [Messier 98] situated on the northern wing of Virgo, & near the same star, no. 6, of Comae Berenices. The nebula is between two stars of seventh & of eighth magnitude."

The annotated negative image in Figure 5.74 is quite crowded, as there are a large number of galaxies in this busy region and the field of view is relatively large

Fig. 5.73 Image of Messier 99, a spiral galaxy in Coma Berenics

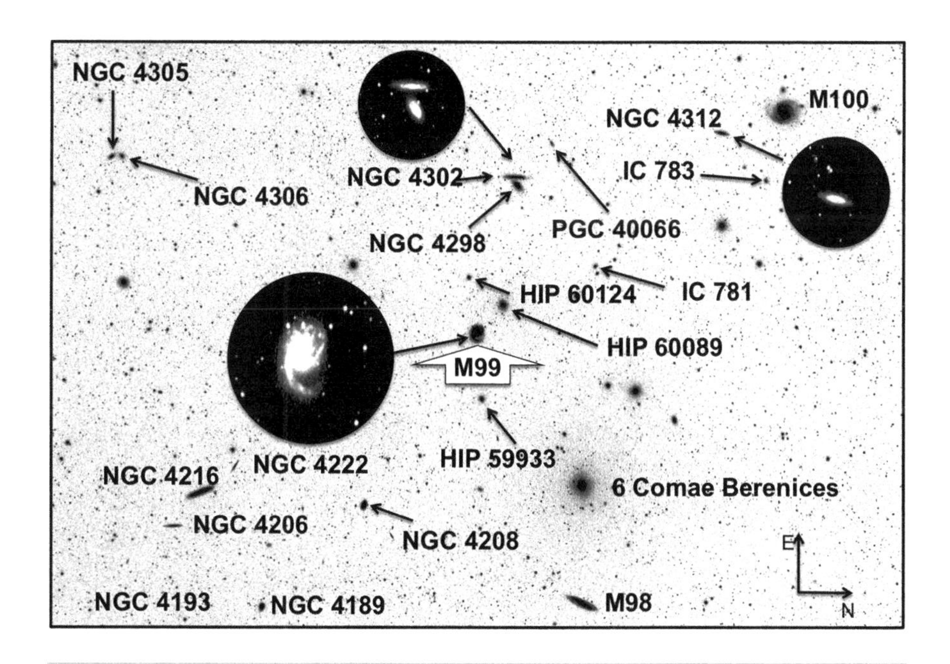

Fig. 5.74 Negative image of Messier 99

at 3° 54' 12" x 2° 36' 08". The enlarged image of Messier 99 is included here. Messier 99 has a major axis of 5.4 arcmin and a minor axis of 4.8 arcmin. At a distance of 60 million ly, this is equivalent to 94,000 x 84,000 ly. Messier referred to M99 being "between two stars of seventh and eighth magnitude." I believe these are HIP 60089 (Messier' magnitude 7 star), to the northeast of M99 and HIP 59933 (Messier's magnitude 8 star), just west of M99. HIP 60089 is a magnitude 6.52 star of spectral type K0 lying at a distance of 751.51 ly, separated from M99 by 10' 33". HIP 59933 is a magnitude 8.17 star of spectral type F8 lying at a distance of 190.74 ly (Hipparcos Catalog Data), separated from M99 by 18' 03".

The position of the image frame is shown on the chart in Figure 5.75, which has north at the top. M99 lies roughly between the two stars 42-Alpha Comae Berenices and 78-Iota Leonis. 42-Alpha Comae Berenices (Diadem or HIP 64241) is a magnitude 4.32 star of spectral type F5 lying at a distance of 46.72 ly. The angular separation between Diadem and M99 is approximately 12° 37'. 78-Iota Leonis (Tsze Tseang or HIP 55642) is a magnitude 4.00 star of spectral type F2 at a distance of 79.05 ly. The angular separation between Tsze Tseang and M99 is roughly 13° 58'.

The image was taken from New Mexico on February 2 at 01h 29m 14s. The Universal Time was 08h 29m 14s. The Right Ascension of M99 is 12h 19m 42s, and the Local Sidereal Time was 10h 16m 52s. So, the spiral galaxy would cross

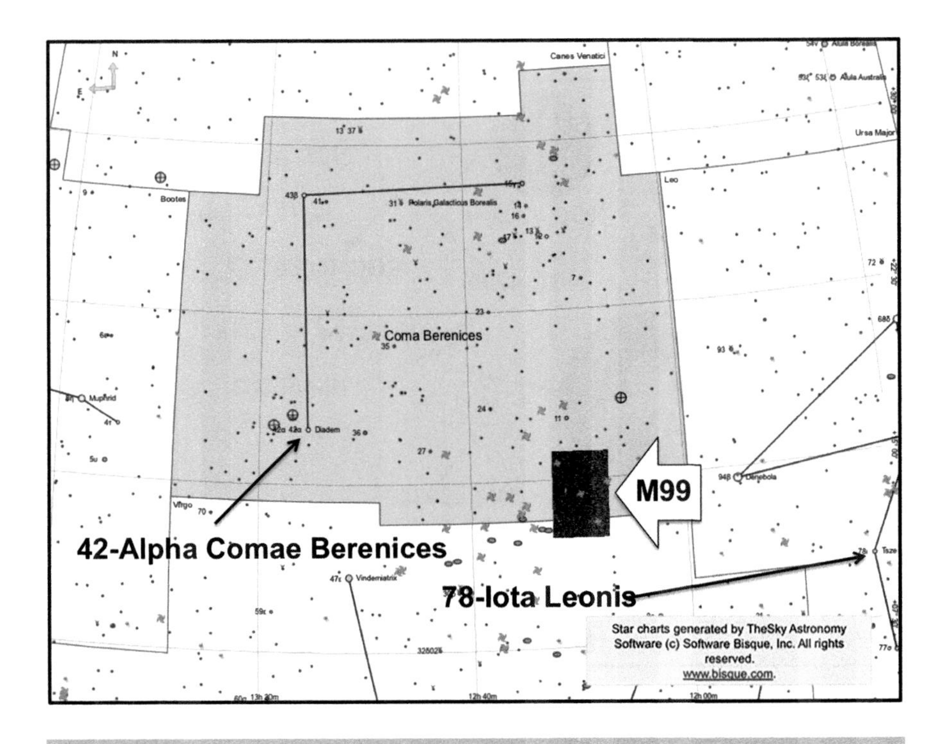

Fig. 5.75 Chart showing the location of M99

the meridian in 2h 2m 50s, giving a meridian transit time of 03h 32m 04s. The telescope used to take the image was a Takahashi FSQ-ED with an aperture of 106 mm and a focal length of 530 mm. The telescope was mounted on a Paramount ME.

Messier 99: Data relating to the image in Figure 5.73

REMOTELY IMAGED FROM MAYHILL, NEW MEXICO
NGC 4254
OBJECT TYPE: Spiral Galaxy

RA: 12h 19m 42s
DEC: +14° 19' 15"
ALTITUDE: +56° 32' 43"
AZIMUTH: 116° 10' 09"
FIELD OF VIEW: 3° 54' 12" x 2° 36' 08"
OBJECT SIZE: 5.4’ X 4.8’
POSITION ANGLE: 89° 54' from North
EXPOSURE TIME: 300 s
DATE: 2nd February
LOCAL TIME: 01h 29m 14s
UNIVERSAL TIME: 08h 29m 14s
SCALE: 3.51 arcsec/pixel
MOON PHASE: 29.98% (waxing)

Telescope Optics
OTA: Takahashi FSQ - ED
Optical Design: Petzval Apochromatic Astrograph
Aperture: 106 mm
Focal Length: 530 mm
F/Ratio: f/5.0
Guiding: External
Mount: Paramount PME

Instrument Package
CCD: SBIG STL-11000M
Pixel Size: 9μm square
Sensor: Frontlit
Cooling: -15°C
Array: 4008 x 2672 pixels
FOV: 188.8 x 233.7 arcmin
Position Angle: 092.7°

Location
Observatory: New Mexico Skies
UTC Minus 7.00 (Daylight savings time is observed)
Minimum Target Elevation: Approx. 25 – 45 Degrees
(N or S) 32° 54' Decimal: 32.9 North
(W or E) 105° 31' Decimal: 105.5 West
Elevation: 2225 meters (7298 ft)

Messier 100

Messier 100 (NGC 4321), shown in Figure 5.76, is a spiral galaxy located in the constellation of Coma Berenices and is a member of the Coma-Virgo group of galaxies. I decided to reuse the same image for Messier 100 as for M99, as the galaxy is included in that frame. The M100 image you see is cropped and enlarged from the full image for M99. North is to the right and east is up in the image. This object is another find for Pierre Méchain, who discovered it on March 15, 1781, the same night he found M98 and M99. Méchain passed the details to Messier, who observed it on April 13, 1781. He noted: Nebula without star, of the same light as the preceding [Messier 99] situated in the ear of Virgo. Seen by M. Méchain on March 15, 1781. The three nebulae, M100, M99 & M98, are very difficult to recognize, because of the faintness of their light: one can observe them only in good weather, & near their passage of the Meridian."

Fig. 5.76 Image of Messier 100, a spiral galaxy in Coma Berenices

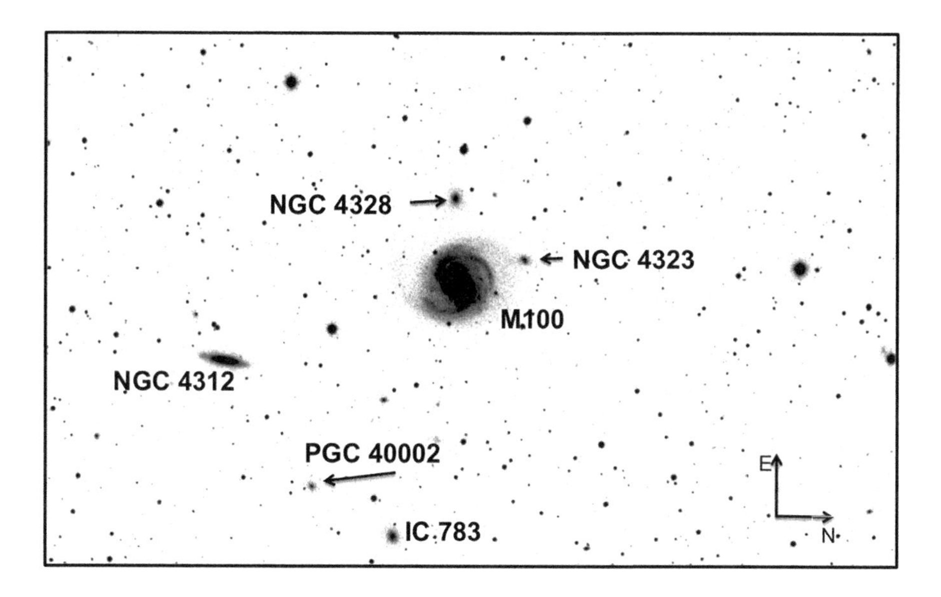

Fig. 5.77 Negative image of Messier 100

The negative image of the cropped original in Figure 5.77 shows M100 at the center, with some surrounding objects of interest. M100 has dimensions 7' x 6' and is at a distance of 60 million ly (SEDS Data). These values give it an actual size of around 122,000 x 105,000 ly. To the southwest of M100 is the spiral galaxy NGC4312 with dimensions 5' x 1.2'. NGC 4328 is an elliptical galaxy with dimensions 1.3' x 0.9' and a magnitude of 13.3. NGC 4323 is a spiral galaxy with dimensions of 1.3 x 0.8 and a magnitude of 13.8. The galaxy IC 783 has dimensions of 1.2' x 1.1' with a magnitude of 14.76. The faint galaxy PGC 40002 is 1.0' x 0.9' at magnitude 15.06.

The chart in Figure 5.78 shows the position of M100 in the constellation of Coma Berenices within the original M99 plate-solved image with north at the top. Using the same reference stars as with M99, the angular separation between 42–Alpha Comae Berenices (Diadem) and Messier 100 is roughly 11° 23' and between 78-Iota Leonis (Tsze Tseang) and M100 is 15° 24'. When the image was taken, the azimuth of M100 was 113° 03' 11", which lies between the east at 90° azimuth and the south at 180° azimuth angle. The meridian was running down the east side of the head of Leo when the image was taken.

The image was taken from New Mexico on the morning of February 2 at 01h 29m 14s. The Universal Time was 08h 29m 14s. The Right Ascension of M100 is 12h 23m 47s, and the Local Sidereal Time was 10h 16m 52s. The spiral galaxy would therefore cross the meridian in 2h 6m 55s with a meridian transit time of 03h 36m 09s. The telescope used to take the image was a Takahashi FSQ-ED with an aperture of 106 mm and a focal length of 530 mm. The telescope was mounted on a Paramount ME.

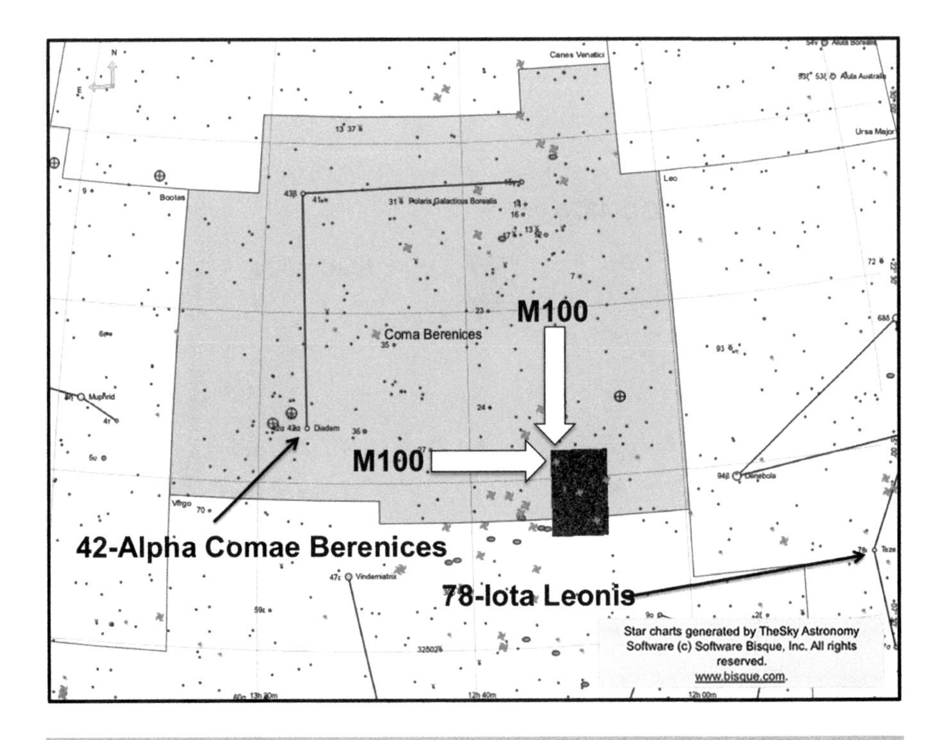

Fig. 5.78 Chart showing the location of M100

Messier 100: Data relating to the image in Figure 5.76

REMOTELY IMAGED FROM MAYHILL, NEW MEXICO
NGC 4321
OBJECT TYPE: Spiral Galaxy

RA: 12h 23m 47s
DEC: +15° 43' 35"
ALTITUDE: +56° 37' 23"
AZIMUTH: 113° 03' 11"
FIELD OF VIEW: 3° 54' 12" x 2° 36' 08"
OBJECT SIZE: 7’ X 6’
POSITION ANGLE: 89° 54' from North
EXPOSURE TIME: 300 s
DATE: 2nd February
LOCAL TIME: 01h 29m 14s
UNIVERSAL TIME: 08h 29m 14s
SCALE: 3.51 arcsec/pixel
MOON PHASE: 29.98% (waxing)

Telescope Optics
OTA: Takahashi FSQ - ED
Optical Design: Petzval Apochromatic Astrograph
Aperture: 106 mm
Focal Length: 530 mm
F/Ratio: f/5.0
Guiding: External
Mount: Paramount PME

Instrument Package
CCD: SBIG STL-11000M
Pixel Size: 9μm square
Sensor: Frontlit
Cooling: -15°C
Array: 4008 x 2672 pixels
FOV: 188.8 x 233.7 arcmin
Position Angle: 092.7°

Location
Observatory: New Mexico Skies
UTC Minus 7.00 (Daylight savings time is observed)
Minimum Target Elevation: Approx. 25 – 45 Degrees
(N or S) 32° 54' Decimal: 32.9 North
(W or E) 105° 31' Decimal: 105.5 West
Elevation: 2225 meters (7298 ft)

Messier 101

Messier 101 (NGC 5457), shown in Figure 5.79, is a spectacular-face on spiral galaxy in the constellation of Ursa Major. The original image had a field of view of 3° 54' 00" x 2° 36' 00" but, in order to have a closer look, an enlarged view of the galaxy is shown here with an approximate field of view of 0° 53' 30" x 0° 33' 30". Pierre Méchain discovered M101 on March 27, 1781. Messier commented: "Nebula without star, very obscure & pretty large, of 6 or 7 arcmin in diameter, between the left hand of Bootes & the tail of the great Bear. It is difficult to distinguish when one lights the graticule wires." W. H. Smyth described the galaxy as "A pale white nebula, in the nebulous field. It is 5d north north east of Alkaid, and a similar distance east half south from Mizar." M101 is usually referred to as the Pinwheel Galaxy.

The negative image in Figure 5.80 shows the full-size original image. There are hundreds of galaxies in the frame area, the brighter of which have been located and identified. Messier 101 has an apparent size of 22 arcmin and is believed to be at a

Fig. 5.79 Image of Messier 101, a spiral galaxy in Ursa Major

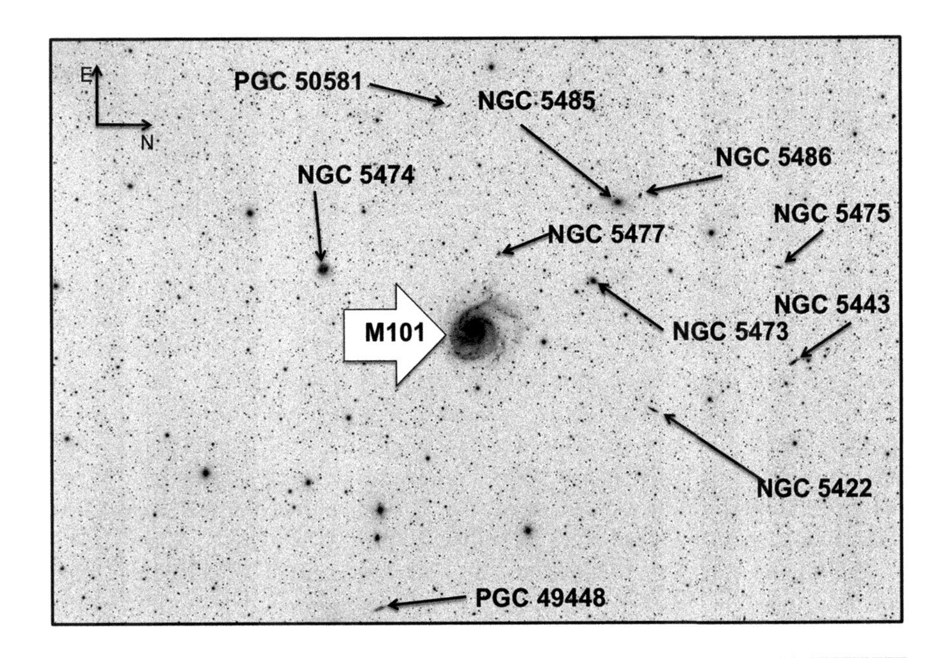

Fig. 5.80 Negative image of Messier 101

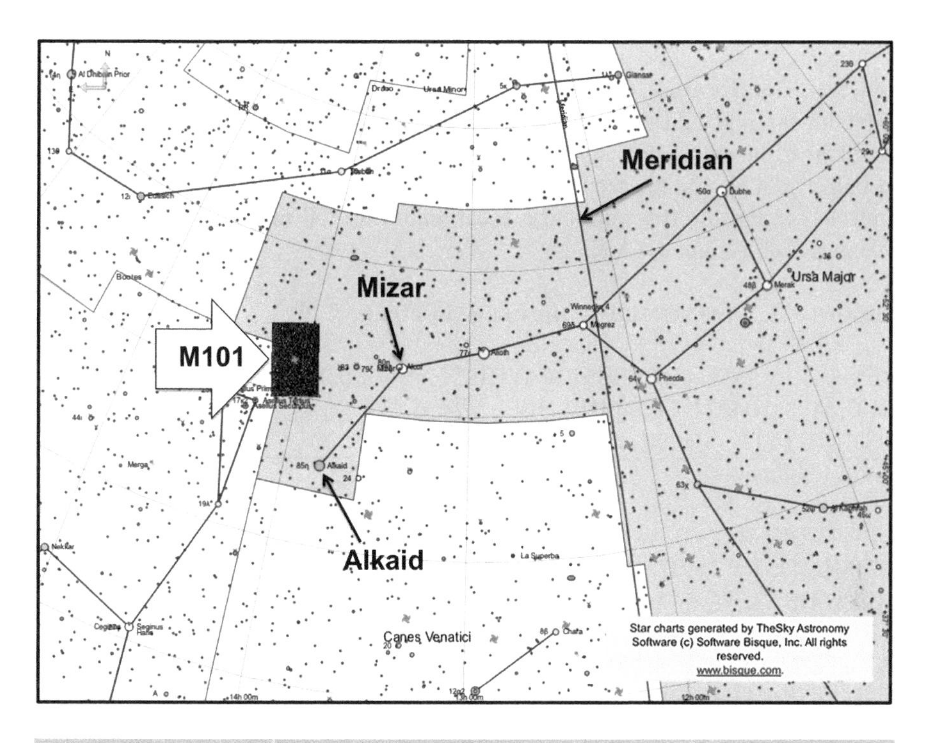

Fig. 5.81 Chart showing the location of M101

distance of 27 million ly (SEDS Data). This is quite close for such a large apparent size, and my calculation using these values results in an actual diameter for Messier 101 of 173,000 ly. This is clearly large in comparison with our own Galaxy, which has a major axis of around 100,000 ly. M101 is the "leader" of the M101 galaxy group of 9 or so galaxies. Some of the galaxies in the M101 group are within the field of view of the original full-size image, shown as a negative, including NGC 5474 and NGC 5477.

The chart in Figure 5.81 shows the position of M101 at the eastern end of Ursa Major, close to its border with Bootes. The plate-solved image is of the full frame, not just the enlarged image of Messier 101. Two stars that astronomers are usually quite familiar with are Mizar and Alkaid in the handle of the Big Dipper asterism. W.H. Smyth mentioned these stars in his description of the location of M101. Mizar is 79-Zeta Ursae Majoris and has its companion star Alcor, which is 80g Ursae Majoris. There is an angular separation of 5° 39' between Mizar and Messier 101. Alkaid is 85-Eta Ursae Majoris (HIP 67301) of magnitude 1.85. The angular separation between Alkaid and M101 is 5° 34'. Smyth was correct when he said the distances were similar. On the chart you can see the meridian passing close to the star Megrez (just to the west), which is at the top of the bowl of the Dipper.

The image was taken from New Mexico on the morning of February 1 at 03h 26m 36s. The Universal Time was 10h 26m 36s. The Right Ascension of M101 is 14h 03m 48s, and the Local Sidereal Time was 12h 10m 37s. The spiral galaxy had yet to cross the meridian and would do so in 1h 53m 11s at a meridian transit time of 05h 19m 47s. The telescope used to take the image was a Takahashi FSQ-ED with an aperture of 106 mm and a focal length of 530 mm. The telescope was mounted on a Paramount ME.

Messier 101: Data relating to the image in Figure 5.79

REMOTELY IMAGED FROM MAYHILL, NEW MEXICO
NGC 5457 (NORTHERN PINWHEEL GALAXY)
OBJECT TYPE: Spiral Galaxy

RA: 14h 03m 48s
DEC: +54° 15' 51"
ALTITUDE: +60° 40' 34"
AZIMUTH: 34° 25' 27"
FIELD OF VIEW: 3° 54' 00" x 2° 36' 00"
OBJECT SIZE: 22'
POSITION ANGLE: 90° 54' from North
EXPOSURE TIME: 300 s
DATE: 1st February
LOCAL TIME: 03h 26m 36s
UNIVERSAL TIME: 10h 26m 36s
SCALE: 3.51 arcsec/pixel
MOON PHASE: 21.07% (waxing)

Telescope Optics
OTA: Takahashi FSQ - ED
Optical Design: Petzval Apochromatic Astrograph
Aperture: 106 mm
Focal Length: 530 mm
F/Ratio: f/5.0
Guiding: External
Mount: Paramount PME

Instrument Package
CCD: SBIG STL-11000M
Pixel Size: 9μm square
Sensor: Frontlit
Cooling: -15°C
Array: 4008 x 2672 pixels
FOV: 188.8 x 233.7 arcmin
Position Angle: 092.7°

Location
Observatory: New Mexico Skies
UTC Minus 7.00 (Daylight savings time is observed)
Minimum Target Elevation: Approx. 25 – 45 Degrees
(N or S) 32° 54' Decimal: 32.9 North
(W or E) 105° 31' Decimal: 105.5 West
Elevation: 2225 meters (7298 ft)

Messier 102

There have historically been two contenders for the position of Messier 102 on the list. Pierre Méchain observed an object that he passed to Messier, who published it in the 1784 *Connoissance des Temps* before he had observed it. Later on, Méchain reported that the observation was a mistake and was in fact a repetition of M101. Later arguments were made proposing that it was not an error and in fact referred to the lenticular galaxy NGC 5866. For the sake of being able to image another object I have gone with the second option, NGC 5866, as shown in Figure 5.82, in the constellation of Draco. A handwritten reference to M102 from Messier reads: "Nebula between the stars Omicron Bootis & Iota Draconis: it is very faint." Note that the Greek letters Omicron (ο) and Theta (θ) are very similar, and a mistake could easily have been made.

Fig. 5.82 Image of Messier 102, a lenticular galaxy NGC 5866

The negative image in Figure 5.83 identifies NGC 5866 and some surrounding objects. There is a star on either side of the galaxy with magnitude 11.36 and 12.11, as shown. The bright star to the southwest of NGC 5866 is HIP 73837 of magnitude 7.69 and spectral type F8 at a distance of 487.53 ly. NGC 5866 itself has the dimension 5.2' x 2.3' and is at a distance of 45 million ly. Based on this distance, the actual dimensions of NGC 5866 are 68,000 x 30,000 ly. To the south of NGC 5866 is the spiral galaxy NGC 5870, which has dimensions of 1.2' x 09'. The small galaxy PGC 166188 is a mere 0.7' x 0.6', and even smaller is PGC 2521141 at 24" x 18". The star Tycho 3864:351 is magnitude 9.22.

The stars mentioned earlier, Theta Bootis and Iota Draconis, are labeled on the chart in Figure 5.84. Drawing a line from one to the other roughly passes through NGC 5866/M102. Theta Bootis is called Asellus Primus, or HIP 70497, and is magnitude 4.04 of spectral type F7, lying at a distance of 47.52 ly. It is situated 7° 15' from NGC 5866. Iota Draconis is called Edasich, or HIP 75458, and is magnitude 3.29 of spectral type K2 at a distance of 102.18 ly. Edasich has an angular separation of 3° 59'.

The image was taken on April 15 from New Mexico at 02h 01m 47s. The corresponding Universal Time was 08h 01m 47s. The Right Ascension of M102 (NGC 5866) is 15h 06m 59s, and the Local Sidereal Time was 14h 33m 12s. M102 would therefore cross the meridian in 33m 47s at a meridian transit time of 02h 35m 34s. The telescope used to take the image of M92 was a Planewave Corrected Dall-Kirkham Astrograph with an aperture of 510 mm and a focal length of 2,280 mm. The mount used was a Planewave Ascension 200HR.

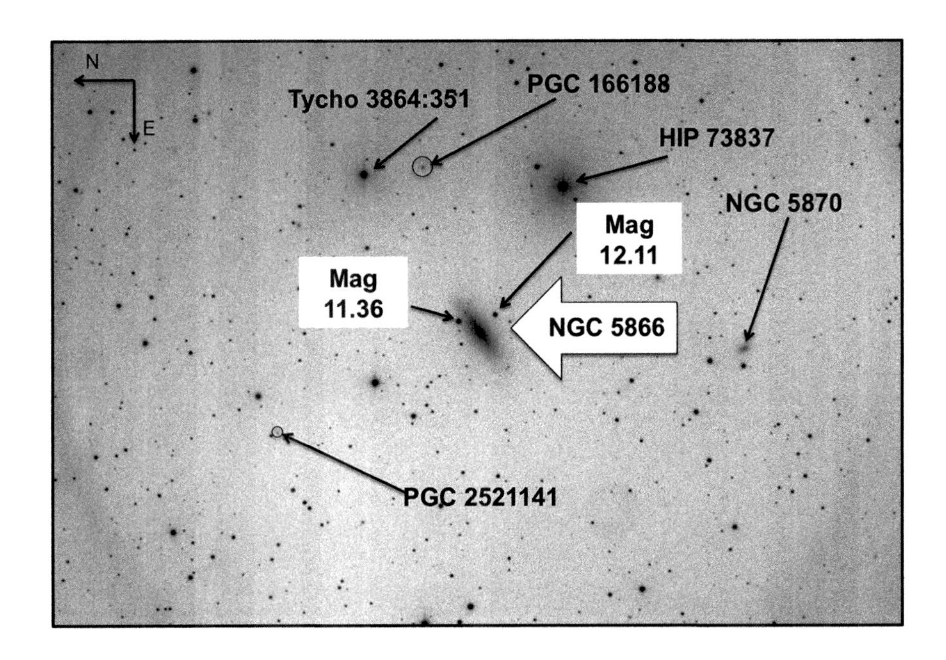

Fig. 5.83 Negative image of M102

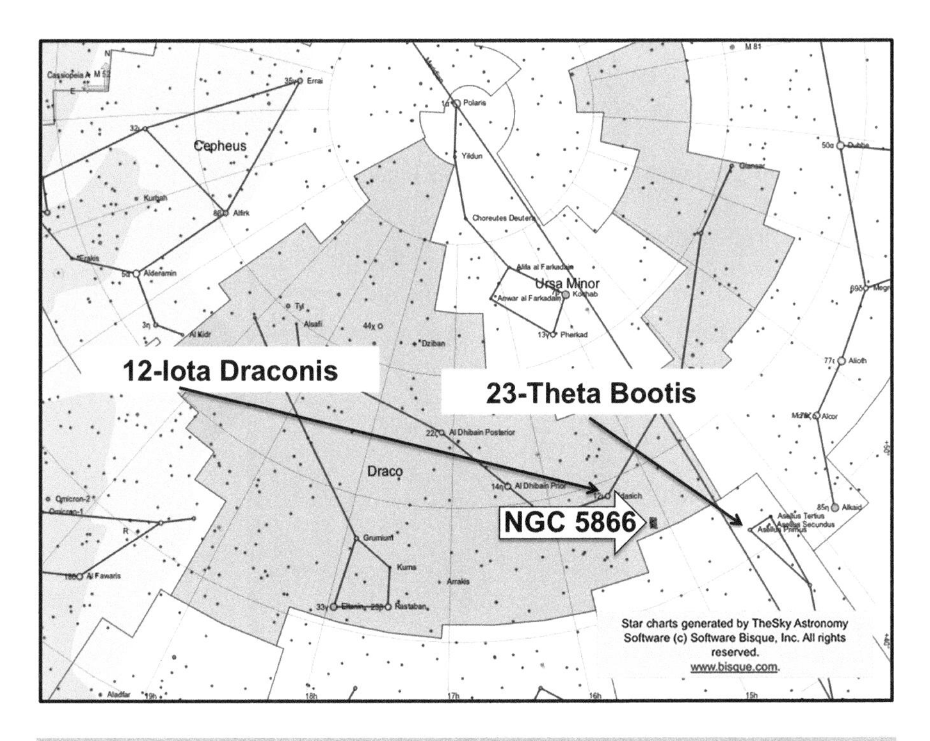

Fig. 5.84 Chart showing the location of M102

Messier 102: Data relating to the image in Figure 5.82

REMOTELY IMAGED FROM MAYHILL, NEW MEXICO
NGC 5866
OBJECT TYPE: Lenticular Galaxy

RA: 15h 06m 59s
DEC: +55° 41' 50"
ALTITUDE: +66° 20' 30"
AZIMUTH: 11° 54' 09"
FIELD OF VIEW: 0° 54' 47" x 0° 36' 31"
OBJECT SIZE: 5.2' x 2.3'
POSITION ANGLE: 271° 20' from North
EXPOSURE TIME: 300 s
DATE: 15th April
LOCAL TIME: 02h 01m 47s
UNIVERSAL TIME: 08h 01m 47s
MOON PHASE: 85.20% (waning)

Telescope Optics
OTA: Planewave 20" CDK
Optical Design: Corrected Dall-Kirkham Astrograph
Aperture: 510 mm
Focal Length: 2280 mm (0.66 Focal Reducer Fitted)
F/Ratio: f/4.5
Guiding: Active Guiding Disabled
Mount: Planewave Ascension 200HR

Instrument Package
FLI Proline PL11002M CCD
Pixel Size: 9μm square
Resolution: 0.81 arcsec/pixel
Sensor:
Cooling: -30°C default
Array: 4008 x 2672 (10.7 Megapixels)

Location
Observatory: New Mexico Skies
UTC Minus 7.00 (Daylight savings time is observed)
Minimum Target Elevation: Approx. 25 – 45 Degrees
(N or S) 32° 54' Decimal: 32.9 North
(W or E) 105° 31' Decimal: 105.5 West
Elevation: 2225 meters (7298 ft)

Messier 103

Messier 103 (NGC 581), shown in Figure 5.85, is an open cluster located in the constellation of Cassiopeia. The image was taken when M103 was at quite a low altitude -- it was not the best time of year to take an image of this interesting open cluster, as it would transit during daylight hours. Pierre Méchain discovered the cluster in 1781. M103 has an angular size of 6 arcmin and lies at a distance of 8,500 ly. This is equivalent to M103 having an actual size of 14.8 ly. The cluster contains a multiple star, Struve 131, and also a red giant star, Tycho 4031:418. W. H. Smyth noted, "My attention was first drawn to this object, by seeing it among Stuve's acervi [list of double stars]; but I soon found that it was also the 103 which Messier describes so vaguely, as being between Delta and Epsilon Cassiopeiae, whereas it is pretty close to Delta, on the Lady's knee."

The bright star on the negative image in Figure 5.86 is a triple star Struve 131. This cannot be a member of the cluster if the Hipparcos satellite data is correct, which gives it a distance of 1,337 ly – much closer to us. The bright star at the bottom is magnitude 8.17 and lies at a distance of 393 ly (Hipparcos) – closer still than

Fig. 5.85 Image of Messier 103, an open cluster in Cassiopeia

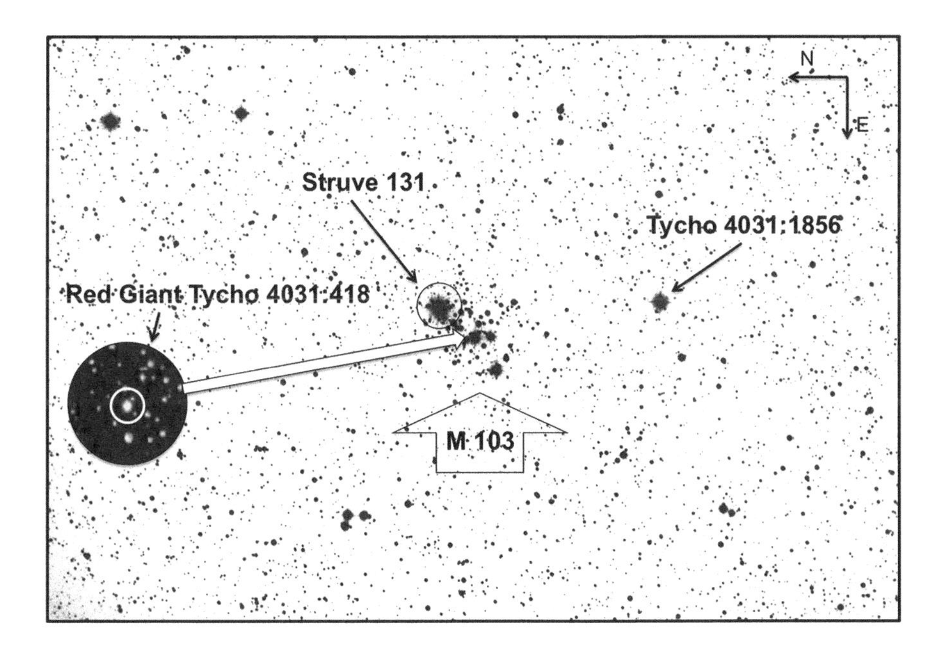

Fig. 5.86 Negative image of Messier 103

the triple star, and not a cluster member. The central star Tycho 4031:418 is magnitude 8.47 and is a red giant. The reason we can tell the type of star is through the data given in the Tycho catalogue for the star, part of the Hipparcos mission. This data includes B and V stellar magnitudes. This means that a star's brightness was measured using a V filter (Green), which approximates the visual response, and a B filter (Blue). The point of using the filters is that if you subtract the V magnitude (V) from the B magnitude (B), you get a measure of the colour – and thus the temperature – of a star. Looking at Tycho 4031:418 again, Tycho gives V = 8.69 and B = 10.89, so B–V = 2.2. This implies a cool, orange star. A hot blue star would have a negative B-V value. Most of the remote telescopes have a filter wheel which often incorporate a B filter and a V filter. All you have to do is to take an image using the B filter, then a corresponding one using the V filter, and then measure the magnitudes of stars through the different filters. It is important to note that intervening interstellar material can affect the B and V readings which can lead to inaccurate results if not taken into account. The enlarged view of the red star in color was taken by the 90-mm single-shot color telescope at Siding Spring.

The chart in Figure 5.87 shows the location of Messier 103 in the constellation of Cassiopeia. Smyth noted that Messier had vaguely referred to M103 lying between Delta and Epsilon Cassiopeia. The star 37-Delta Cassiopeiae is called Ruchbah (HIP 6686) and has a magnitude of 2.66 with spectral type A5, lying at a distance of 99.41 ly. There is an angular separation between Ruchbah and M103 of

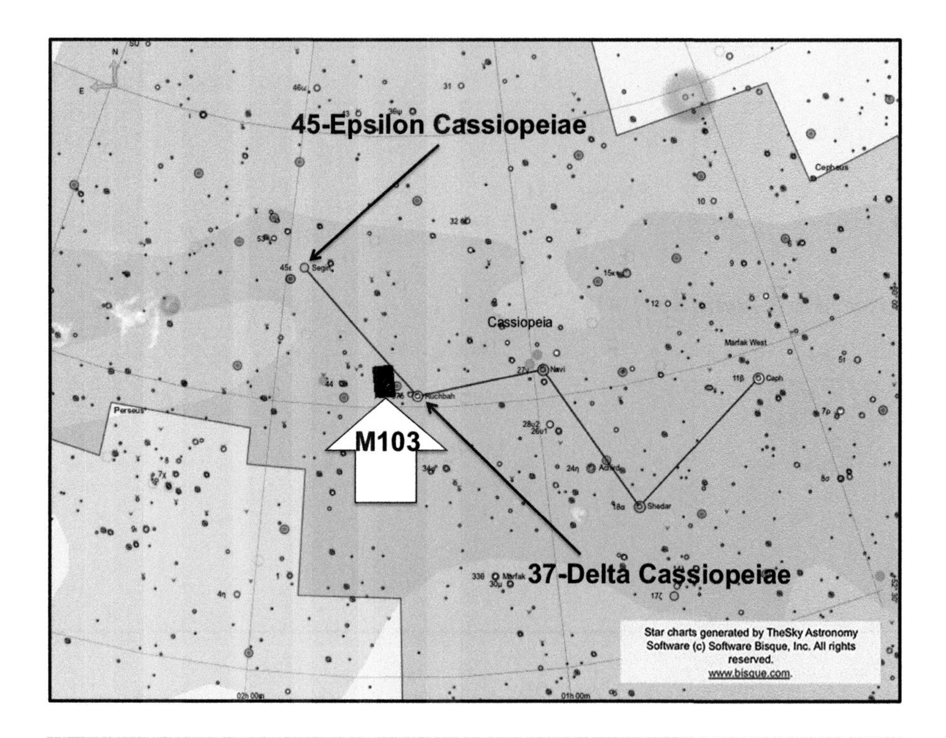

Fig. 5.87 Chart showing the location of M103

1° 1'. The star 45-Epsilon Cassiopeiae is called Segin (HIP 8886) and has a magnitude of 3.35 with spectral type B2, lying at a distance of 441.95 ly. There is an angular separation between Segin and M103 of 3° 53'.

The image was taken on June 13 from New Mexico at 03h 02m 13s. The corresponding Universal Time was 09h 02m 13s. The Right Ascension of M103 (NGC581) is 01h 34m 31s, and the Local Sidereal Time was 19h 26m 25s. Thus, M103 would cross the meridian in 6h 8m 06s, providing a meridian transit time of 09h 10m 19s. The telescope used to take the image of M92 was a Planewave 20" Corrected Dall-Kirkham Astrograph with an aperture of 510 mm and a focal length of 2,280 mm. The mount used was a Planewave Ascension 200HR.

Messier 103: Data relating to the image in Figure 5.85

REMOTELY IMAGED FROM MAYHILL, NEW MEXICO
NGC 581
OBJECT TYPE: Open Cluster

RA: 01h 34m 31s
DEC: +60° 44' 27"
ALTITUDE: +27° 15' 01"
AZIMUTH: 33° 19' 41"
FIELD OF VIEW: 0° 54' 47" x 0° 36' 31"
OBJECT SIZE: 6'
POSITION ANGLE: 271° 12' from North
EXPOSURE TIME: 300 s
DATE: 13th June
LOCAL TIME: 03h 02m 13s
UNIVERSAL TIME: 09h 02m 13s
MOON PHASE: 87.12% (waning)

Telescope Optics
OTA: Planewave 20" CDK
Optical Design: Corrected Dall-Kirkham Astrograph
Aperture: 510 mm
Focal Length: 2280 mm (0.66 Focal Reducer Fitted)
F/Ratio: f/4.5
Guiding: Active Guiding Disabled
Mount: Planewave Ascension 200HR

Instrument Package
FLI Proline PL11002M CCD
Pixel Size: 9μm square
Resolution: 0.81 arcsec/pixel
Cooling: -30°C default
Array: 4008 x 2672 (10.7 Megapixels)

Location
Observatory: New Mexico Skies
UTC Minus 7.00 (Daylight savings time is observed)
Minimum Target Elevation: Approx 25 – 45 Degrees
(N or S) 32° 54' Decimal: 32.9 North
(W or E) 105° 31' Decimal: 105.5 West
Elevation: 2225 meters (7298 ft)

Messier 104

Spiral galaxy Messier 104 (NGC 4594) in the constellation of Virgo, shown in Figure 5.88, is better known as the Sombrero Galaxy due to its large central bulge and edge-on perspective from our viewpoint, together with a dark dust lane bisecting the upper and lower sections. Pierre Méchain discovered it on May 11, 1781. Méchain noted: "I discovered a nebula above the Raven [Corvus] which did not appear to me to contain any single star. It is of a faint light and difficult to find if

Fig. 5.88 Image of Messier 104, a spiral galaxy in Virgo

the micrometer wires are illuminated. I have compared its position on this day and the following with Spica in the Virgin and from this derived its right ascension and declination." French astronomer Camille Flammarion managed to obtain Charles Messier's handwritten notes and in 1921 added M104 as the first new object on Messier's list.

The negative image in Figure 5.89 includes a positive enlargement of M104 from the original image. The brightest star in the frame is magnitude 8.77 (center right on the western edge of the frame). This is Tycho 5531:1324, which lies at a distance of 160.67 ly. M104 itself is given a distance of 30 million ly to 50 million ly by different sources. The major axis of M104 is 9 arcmin, which would correspond to an actual size of 131,000 ly at the 50 million ly distance, or 79,000 ly at the 30 million ly distance – quite a difference. The minor axis is 4', which would give an actual size of 58,000 or 35,000 ly for the two distances quoted. M104 is said to have about 2000 globular clusters, many more than in our own Galaxy. There is a small group of galaxies known as the M104 group.

The chart in Figure 5.90 shows the position of M104 within the embedded image, which straddles the constellations of Virgo and Corvus (the magnitude 9.26 star shown next to the enlarged M104 in the negative image is HIP 61823 is in the

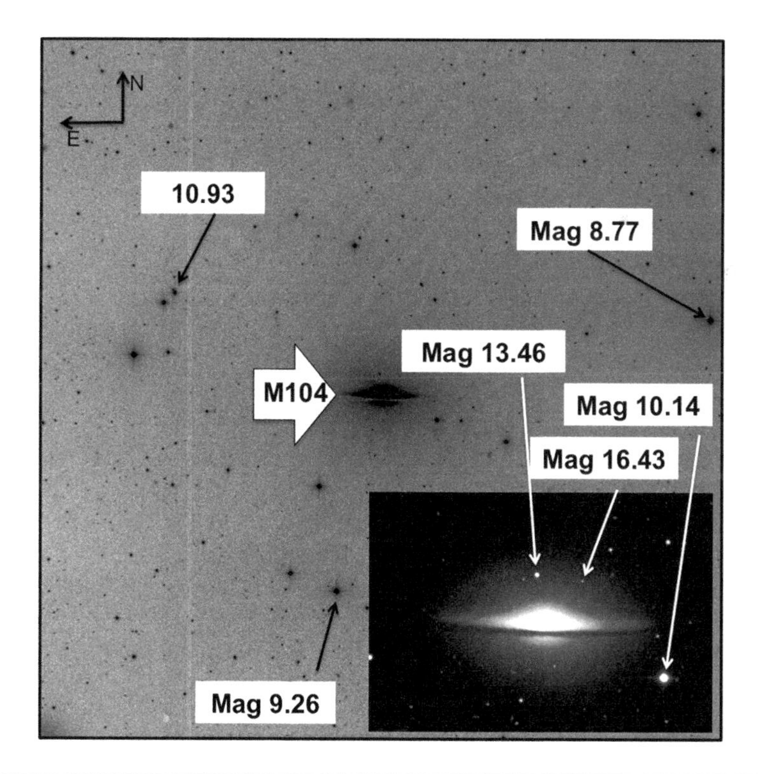

Fig. 5.89 Negative image of Messier 104

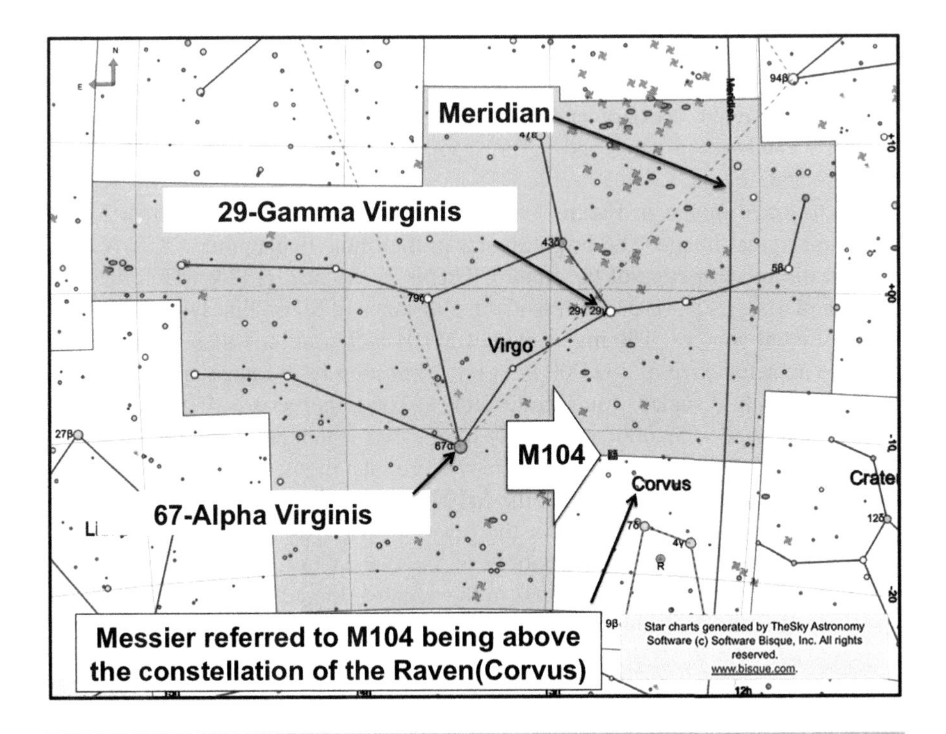

Fig. 5.90 Chart showing the location of M104

constellation of Corvus). Two reference stars are labeled on the chart, 67-Alpha Virginis and 29-Gamma Virginis. The star 67-Alpha Virginis is better known as Spica (HIP 65474), which has a magnitude of 0.98 and is of spectral type B1, lying at a distance of 262.18 ly. There is an angular separation of 11° 6' between Spica and M104. The star 29-Gamma Virginis is Porrima (HIP 61941), which has a magnitude of 2.74 and is of spectral type F0, lying at a distance of 38.58 ly. The meridian has been pointed out on the chart to the west of M104.

The image was taken on February 21 from Siding Spring at 03h 09m 02s. The corresponding Universal Time was 16h 09m 02s (February 20). The Right Ascension of M104 is 12h 40m 54s, and the Local Sidereal Time was 12h 08m 10s. So, M104 will cross the meridian in 32m 44s, making the meridian transit time 03h 41m 46s. The telescope used to take the image of M104 was a Planewave 17" Corrected Dall-Kirkham Astrograph with an aperture of 431 mm and a focal length of 2,912 mm. The mount used was a Planewave Ascension 200HR.

Messier 104: Data relating to the image in Figure 5.88

REMOTELY IMAGED FROM NEW SOUTH WALES, AUSTRALIA
NGC 4594 (SOMBRERO GALAXY)
OBJECT TYPE: Spiral Galaxy

RA: 12h 40m 54s
DEC: -11° 43' 00"
ALTITUDE: +69° 03' 53"
AZIMUTH: 22° 57' 12"
FIELD OF VIEW: 0° 43' 54" x 0° 43' 54"
OBJECT SIZE: 9' x 4'
POSITION ANGLE: 90° 47' from North
EXPOSURE TIME: 300 s
DATE: 21st February
LOCAL TIME: 03h 09m 02s
UNIVERSAL TIME: 16h 09m 02s (20th February)
SCALE: 0.64 arcsec/pixel
MOON PHASE: 32.88% (waning)

Telescope Optics
OTA: Planewave 17" CDK
Optical Design: Corrected Dall-Kirkham Astrograph
Aperture: 431 mm
Focal Length: 2912 mm
F/Ratio: f/6.8
Guiding: Active Guiding Disabled
Mount: Planewave Ascension 200HR

Instrument Package
CCD: FLI Proline 16803
Pixel Size: 9μm square
Sensor: KAF 16803
Cooling: -35°C default
Array: 4096 x 4096 pixels
FOV: 43.2 x 43.2 arcmin

Location
Observatory: Siding Spring
UTC +10.00 (Australia Daylight savings time is observed)
31° 16' 24" South
149 03' 52" East
Elevation: 1122 meters (3681 ft)

Messier 105

Messier 105, shown in Figure 5.91, is an elliptical galaxy located in the constellation of Leo. The area is swarming with galaxies, as can readily be seen in the monochromatic image, which also includes M95 and M96. It appears to have a major axis of around 5 arcmin and a minor axis of around 4 arcmin. It is believed to be at a distance of around 38 million ly and was discovered by Pierre Méchain in 1781. The designation of the galaxy as M105 was proposed by 20th century astronomer Helen Sawyer Hogg. She discovered notes in a letter from Méchain and wrote the following: "He also lists four other nebulae which he has discovered, and these should logically be given Messier numbers as follows: NGC 4594 as M 104; NGC 3379 as M 105." The proposals were adopted in 1947. This galaxy is known to contain a supermassive black hole with a mass hundreds of millions of times the mass of the Sun!

The negative image in Figure 5.92 shows the elliptical galaxy M105 and also identifies spiral galaxies M95 and M96. There are however some other interesting galaxies in the same field. NGC 3371 is also identified as NGC 3384 in the NGC Catalog and is an elliptical galaxy. It has a major axis of 5.4' and a minor axis of 2.7'. The distance between M105 and NGC 3371 is just over 7' center to center, and there is a position angle of roughly 68° between north and NGC 3371 measured from M105 towards the east. Note that north is to the right on the image and east is

Fig. 5.91 Image of Messier 105, an elliptical galaxy in Leo

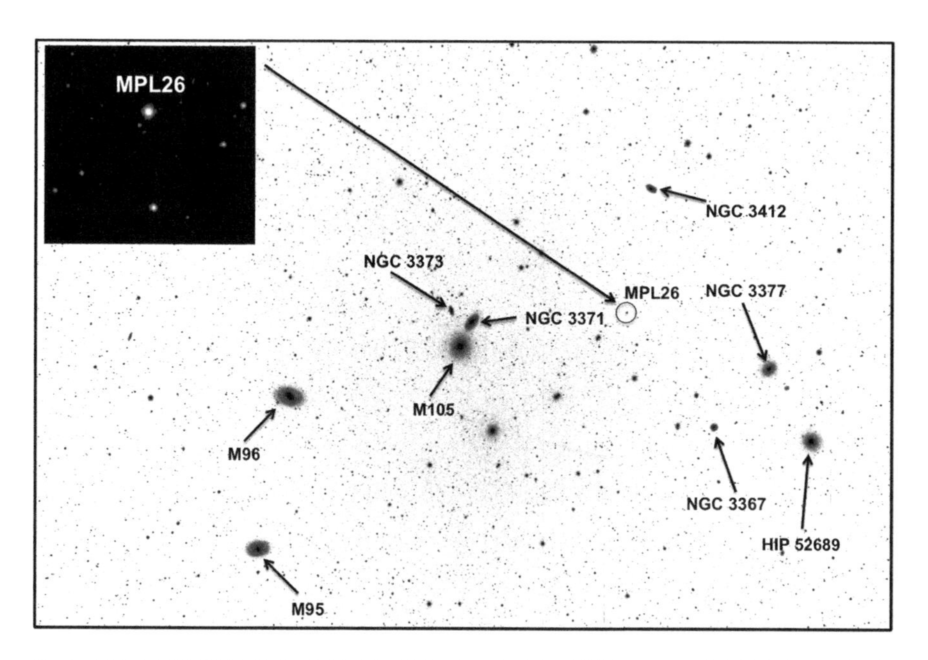

Fig. 5.92 Negative image of M105

up. Notice also the much smaller galaxy NGC 3373, which has a major axis of 2.9' and a minor axis of 1.3'. It lies 10' from M105 at an angle of about 108°. NGC 3377 is an elliptical galaxy with a major axis of 3.9' and a minor axis of 1.9'. It is believed to be about 26 million ly distant. At that distance, the 3.9' axis would correspond to a length of around 30,000 ly. NGC 3367 is a spiral galaxy that is given a size of 2.9' X 2.8' by the Revised New General Catalog (RNGC). NGC 3412 is a spiral galaxy with dimensions 3.7' X 2.2' (RNGC).

The chart in Figure 5.93 with north at the top shows the position of M105, centrally located within the constellation of Leo. The solved image is next to the star HIP 52911, which is 53 l Leonis. The image itself contains the star HIP 52689, which is 52 k Leonis. When examining the image, I spotted the asteroid -- minor planet MPL 26 Prosperpina which just happened to be passing through Leo at the time. This is marked on the negative image in Figure 5.92. It was moving in a northwesterly direction at a rate of roughly 45 arcsec per hour. Prosperpina had a magnitude of 11.31 at the time the image was taken. Prosperpina is a main belt asteroid discovered in 1853 by R. Luther. If you were to monitor the brightness of this minor planet over a 13-hour period, it would complete a full cycle of brightness variation, meaning that it has rotated once in that period.

The image was taken on February 1 at 01h 36m 01s from the telescope site in New Mexico. The Universal Time was 08h 36m 01s. The Right Ascension of M105 is 10h 48m 44s, and the Local Sidereal Time was 10h 19m 44s. The difference of

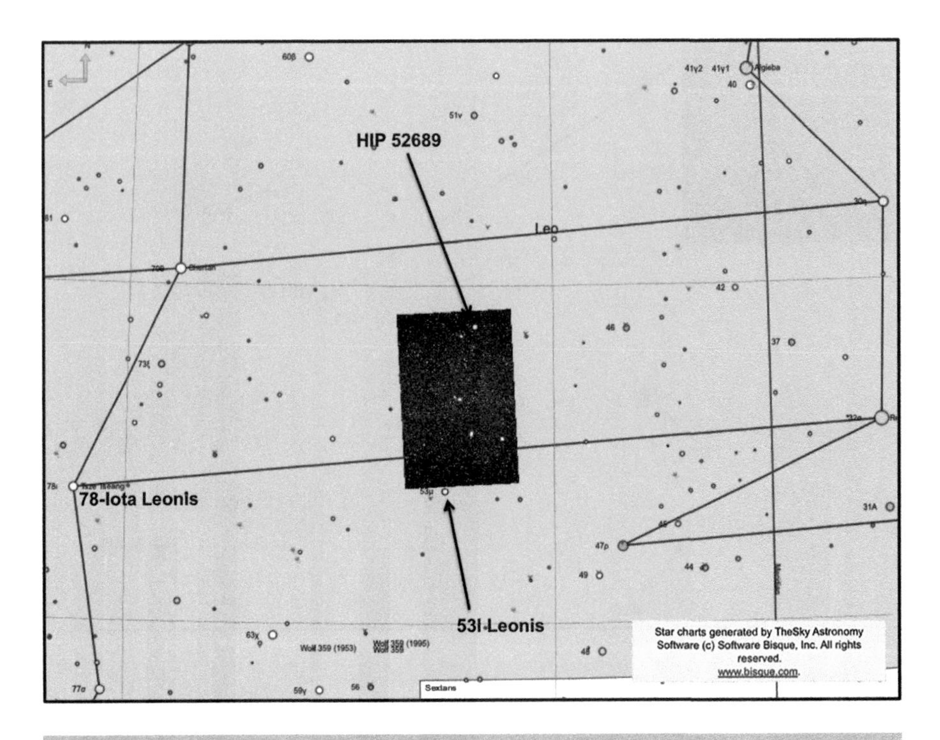

Fig. 5.93 Chart showing the location of M105

29m indicates that this is how long it would take before M105 reached the meridian. Transit would therefore occur therefore at 02h 05m 0s. The altitude of M105 was +68° 38' 48" when the image was taken. It would reach +69° 42' 18" at the meridian. The azimuth of M105 was 160° 12' 54"when the image was taken. The telescope used was a 106-mm Petzval Apochromat Astrograph.

Messier 105: Data relating to the image in Figure 5.91

REMOTELY IMAGED FROM MAYHILL, NEW MEXICO
OBJECT TYPE: Elliptical Galaxy

RA: 10h 48m 44s
DEC: +12° 29' 18"
ALTITUDE: +68° 38' 48"
AZIMUTH: 160° 12' 54"
FIELD OF VIEW: 3° 53' 24" x 2° 35' 36"
OBJECT SIZE: 2'
POSITION ANGLE: 92° 43' from North
EXPOSURE TIME: 300 s

DATE: 1st February
LOCAL TIME: 01h 36m 01s
UNIVERSAL TIME: 08h 36m 01s
SCALE: 3.49 arcsec/pixel
MOON PHASE: 20.36% (waxing)

Telescope Optics
OTA: Takahashi FSQ Fluorite
Optical Design: Petzval Apochromat Astrograph
Aperture: 150 mm
Focal Length: 106 mm
F/Ratio: f/5.0
Guiding: External
Mount: Paramount GTS-1100S

Instrument Package
SBIG STL-11000M
A/D Gain: 2.2e-/ADU
Pixel Size: 9um square
Resolution: 3.5 arcsec/pixel
Sensor: Frontlit
Cooling: -15°C default
Array: 4008 x 2672 (10.7 Megapixels)
FOV: 155.8 x 233.7 arcmin

Location
Observatory: New Mexico Skies
UTC Minus 7.00 (Daylight savings time is observed)
Minimum Target Elevation: Approx. 25 – 45 Degrees
(N or S) 32° 54' Decimal: 32.9 North
(W or E) 105° 31' Decimal: 105.5 West
Elevation: 2225 meters (7298 ft)

Messier 106

Messier 106 (NGC 4258), shown in Figure 5.94, is a spiral galaxy located in the constellation of Canes Venatici. The wide-angle 5-minute monochrome image shown is almost 4° across, so the galaxy is quite large with a major axis of 19' of arc and a minor axis of 8'. This object was added to Messier's list in 1947 by astronomer Helen Sawyer Hogg but was originally discovered by Messier's colleague Pierre Méchain in 1781. M105 and M107 were added in similar fashion. In a letter concerning his discovery, Méchain wrote, "In July 1781 I found another

Fig. 5.94 Image of Messier 106, a spiral galaxy in Canes Venatici

nebula close to the Great Bear near the star No. 3 of the Hunting Dogs and 1 degree more south." The galaxy is believed to lie at a distance of 25 million ly.

The negative of the M106 image in Figure 5.95 is used to highlight the fact that there are many more galaxies in the frame. The detectable ones are indicated on the image. NGC 4217 is also known as PGC 39241,where PGC stands for Principal Galaxy Catalog and includes many of the fainter galaxies. This spiral galaxy has a major axis of 5.5' and a minor axis of 1.6'. Spiral galaxy NGC 4220, also known as PGC 39285, has a major axis of 3.3' and a minor axis of 1'. NGC 4242, also known as PGC 39423, is another spiral galaxy with a major axis of 3.8' and a minor axis of 2.7'. The galaxy PGC 39864 is much smaller with a major axis of 1' and a minor axis of 0.7. These are the brighter galaxies – there are many more in the frame that are too faint to be seen in the image. The constellation of Canes Venatici contains many galaxies and searching the PGC Catalog indicated that there were in excess of 20,000 galaxies in this constellation.

The chart in Figure 5.96 shows the solved image accurately placed in the constellation. Note that the chart has north to the right, with east at the top. In the report of his discovery, Méchain mentioned the object being placed "near star No. 3 of the Hunting Dogs." This refers to its Flamsteed Catalog designation. This star is arrowed on the chart as "3." It is more clearly seen in the negative image with its Hipparcos Catalog depiction as the magnitude 5.28 cool M-type star HIP 60122.

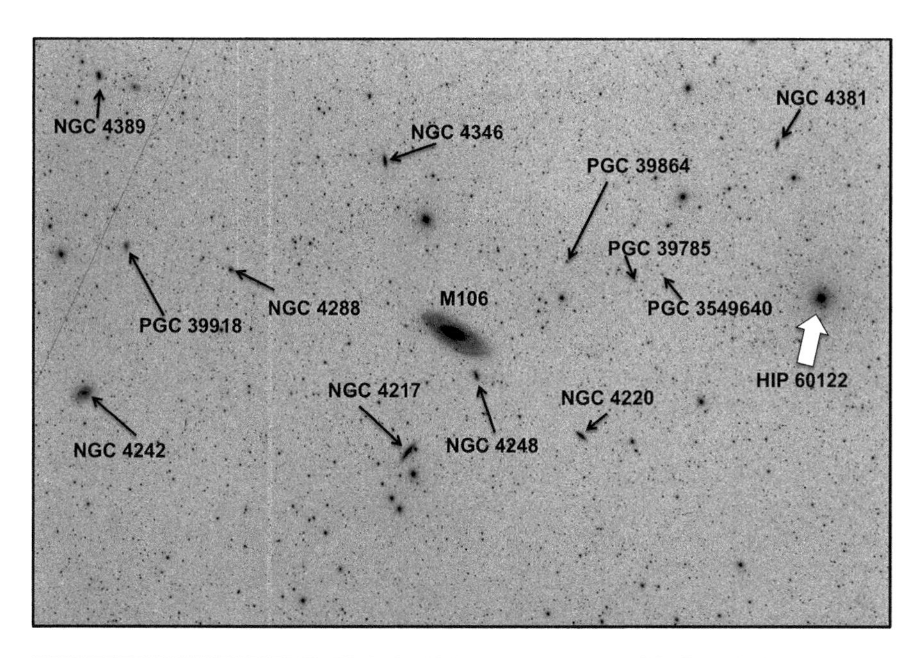

Fig. 5.95 Negative image of Messier 106

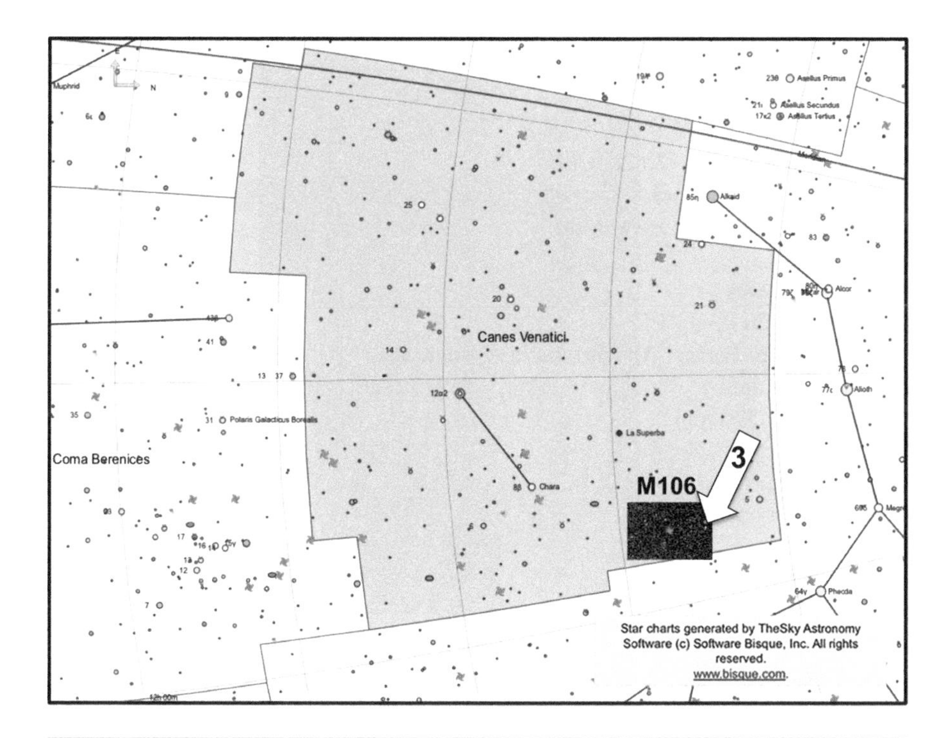

Fig. 5.96 Chart showing the location of M106

The reference to the Big Bear (Ursa Major) indicates the object's proximity to that constellation. M106 lies just under the handle of the Big Dipper.

The image was taken on February 1 at 05h 20m 12s from New Mexico. The Universal Time was 12h 20m 12s. M106 has a Right Ascension of 12h 19m 48s, and the Local Sidereal Time was 14h 04m 32s. Thus, M106 crossed the meridian some time earlier. The time difference is 1h 44m 44s, so meridian transit will have been at 03h 35m 28s after midnight. The azimuth value of 313° 44' 38" shows that M106 was 133° 44m 38s to the west of the meridian when the image was taken. The telescope used was a wide-angle instrument with an aperture of 106 mm on a Paramount ME mount.

Messier 106: Data relating to the image in Figure 5.94

REMOTELY IMAGED FROM MAYHILL, NEW MEXICO
NGC 4258
OBJECT TYPE: Spiral Galaxy

RA: 12h 19m 48s
DEC: +47° 12' 31"
ALTITUDE: +65° 29' 22"
AZIMUTH: 313° 44' 38"
FIELD OF VIEW: 3° 54' 00" x 2° 36' 00"
OBJECT SIZE: 19' X 8'
POSITION ANGLE: 89° 30' from North
EXPOSURE TIME: 300 s
DATE: 1st February
LOCAL TIME: 05h 20m 12s
UNIVERSAL TIME: 12h 20m 12s
SCALE: 3.5 arcsec/pixel
MOON PHASE: 21.78% (waxing)

Telescope Optics
OTA: Takahashi FSQ - ED
Optical Design: Petzval Apochromatic Astrograph
Aperture: 106 mm
Focal Length: 530 mm
F/Ratio: f/5.0
Guiding: External
Mount: Paramount PME

Instrument Package
CCD: SBIG STL-11000M
Pixel Size: 9µm square
Sensor: Frontlit
Cooling: -15°C

Array: 4008 x 2672 pixels
FOV: 188.8 x 233.7 arcmin
Position Angle: 092.7°

Location
Observatory: New Mexico Skies
UTC Minus 7.00 (Daylight savings time is observed)
Minimum Target Elevation: Approx 25 – 45 Degrees
(N or S) 32° 54' Decimal: 32.9 North
(W or E) 105° 31' Decimal: 105.5 West
Elevation: 2225 meters (7298 ft)

Messier 107

Messier 107 (NGC 6171), shown in Figure 5.97, is a globular cluster in the constellation of Ophiuchus. The monochrome image has a wide field, so the object is fairly small at this scale. The image has a size of almost 4° x 2.5°. The cluster itself has an apparent diameter of 13 arcmin. M107 was a later addition to Messier's original list. It was discovered by Pierre Méchain in 1782. In a letter to fellow mathematician and astronomer Bernoulli, Méchain wrote, "In April 1782 I discovered a small

Fig. 5.97 Image of Messier 107, a globular cluster in Ophiuchus

nebula in the left flank of Ophiuchus between the stars Zeta and Phi, the position of which I have not yet observed any closer." In 1837, W. H. Smyth noticed, "A large but pale granulated cluster of small stars, on the Serpent-bearer's right leg. There are five telescopic stars around it, so placed as to form a crucifix, when the cluster is high in the field; but the region immediately beyond is a comparative desert. After long gazing, this object becomes more compressed in the centre, and perplexes the mind by so wonderful an aggregation. It was discovered by WH in May 1793, and was registered 5' or 6' in diameter. The mean place was obtained by differentiation with Zeta Ophiuchi, from which it is distant 3 degrees to the south-southwest, in the line between Beta Scorpii and Beta Ophiuchi."

The negative image in Figure 5.98 shows M107 and identifies some surrounding stars. The inset is an expanded positive image of M107 from the original. To give an idea of the magnitudes of the stars surrounding the globular cluster, the relatively bright magnitude 10.84 star UCAC4 386:72712 is identified. To the left of this star there is a little string of three fainter stars. The star on the right is UCAC4 386:72709 with a magnitude of 14.17. Many of the outer cluster stars that can be differentiated from the cluster background are of similar magnitude or fainter than this. M107 lies at a distance of 20,900 ly and has a diameter of around 80 ly.

The chart in Figure 5.99 shows the large field of the image taken of M107. Méchain referred to the location of M107 in Ophiuchus between Zeta and Phi. Zeta is the star "Han," shown in the chart to the right of M107. This is HIP 81377, a

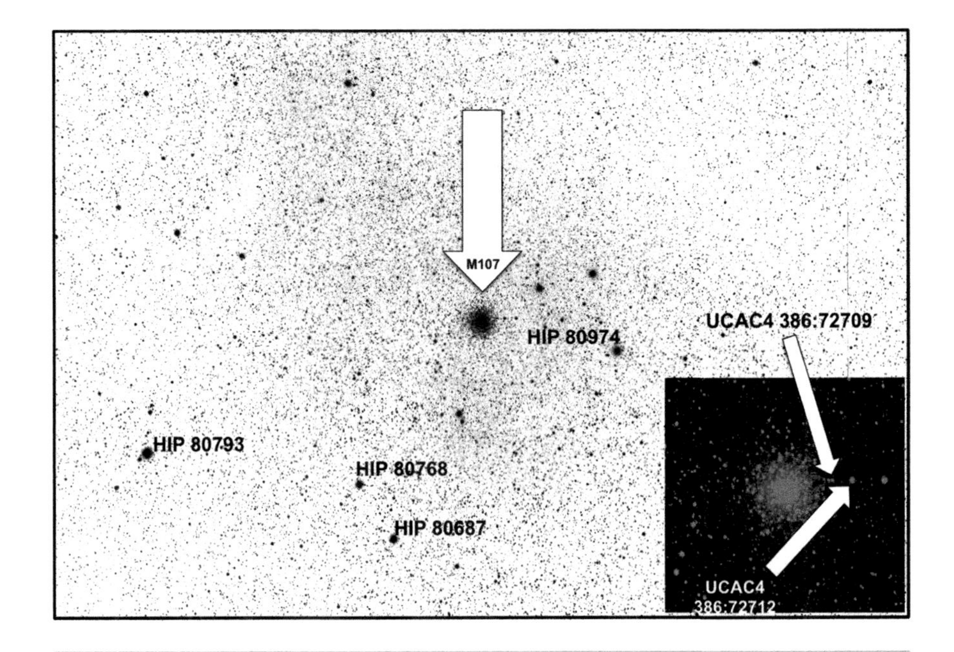

Fig. 5.98 Negative image of Messier 107

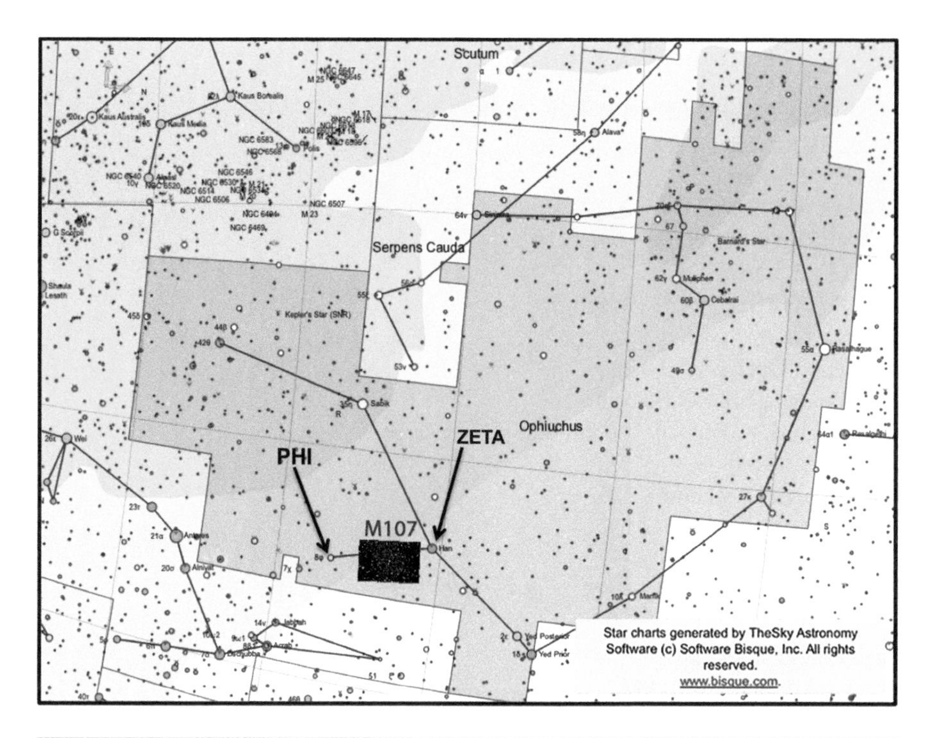

Fig. 5.99 Chart showing the location of M107

magnitude 2.54 star that is at a distance of 458.09 ly. To the left of M107 is the star HIP 80894, which is the magnitude 4.29 star Phi, lying at a distance of 210.02 ly. The plate-solved rectangular image lies between these stars.

The M107 image was taken at 02h 04m 21s on April 25 in New Mexico. The Universal Time was 08h 04m 21s. M107 has a Right Ascension of 16h 33m 30s, and the Local Sidereal Time was 13h 12m 59s. This gives a time difference of 3h 20m 31s, which, if added to the local time, gives a meridian transit time of 05h 24m 52s. When the image was taken, the altitude of M107 was only +23° 43' 34" and the azimuth 125° 15' 31". At transit, M107 would reach an altitude of +44° 07' 45". The telescope used was a 0.1m f/5 astrograph.

Messier 107: Data relating to the image in Figure 5.97

REMOTELY IMAGED FROM MAYHILL, NEW MEXICO
NGC 6171
OBJECT TYPE: Globular Cluster

RA: 16h 33m 30s DEC: -13° 05' 15"
ALTITUDE: +23° 43' 34"
AZIMUTH: 125° 15' 31"

FIELD OF VIEW: 3° 53' 44" x 2° 35' 49"
OBJECT SIZE: 13'
POSITION ANGLE: 93° 14' from North
EXPOSURE TIME: 300 s
DATE: 25th March
LOCAL TIME: 02h 04m 21s
UNIVERSAL TIME: 08h 04m 21s
SCALE: 3.5 arcsec/pixel
MOON PHASE: 9.60% (waning)

Telescope Optics
OTA: Takahashi FSQ Fluorite
Optical Design: Petzval Apochromatic Astrograph
Aperture: 106 mm
Focal Length: 530 mm
F/Ratio: f/5.0
Guiding: External
Mount: Paramount GT- 1100S

Instrument Package
CCD: SBIG STL – 1100M
Pixel Size: 9µm square
Sensor: Frontlit
Cooling: -15°C default
Array: 4008 x 2672 pixels
FOV: 155.8 x 233.7 arcmin

Location
Observatory: New Mexico Skies
UTC Minus 7.00 (Daylight savings time is observed)
Minimum Target Elevation: Approx. 25 – 45 Degrees
(N or S) 32° 54' Decimal: 32.9 North
(W or E) 105° 31' Decimal: 105.5 West
Elevation: 2225 meters (7298 ft)

Messier 108

Messier 108 (NGC 3556), shown in Figure 5.100, is a spiral galaxy in the constellation of Ursa Major. It is very dusty with obscuring material giving it a mottled appearance. It lies very close to the Owl Nebula (Messier 97) and can be seen in the same field of view through an eyepiece. The field of view in the image is too small to include M97, but if you look back to the image shown for M97, Messier

Fig. 5.100 Image of Messier 108, a spiral galaxy in Ursa Major

108 can be seen at the bottom right. Pierre Méchain discovered the galaxy in February 1781. Later, Messier commented, "Near this nebula [M97] he [Méchain] has seen another one, which has not yet been determined [M108] and also a third which is near Gamma of the Great Bear. [M109]."

The negative image in Figure 5.101 shows the near edge on M108 within the plate-solved image and also identifies a number of other galaxies in the frame. Messier 108 has a major axis of 8' of arc and lies at a distance of 45 million ly. This gives the major axis an actual size of 105,000 ly. The minor axis is about 2', which is equivalent to 26, 000 ly. The bright star to the west of M108 (at the bottom of the image) is HIP 54584, which is a magnitude 8.91 star of spectral type F0 lying at a distance of 1,065.87 ly. The star Tycho 3827:533 to the south of M108 is a magnitude 10.56 star at a distance of 51.53 ly. The galaxy PGC 34128 labeled on the image has dimensions 0.6' x 0.3'. This is quite faint at magnitude 16.16. The 24-inch aperture of the telescope was able to capture a number of these faint objects. I leave the other galaxies for the reader to research.

Messier 108 lies just under the westerly end of the bottom of the bowl of the Big Dipper asterism in the constellation of Ursa Major. The nearby star 48-Beta Ursae

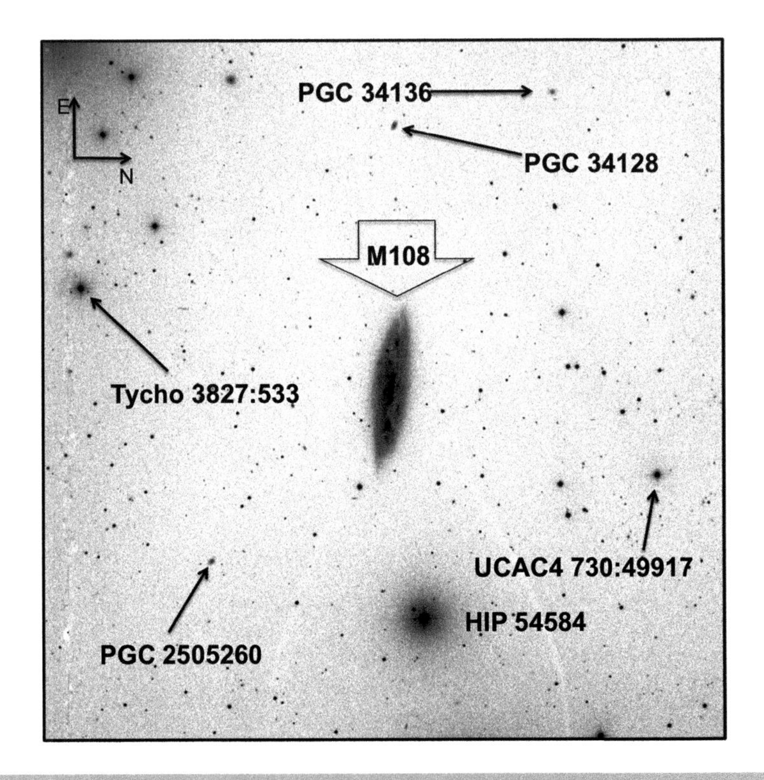

Fig. 5.101 Negative image of Messier 108

Majoris is labeled on the chart in Figure 5.102. This is Merak (HIP 53910), a star of magnitude 2.34 and spectral type A1, lying at a distance of 79.41 ly. This is separated from M108 by an angle of 1° 31'. The star 64-Gamma Ursae Majoris is Phecda (HIP 58001), slightly fainter than Merak at magnitude 2.41 and of spectral type A0. It lies at a distance of 83.65 ly. There is an angular separation of 6° 26' between Phecda and Messier 108. The meridian line cannot be seen on the zoomed-in view of the chart.

The image was taken from California at local time 22h 01m 05s on June 11. Universal Time was 05h 01m 05s (June 12). M108 has a Right Ascension of 11h 12m 28s, and the Local Sidereal Time was 14h 26m 21s. The LST is higher than the Right Ascension, so M108 had already crossed the meridian. The difference in these times is 03h 13m 53s, giving M108 a transit time of 18h 47m 12s. The azimuth angle is 315° 37' 00". The image was taken when the altitude of the galaxy was +52° 46' 36". At meridian transit, M108 had an altitude of +71° 29' 14". The image was taken with the excellent 24-inch Corrected Dall-Kirkham Astrograph. An exposure with a smaller telescope from New Mexico had been taken previously, but it did not bring out the same detail.

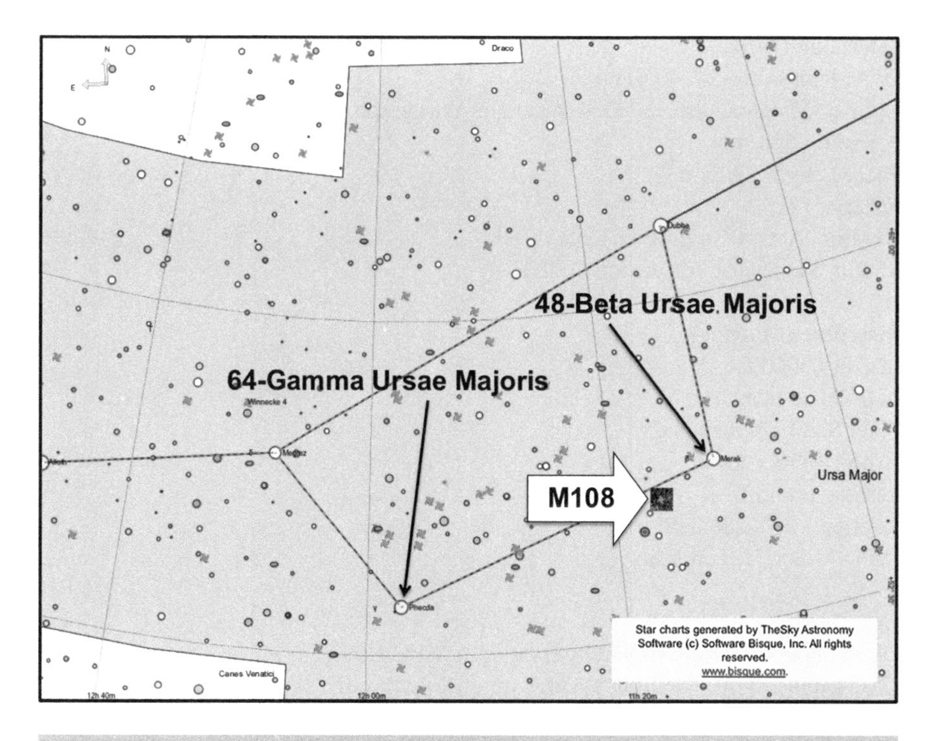

Fig. 5.102 Chart showing the location of M108

Messier 108: Data relating to the image in Figure 5.100

REMOTELY IMAGED FROM AUBERRY, CALIFORNIA, USA
NGC 3556
OBJECT TYPE: Spiral Galaxy

RA: 11h 12m 28s
DEC: +55° 34' 58"
ALTITUDE: +52° 46' 36"
AZIMUTH: 315° 37' 00"
FIELD OF VIEW: 32’ 08” x 32’ 08”
OBJECT SIZE: 8’ X 1’
POSITION ANGLE: 87° 55’ from North
EXPOSURE TIME: 300 s
DATE: 11th June
LOCAL TIME: 22h 01m 05s
UNIVERSAL TIME: 05h 01m 05s (12th June)
SCALE: 0.63 arcsec/pixel
MOON PHASE: 93.64% (waning)

Telescope Optics
OTA: Planewave 24" (0.61m)
Optical Design: Corrected Dall-Kirkham Astrograph
Aperture: 610 mm
Focal Length: 3962 mm
F/Ratio: f/6.5
Guiding: Active Guiding Disabled
Mount: Planewave Ascension 200HR

Instrument Package
FLI-PL09000 CCD
A/D Gain: e-/ADU
Pixel Size: 12-μm square
Array: 3056 x 3956
Sensor: Frontlit
Cooling: -35°C default
FOV: 31.8 x 31.8 arcmin

Location
Observatory: Sierra Remote Observatory – MPC U69
UTC Minus -8.00 (Daylight savings time is observed)
Minimum Target Elevation: Approx 25 Degrees
37.07°N, 119.4W
Elevation: 1405 metres (4610ft)

Messier 109

Messier 109 (NGC 3992), shown in Figure 5.103, is a barred spiral galaxy in the constellation of Ursa Major, just outside and to the east of the bottom of the bowl of the big Dipper asterism. On February 16, 1781, Pierre Méchain observed M97 and two other nearby galaxies and passed the information on to his friend and colleague Charles Messier. Messier observed M97 on March 24, 1781 and noted, "Near this nebula (M97) Méchain has seen another one, which has not yet been determined [M108] and also a third which is near Gamma of the Great Bear [M109]." It was not until 1953 that the American astronomer Owen Gingerich added both of these objects (M108 and M109) to the Messier list.

The shape of the lowercase Greek letter Theta (θ) is like a zero (0) with a bar drawn across it. M109 is similar in that it is a spiral galaxy with a bar drawn through it. The apparent size of the major axis of M109 is 7' of arc. The distance to M109 is given as 55 million ly by SEDS, but other sources give 60 million ly and 83 million ly. At 55 million ly, M109 would have an actual major axis size of 112,000 ly, and at 83 million ly distant, a size of 170,000 ly. Similarly, the minor axis of 4'

Fig. 5.103 Image of Messier 109, a barred spiral galaxy in Ursa Major

would give a range of 64,000 ly to 97,000 ly. The negative image in Figure 5.104 shows the location of a number of galaxies in the field of view, ranging from PGC 37553 with an apparent magnitude of 13.92 to PGC 2445913 with an an apparent magnitude of 17.41.

The chart in Figure 5.105 shows the position of M109 in Ursa Major "near Gamma of the Great Bear," as Messier put it. This star is 64-Gamma Ursae Majoris, better known as Phecda (HIP 58001), which has a magnitude of 2.41 and spectral type A0, lying at a distance of 83.65 ly. There is an angular separation of just less than 39' between Phecda and Messier 109. M109 lies on a line drawn between Phecda and 5 Canum Venaticorum (HIP 60485), as identified on the chart. HIP 60485 is a star of magnitude 4.76 and spectral type G7, lying at a distance of 392.96 ly. There is an angular separation of 4° 24' between 5 Canum Venaticorum and Messier 109.

The image was taken on February 1 at 05h 40m 09s local time in New Mexico. This corresponded to a Universal Time of 12h 40m 09s. The Right Ascension of M109 is 11h 58m 29s, and the Local Sidereal Time was 14h 24m 32s. The LST is higher than the RA, so M109 must have crossed the meridian. Subtracting one from

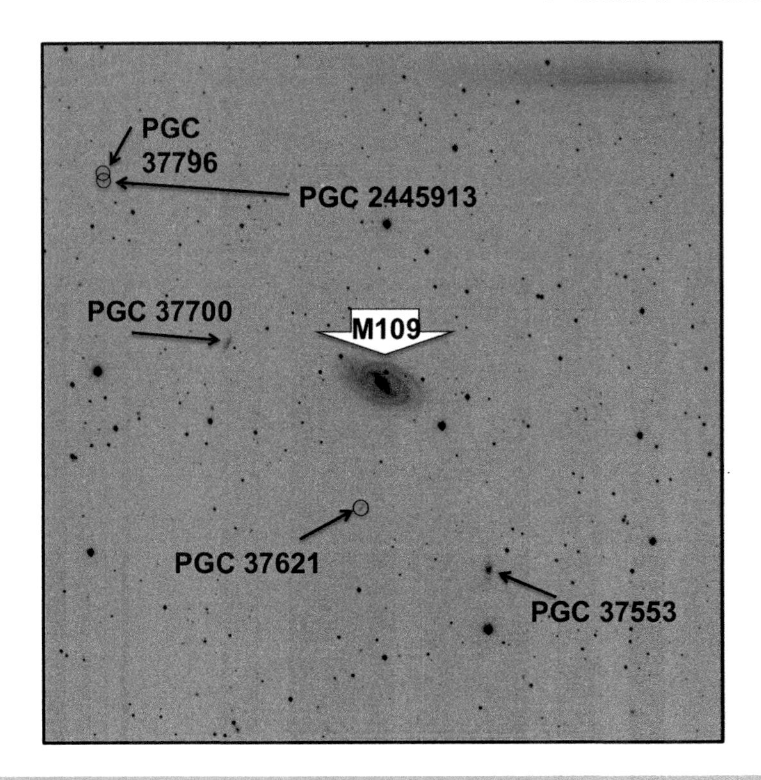

Fig. 5.104 Negative image of Messier 109

the other gives a difference of 2h 26m 3s, indicating a meridian transit time of 03h 14m 06s. The telescope used to take the image of M109 was a Takahashi TOA-150 apochromatic refractor with an aperture of 150 mm.

Messier 109: Data relating to the image in Figure 5.103

REMOTELY IMAGED FROM MAYHILL, NEW MEXICO
NGC 3992
OBJECT TYPE: Barred Spiral Galaxy

RA: 11h 58m 29s DEC: +53° 16' 29"
ALTITUDE: +56° 56' 03"
AZIMUTH: 319° 17' 50"
FIELD OF VIEW: 47' 6"' X 47' 6"
OBJECT SIZE: 7' X 4'
POSITION ANGLE: 358° 17' from North
EXPOSURE TIME: 300 s
DATE: 1st February

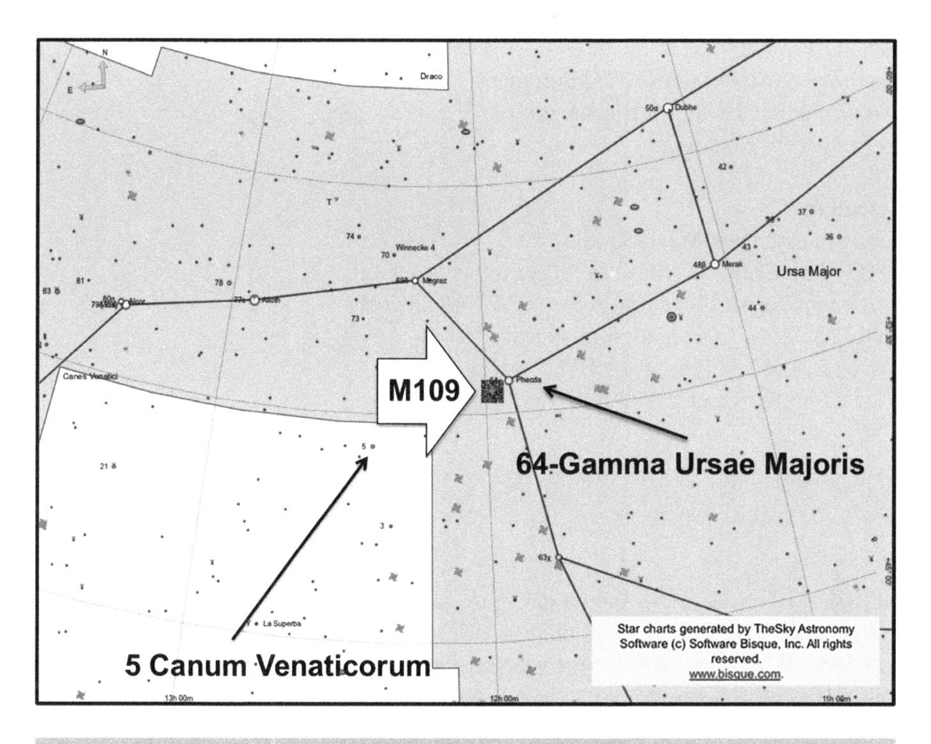

Fig. 5.105 Chart showing the location of M109

LOCAL TIME: 05h 40m 09s
UNIVERSAL TIME: 12h 40m 09s
SCALE: 1.38 arcsec/pixel
MOON PHASE: 21.90% (waxing)

Telescope Optics
OTA: Takahashi TOA-150
Optical Design: Apochromatic Refractor
Aperture: 150 mm
Focal Length: 1095 mm
F/Ratio: f/7.3
Guiding: Internal
Mount: Paramount GTS

Instrument Package
SBIG ST-4000M One-Shot Color CCD
A/D Gain: 0.6e-/ADU
Pixel Size: 7.4um square
Resolution: 1.45 arcsec/pixel

Sensor: Frontlit
Cooling: -20°C Winter (-10°C Summer)
Array: 2048 x 2048 (8.3 Megapixels)
FOV: 49.6 x49.6 arcmin

Location
Observatory: New Mexico Skies
UTC Minus 7.00 (Daylight savings time is observed)
Minimum Target Elevation: Approx 25 – 45 Degrees
(N or S) 32° 54' Decimal: 32.9 North
(W or E) 105° 31' Decimal: 105.5 West
Elevation: 2225 meters (7298 ft)

Messier 110

Messier 110 (NGC 205), shown in Figure 5.106, is a dwarf elliptical galaxy and a companion to M31 in the constellation of Andromeda. The 180-second image captured M110, M31 and M32 in the frame. Charles Messier discovered M110 in 1773 and made the following note: "On August 10 I examined, under a very good sky, the beautiful nebula of the girdle of Andromeda [M31] with my achromatic refractor, which I had made to magnify 68 times, for creating a drawing like the one of that in Orion [M42]. I saw that nebula which Citizen Legentil discovered on October 29, 1749 [M32]. I also saw a new, fainter one, placed north of the great nebula, which was distant from it about 35' in right ascension and 24' in declination. It appeared to me amazing that this faint nebula has escaped the discovery by the astronomers and myself, since the discovery of the great nebula by Simon Marius in 1612, because when observing the great nebula, the small is located in the same field of view of the telescope. I will give a drawing of that remarkable nebula in the girdle of Andromeda, with the two small, which accompany it."

The negative image in Figure 5.107 shows M110, M31 and M32. An enlarged positive image of M110 has been superimposed onto the negative image. North is to the right and east is up on the images. Messier pointed out that M110 was to the north of M31, but you can see that it is actually to the northwest of the center of M31. M110 has an apparent major axis size of 17' of arc and minor axis size of 10'. M110 is believed to be at the same distance as M31, so the apparent difference in size in the image actually represents the real difference. In other words, M110 is dwarfed by M31. The distance to these galaxies is 2.9 million ly, so Messier 110 is 14,000 x 8,500 ly (SEDS). Eight globular clusters have been observed in M110, one of which was spotted through the 106-mm aperture telescope. This is the brightest M110 globular at magnitude 15 and is known as G73. Its location is shown on the negative image and on the enlarged positive view of Messier 110.

Fig. 5.106 Image of Messier 110, a dwarf elliptical galaxy in Andromeda

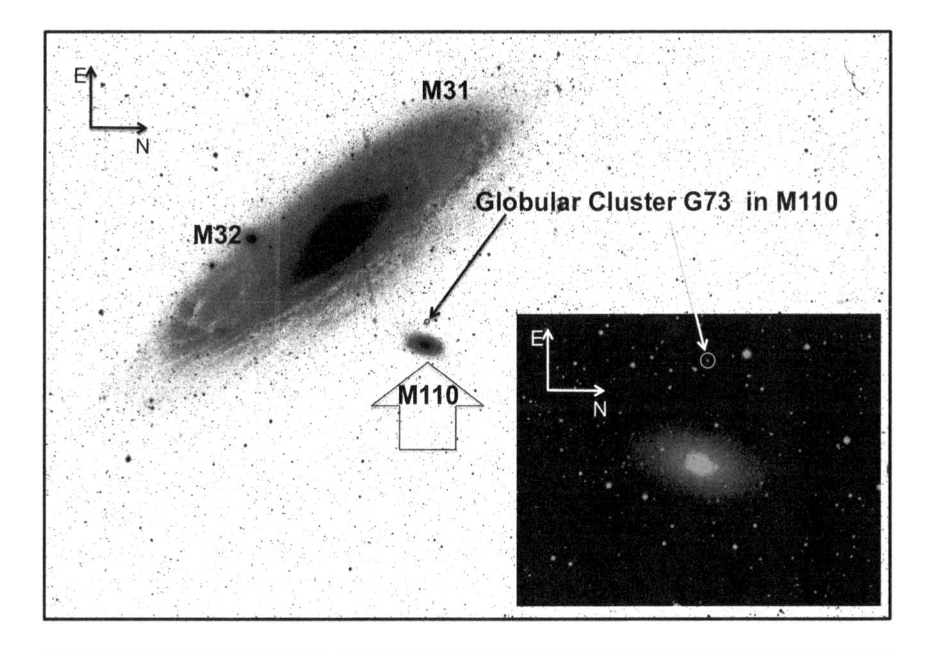

Fig. 5.107 Negative image of Messier 110

The chart in Figure 5.108 shows the plate-solved image of M110 embedded in the constellation of Andromeda. The image has a field of view of 3° 53' 16" x 2° 35' 31", so it is quite a chunk of the constellation. M110 lies roughly on a line drawn between the two stars 35-Nu Andromedae and 19-Kappa Andromedae. 35-Nu Andromedae is HIP 3881 at magnitude 4.53 and spectral type B5, lying at a distance of 679.49 ly. There is an angular separation between 35-Nu Andromedae and Messier 110 of roughly 1° 52'. 19-Kappa Andromedae is HIP 116805 with a magnitude of 4.15 and spectral type B9, lying at a distance of 169.70 ly. The angular separation between M110 and 19-Kappa Andromedae is in the region of 11° 15'. The meridian line actually passes through the image, as can be seen in Figure 5.108, showing that M110 crossed the meridian recently.

The image was taken on September 24 at 01h 34m 01s local time in New Mexico. This corresponded to a Universal Time of 07h 34m 01s. The Right Ascension of M110 is 00h 41m 19s, and the Local Sidereal Time was 00h 45m 01s. The LST is slightly higher than the RA, so M31 must have crossed the meridian from east to west already -- in fact, just a few minutes ago. Subtracting one from the other gives a difference of 0h 03m 42s. Subtracting this from the local time the image was taken gives a meridian transit time of 01h 30m 19s. The telescope used to take the image of M110 was a Takahashi FSQ apochromatic astrograph with an aperture of 106 mm.

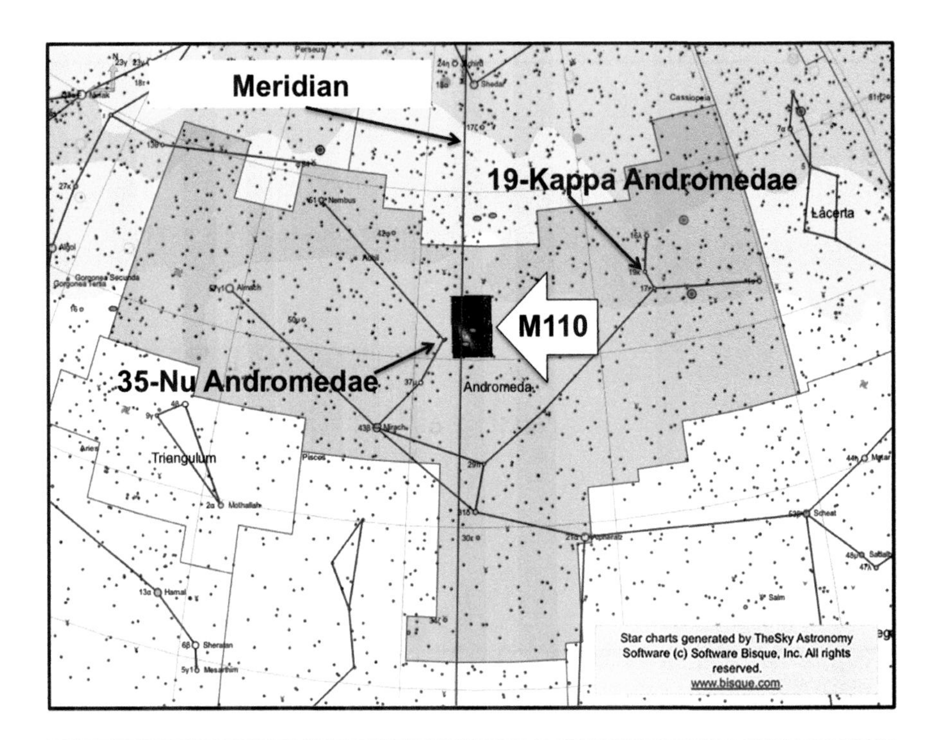

Fig. 5.108 Chart showing the location of M110

Messier 110: Data relating to the image in Figure 5.106

REMOTELY IMAGED FROM MAYHILL, NEW MEXICO
NGC 205
OBJECT TYPE: Dwarf Elliptical Galaxy

RA: 00h 41m 19s
DEC: +41° 46' 38"
ALTITUDE: +80° 58' 34"
AZIMUTH: 355° 35' 16"
FIELD OF VIEW: 3° 53' 16" x 2° 35' 31"
OBJECT SIZE: 17' X 10'
POSITION ANGLE: 92° 16' from North
EXPOSURE TIME: 180s
DATE: 24th September
LOCAL TIME: 01h 34m 01s
UNIVERSAL TIME: 07h 34m 01s
SCALE: 3.49 arcsec/pixel
MOON PHASE: 40.22% (waning)

Telescope Optics
OTA: Takahashi FSQ Fluorite
Optical Design: Petzval Apochromat Astrograph
Aperture: 106 mm
Focal Length: 530 mm
F/Ratio: f/5.0
Guiding: External
Mount: Paramount GTS-1100S

Instrument Package
SBIG STL-11000M
A/D Gain: 2.2e-/ADU
Pixel Size: 9um square
Resolution: 3.5 arcsec/pixel
Sensor: Frontlit
Cooling: -15°C default
Array: 4008 x 2672 (10.7 Megapixels)
FOV: 155.8 x 233.7 arcmin

Location
Observatory: New Mexico Skies
UTC Minus 7.00 (Daylight savings time is observed)
Minimum Target Elevation: Approx. 25 – 45 Degrees
(N or S) 32° 54' Decimal: 32.9 North
(W or E) 105° 31' Decimal: 105.5 West
Elevation: 2225 meters (7298 ft)

Chapter 6

Quick Reference Image Library

Images of all the Messier Objects Messier 1 – Messier 110

L. Adam, *Imaging the Messier Objects Remotely from Your Laptop*, The Patrick Moore Practical Astronomy Series, https://doi.org/10.1007/978-3-319-65385-3_6

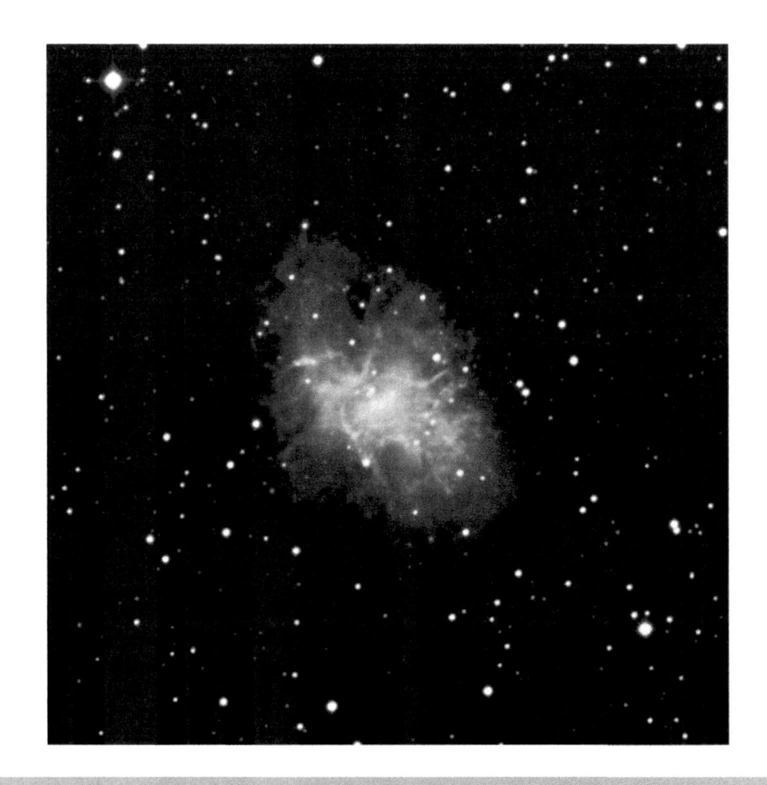

Messier 1 – NGC 1952 - Supernova Remnant - Taurus

Messier 2 – NGC 7089, a globular cluster - Aquarius

Messier 3 – NGC 5272, a globular cluster – Canes Venatici

Messier 4 – NGC 6121, a globular cluster – Scorpius

Messier 5 – NGC 5904, a globular cluster – Serpens Caput

Messier 6 – NGC 6405, an open cluster – Scorpius

Messier 7 – NGC 6475, an open cluster – Scorpius

Messier 8 – NGC 6523 – Star-forming Nebula – Sagittarius

Messier 9 – NGC 6333, a globular cluster – Ophiuchus

Messier 10 – NGC 6254, a globular cluster – Ophiuchus

Messier 11 – NGC 6705, an open cluster – Scutum

Messier 12 – NGC 6218, a globular cluster – Ophiuchus

Messier 13 – NGC 6205, a globular cluster – Hercules

Messier 14 – NGC 6402, a globular cluster – Ophiuchus

Messier 15 – NGC 7078, a globular cluster – Pegasus

Messier 16 – NGC 6611, an open cluster and Nebula – Serpens Cauda

Messier 17 – NGC 6618, an open cluster and Nebula - Sagittarius

Messier 18 – NGC 6613, an open cluster – Sagittarius

Messier 19 – NGC 6273, a globular cluster – Ophiuchus

Messier 20 – NGC 6514 – Bright Nebula – Sagittarius

Messier 21 – NGC 6531, an open cluster – Sagittarius

Messier 22 – NGC 6656, a globular cluster – Sagittarius

Messier 23 – NGC 6494, an open cluster – Sagittarius

Messier 24 – NGC 6603 (Centered on) – Star Cloud – Sagittarius

Messier 25 – IC 4725, an open cluster – Sagittarius

Messier 26 – NGC 6694, an open cluster – Scutum

Messier 27 – NGC 6853, a planetary nebula – Vulpecula

Messier 28 – NGC 6626, a globular cluster – Sagittarius

Messier 29 – NGC 6913, an open cluster – Cygnus

Messier 30 – NGC 7099, a globular cluster – Sagittarius

Messier 31 – NGC 224, a spiral galaxy – Andromeda

Messier 32 – NGC 221 – Dwarf Elliptical Galaxy – Andromeda

Messier 33 – NGC 598, a spiral galaxy – Triangulum

Messier 34 – NGC 1039, an open cluster – Perseus

Messier 35 – NGC 2168, an open cluster – Gemini

Messier 36 – NGC 1960, an open cluster – Auriga

Messier 37 – NGC 2099, an open cluster – Auriga

Messier 38 – NGC 1912, an open cluster – Auriga

Messier 39 – NGC 7092, an open cluster – Cygnus

Messier 40 – Winnecke 4 – Double Star - Ursa Major

Messier 41 – NGC 2287, an open cluster – Canis Major

Messier 42 - NGC 1976 – Star-forming Nebula – Orion

Messier 43 – NGC 1982 – Bright Nebula – Orion

Messier 44 – NGC 2632, an open cluster – Cancer

Messier 45 – Open Cluster - Taurus

Messier 46 – NGC 2437, an open cluster – Puppis

Messier 47 – NGC 2422, an open cluster – Puppis

Messier 48 – NGC 2548, an open cluster – Hydra

Messier 49 – NGC 4472, an elliptical galaxy – Virgo

Messier 50 – NGC 2323, an open cluster – Monoceros

Messier 51 – NGC 5194, a spiral galaxy – Canes Venatici

Messier 52 – NGC 7654 - Open Cluster – Cassiopeia

Messier 53 – NGC 5024, a globular cluster – Coma Berenices

Messier 54 – NGC 6715, a globular cluster – Sagittarius

Messier 55 – NGC 6809, a globular cluster – Sagittarius

Messier 56 – NGC 6779, a globular cluster – Lyra

Messier 57 – NGC 6720, a planetary nebula – Lyra

Messier 58 – NGC 4579, a spiral galaxy – Virgo

Messier 59 – NGC 4621, an elliptical galaxy – Virgo

Messier 60 – NGC 4649, an elliptical galaxy – Virgo

Messier 61 – NGC 4303, a spiral galaxy – Virgo

Messier 62 – NGC 6266, a globular cluster – Ophiuchus

Messier 63 – NGC 5055, a spiral galaxy – Canes Venatici

Messier 64 – NGC 4826, a spiral galaxy – Coma Berenices

Messier 65 – NGC 3623, a spiral galaxy – Leo

Messier 66 – NGC 3627, a spiral galaxy – Leo

Messier 67 – NGC 2682, an open cluster – Cancer

Messier 68 – NGC 4590, a globular cluster – Hydra

Messier 69 – NGC 6637, a globular cluster – Sagittarius

Messier 70 – NGC 6681, a globular cluster – Sagittarius

Messier 71 – NGC 6838, a globular cluster – Sagitta

Messier 72 – NGC 6981, a globular cluster – Aquarius

Messier 73 – NGC 6994, an open cluster – Aquarius

Messier 74 – NGC 628, a spiral galaxy – Pisces

Messier 75 – NGC 6864, a globular cluster – Sagittarius

Messier 76 – NGC 650/651, a planetary nebula – Perseus

Messier 77 – NGC 1068, a spiral galaxy – Cetus

Messier 78 – NGC 2068 – Star-forming Nebula – Orion

Messier 79 – NGC 1904, a globular cluster – Lepus

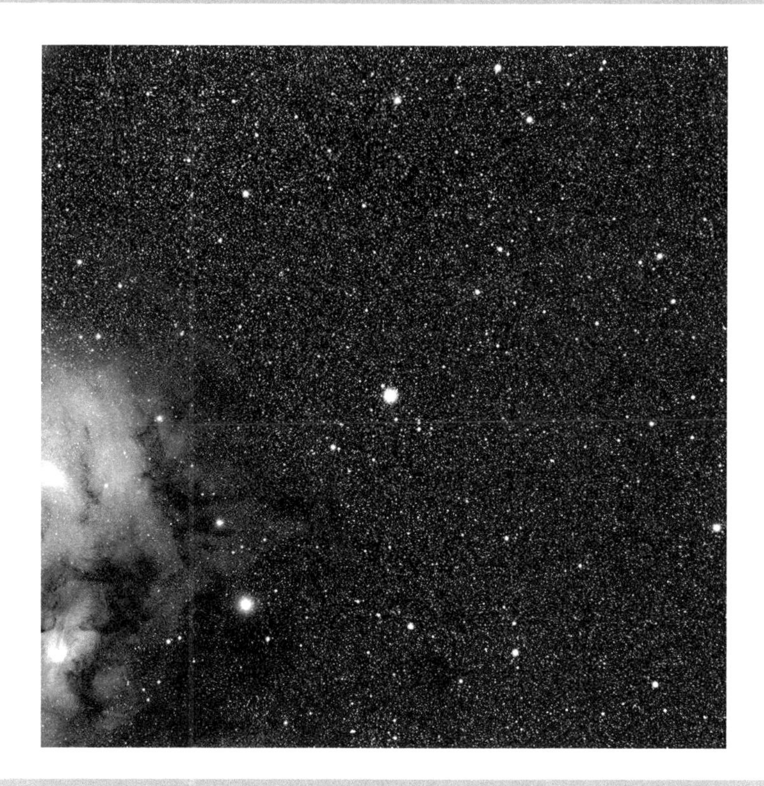

Messier 80 – NGC 6093, a globular cluster – Scorpius

Messier 81 – NGC 3031, a spiral galaxy – Ursa Major

Messier 82 – NGC 3034 – Irregular Galaxy – Ursa Major

Messier 83 – NGC 5236, a spiral galaxy – Hydra

Messier 84 – NGC 4374, an elliptical galaxy – Virgo

Messier 85 – NGC 4382, a lenticular galaxy – Coma Berenices

Messier 86 – NGC 4406 – Elliptical/Lenticular Galaxy – Virgo

Messier 87 – NGC 4486, an elliptical galaxy – Virgo

Messier 88 – NGC 4501, a spiral galaxy – Coma Berenices

Messier 89 – NGC 4552, an elliptical galaxy – Virgo

Messier 90 – NGC 4569, a spiral galaxy – Virgo

Messier 91 – NGC 4548 – Barred Spiral Galaxy – Coma Berenices

Messier 92 – NGC 6341, a globular cluster – Hercules

Messier 93 – NGC 2447, an open cluster – Puppis

Messier 94 – NGC 4736, a spiral galaxy – Canes Venatici

Messier 95 – NGC 3351 – Barred Spiral Galaxy – Leo

Messier 96 – NGC 3368, a spiral galaxy – Leo

Messier 97 – NGC 3587, a planetary nebula – Ursa Major

Messier 98 – NGC 4192, a spiral galaxy – Coma Berenices

Messier 99 – NGC 4254, a spiral galaxy – Coma Berenices

Messier 100 – NGC 4321, a spiral galaxy – Coma Berenices

Messier 101 – NGC 5457, a spiral galaxy – Ursa Major

Messier 102 – NGC 5866, a lenticular galaxy – Draco

Messier 103 – NGC 581, an open cluster – Cassiopeia

Messier 104 – NGC 4594, a spiral galaxy – Virgo

Messier 105 – NGC 3379, an elliptical galaxy – Leo

Messier 106 – NGC 4258, a spiral galaxy – Canes Venatici

Messier 107 – NGC 6171, a globular cluster – Ophiuchus

Messier 108 – NGC 3556, a spiral galaxy – Ursa Major

Messier 109 – NGC 3992 – Barred Spiral Galaxy – Ursa Major

Messier 110 – NGC 205, an elliptical galaxy – Andromeda

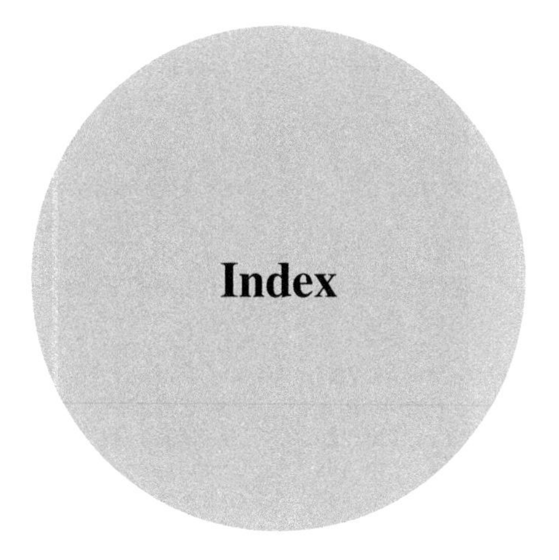

Index

L. Adam, *Imaging the Messier Objects Remotely from Your Laptop*, The Patrick Moore Practical Astronomy Series, https://doi.org/10.1007/978-3-319-65385-3

N

R

S

T

U